Anthology

Test Equipment And Repair Techniques

(From The Pages Of Ham Radio Magazine)

First Printing 2003

Library of Congress Catalog Card Number 2003105141
ISBN 0-943016-26-6

Editor: Lew Ozimek
Layout and Design: Elizabeth Ryan

Published by CQ Communications, Inc.
25 Newbridge Road
Hicksville, NY 11801 USA

Printed in the United States of America

Preface

In October of 1967 two Radio Amateurs, Jim Fisk, W1DTY, and Skip Tenney, W1NLB, discussed what they perceived to be a serious shortfall in published technical literature available to the Ham community. They were aware that a rapid growth had occurred in the electronics and communication industry in general, without a commensurate rate of development in amateur radio. Single sideband had largely replaced A-M, transistors were taking the place of vacuum tubes, and integrated circuits were slowly finding their way into electronic projects. They remembered the past, however, where, when vacuum tubes became practical devices, amateurs were among the first to use them in home construction. The same ham development and growth had occurred when semiconductor devices became available. And yet with the increasing availability of transistors, they did not see the absorption of this challenging technical development by the amateur radio community. They were certain that integrated circuits would likewise languish in engineering laboratories and amateurs also would, in all probability, under utilize other new technology.

They asked each other why this had occurred (or if you prefer, had not occurred)? Why, indeed? To them it was very simple, hams did not have enough good practical information available to use and apply. An awful lot of state-of-the-art design fundamentals existed that had to be understood before anyone could get into state-of-the-art design. If you are an engineer, fine, they reasoned, but not all hams were engineers. They perceived that most hams wanted good explanations of these technical advances and needed practical circuits that they could adapt to special jobs and projects.

A good example of the problem to them was that SSB appeared to be a lot more complex than A-M. Certainly the gear required to generate a SSB signal was more complex than that used in an old A-M rig, but understanding what makes it tick was not. They perceived that the two modes were very closely related, not particularly compatible, but related. To them the problem was that a simple, concise explanation of sideband was lacking. A few articles had been published on the subject in the early fifties, but many of the hams who actually used sideband equipment had never seen the articles. If that happened with single-sideband at the time, then this failing would continue with transistors and with any new technological approaches developed in the future.

With this in mind, these two individuals decided to launch a new magazine for the ham community entitled 'Ham Radio'. In the words of Jim Fisk, the magazine's editor, "Ham Radio is designed to fill this gap (in the availability of data). It is designed to inform. It will be geared to the state of the art- the state of the art in practice. It will be a magazine (that) shows you how to use new devices and old. Although we will encourage the use of solid state (devices), we will not discriminate against (items like) vacuum tubes for the sake of being modern. There are a lot of places where vacuum tubes are still very practical and desirable gadgets."

Skip Tenney, the publisher, confirmed this approach by saying, "A new outlook is necessary. If we continue to work with old ideas and concepts, we can hardly expect to maintain our traditional spot in the electronics world. We are a branch of one of the fastest moving areas of technology. If you have any doubts, look at the developments of the past few years in solid-state techniques or satellite communications. Amateur radio will have to look and act the part if we are to keep up." The two of them obviously agreed on the principles to be used in the development of this new publication.

They officially formed the company to launch this project in January of 1968 and published the first concept issue in February of that year. True to their perception of the needs of amateurs, the magazine's articles ran the gamut from the simple to the complex, but always oriented to a practical approach – the amateur approach. They included articles covering simple projects for the novice and the one-night-a-week experimenters, involved projects for the experienced ham who likes to work in his shop, practical design and theory for the individual who wants to start from scratch, and the 'last word for the (specialized) enthusiast.'

The magazine prospered from February of 1968 until its final issue in June 1990. During that period it provided to an enthusiastic ham following a continuous stream of high quality technical articles and projects. It concentrated on advanced levels of technology and as a result many of the articles published a number of years ago are still of interest and use today! The full spectrum of Amateur Radio activity was included; there was something there for everyone. Many of the topics covered are timeless and will always be of value and interest to the hobbyist. A large number of the construction projects are just as useful today as they were the day they were first published. The technical descriptions presented during the 22 years of Ham Radio's existence, are just as illuminating today as they were back then.

In order to prevent such a wealth of knowledge to languish on the shelves, CQ Magazine decided to review the issues of Ham Radio magazine and to extract those articles that would be of interest to amateurs today. This is one in the series of volumes published by CQ. It is hoped that it will provide a wealth of useful knowledge and bring as much pleasure to the reader as it did to the editorial staff of CQ who worked on the selection and publishing of this Anthology.

L. Ozimek, N2OZ

Chronological Index

Subject Index

Clustered Index By Topic

TROUBLE SHOOTING TECHNIQUES

SPECIAL TEST INSTRUMENT APPLICATIONS

CONSTRUCTION PROJECTS: TEST INSTRUMENTS

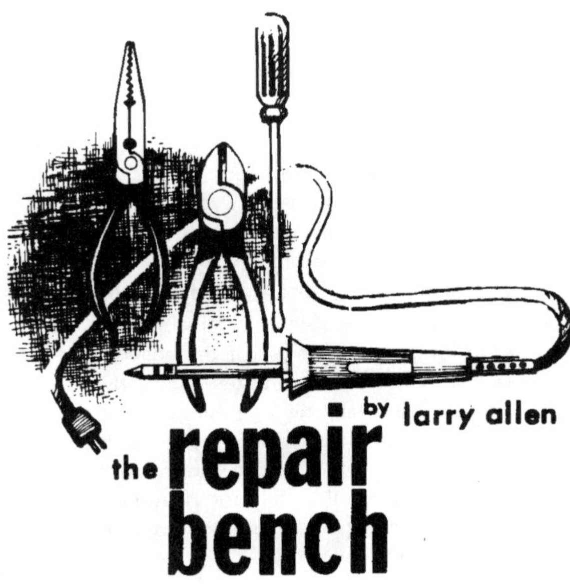

troubleshooting transistor ham gear

Hardly a month goes by that someone doesn't introduce new ham transistor equipment. It's about time we included some transistor troubleshooting information in this column. There are plenty of books on the subject, but it can't hurt to recap briefly some of the more fundamental transistor troubleshooting principles.

First of all, you should understand what a transistor does. In its most common use, a transistor is simply an amplifying device. For all practical purposes, you can see a transistor as a device through which controllable current flows. There are two kinds. In one, current flows from collector to emitter; in the other, from emitter to collector.

In either case, it is sufficient in this preliminary explanation for you to know that current in the transistor flows between these two elements. The phenomenon of amplification takes place because the transistor has a third element—called the **base**—that can control this flow of current.

Take a look at **fig. 1A.** This will acquaint you with the schematic diagram of a PNP transistor, one of the two types (the other is called an NPN). A transistor needs dc operating voltages. For a PNP transistor, they are applied as shown in **fig. 1A.** The emitter is usually grounded or connected to ground through a low-value resistor. Ground is therefore the common connection for all supply voltages.

The collector of a PNP transistor is connected to a strong negative voltage. This alone does not cause a flow of current in a normal transistor, but the possibility is there. It remains for **forward bias** to be applied to the base before collector current can flow. To cause normal current flow in this PNP transistor, a small negative voltage must be applied to the base. This makes the base more negative than the emitter (though still much less negative than the collector). With negative voltage applied to the base of a PNP transistor, the base-emitter junction is said to be **forward** biased, because the current flows easily across the junction from the N-material of the base to the P-material of the emitter. This base current is small, but it releases a large current flow between emitter and collector.

A small current in the base circuit controls a large current in the collector circuit; thus amplification is possible. Suppose a transistor, connected as in **fig. 1A,** has a small audio voltage applied to the base along with the forward dc bias. What the audio voltage does is increase and decrease the bias, which in turn lets more and less collector current flow. Therefore, a tiny signal voltage or current in the base-emitter junction controls large amounts of signal current in the collector circuit. All you have to do is place a load resistor in the collector circuit and the current is converted to a strong voltage, and you have voltage amplification.

fig. 1. Power-supply connections for typical transistors: the PNP takes negative on collector and base (A), the NPN, positive on collector and base.

Fig. 1B shows the power supply connection for an NPN transistor. The collector of this transistor type is connected to a positive voltage. So is the base, but to a lower voltage. Again, any varying voltage, such as an af or rf signal, applied to the base will control the flow of current in the emitter-collector circuit.

Amplification is thus accomplished by either type of transistor; the only difference is in the polarity of dc power-supply voltages applied to operate the transistor. **Figs. 2A** and **2B** show both types of transistor connected in normal amplifying circuits. **Fig. 2A,** using the PNP, is an audio amplifier, as you can see from the values of components. **Fig. 2B,** using an NPN, is an rf amplifier, which you can see from the tuned air-core transformers used to couple the signal in and out. Either polarity of transistor could be used in either circuit, merely by reversing polarity of the power-supply connections.

Normal operating voltages are shown in both schematics of **fig. 2,** as they would appear in a diagram of equipment you might want to troubleshoot. The "fun" of troubleshooting starts when the dc operating voltages on a transistor have changed from normal. One may be high and another low; one may be okay while another is way off; or they may all be wrong. Whatever discrepancy you find in measuring dc voltages at the elements of a transistor, your problem is to figure out what's causing it. The transistor itself could be at fault, or there could be a problem in one of the other parts.

One way to find out if the transistor is faulty would be to remove it from the circuit and check it either with a tester or with your ohmmeter. (I'll tell you how to use your ohmmeter for a quick test later.) The trouble is that most transistors are soldered in, and are not easy to get loose for out-of-circuit testing. As a result, it is better to be able to **interpret** incorrect voltages. It really isn't too difficult, if you stop to *analyze the direction of current flow* in the transistor, whatever is polarity.

Remember that current always flows from the negative power-supply terminal toward the positive one. Therefore, if the collector of a PNP transistor is connected to negative voltage, it can mean only one thing: current in that transistor flows from collector to emitter. Conversely, if you're dealing with an NPN transistor, the collector is positive, and the direction of current flow is from *emitter toward collector.*

fig. 2. Use of the transistor in typical circuits: PNP in an audio circuit (A), NPN in an rf circuit (B).

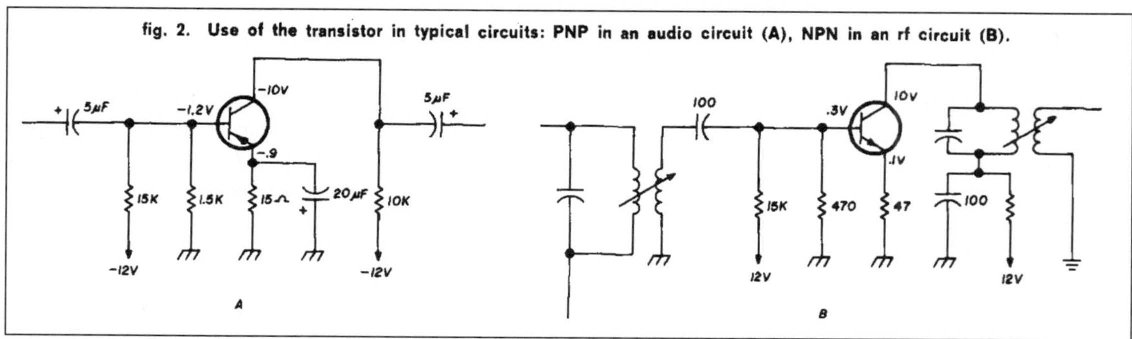

In most cases, however, knowing the direction of flow is not really as important as figuring out whether the voltage at the collector has increased or decreased. If it has increased—that is, if it's **closer to** the power supply voltage—you can reason quite easily that there must be less current flowing through the load resistor and consequently less voltage drop across it. Since the load resistor is in series with the collector circuit, it stands to reason that less current is flowing through the transistor, too.

If the power-supply voltage is negative, say −12 volts, the normal collector voltage may be about −10 volts. Suppose when you measure the voltage you find it to be −11.6 on the collector. This can mean only one thing: there is less collector current, signified by less **drop** across the load resistor. You can also be sure the cause is not the load resistor having increased in value, because that would cause a larger voltage drop and the voltage at the collector would be **less** than the normal −10 volts.

Exactly the same reasoning follows if you're dealing with an NPN transistor, where the collector voltage is positive. If the collector voltage shifts nearer to the power-supply voltage, whatever its polarity, it is a sign of reduced collector current. There is an outside possibility that the load resistor has lowered in value and therefore does not drop as much voltage. In that case, collector current would be unusually high, yet you'd still find a higher voltage at the collector. This seldom happens.

How many things can cause abnormal collector voltage? Or, phrased another way, depending on your conclusion from the first collector-voltage measurement, what makes a transistor draw less current than normal?

Two things: a faulty transistor, or reduced bias voltage (which causes reduced bias current). The way to find out which is to measure the bias voltage. If it is much lower than it should be, then chances are the bias voltage is the cause of reduced current; the transistor may be okay. If bias is normal, yet current is very low through the transistor, the transistor is probably defective.

Once you suspect by this method of reasoning that the transistor is okay, you might as well go ahead and check the other parts. Measure resistors to see if they've

fig. 3. Using ohmmeter readings to check the quality of a transistor. Both types should show high readings in both directions between the emitter and collector terminals.

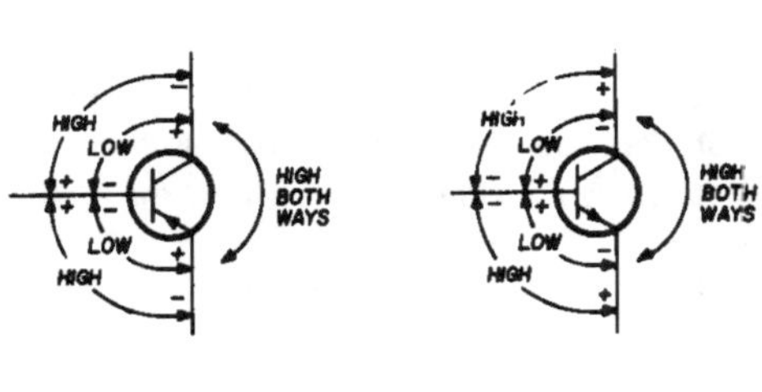

changed value, and check capacitors to see if they are leaky. Your ohmmeter is handy for both these tests.

If you decide to suspect the transistor, there are a couple more checks you can make before you disconnect the transistor for external testing. Clip your voltmeter—a vtvm is best—to measure collector voltage. If it reads at least slightly below the power-supply voltage it is connected to, proceed with this test. With a jumper lead, short the **base** of the transistor to the **emitter.** CAUTION: Be sure it is the emitter you short the base to; if you short it to the collector, you'll burn up the transistor. Watch the collector voltage as you make the jumper connection. If the transistor is operating normally, the collector voltage will jump upward, and read virtually the same as its power-supply source. If it doesn't, either the base element is open inside the transistor or the collector junction is leaky. Either condition signifies a defective transistor.

Now let's go back and see what we'd have done if the voltage on the collector were too low. Again keep in mind that this depends not at all on which polarity of transistor is involved; just consider "higher" voltage as being **closer to** the supply value and "lower" as being **further from** the supply value.

If the voltage at the collector is low, it means that either the load resistor has increased in value (unlikely) or more current is being drawn through it. If the latter, then more current than normal is flowing in the transistor collector. This in turn may be caused either by a faulty transistor that allows too much current to flow, or by too much bias current flowing between base and emitter.

Your voltmeter will tell you whether the bias voltage is too high. If it is, track down the trouble in the resistive network that develops the bias. You'll find any of several different resistor arrangements, but all of them are simple voltage dividers. You should have no trouble checking the resistors with your ohmmeter.

If bias is okay, the trouble is likely inside the transistor. Either it is leaky, allowing too much current to flow, or some trouble in the base-emitter junction is not letting the bias control the transistor as it should. Again, you can make the "jumper" test mentioned earlier: while measuring the collector voltage, connect the base terminal to the emitter terminal. The abnormally low voltage at the collector should suddenly jump up to the power-

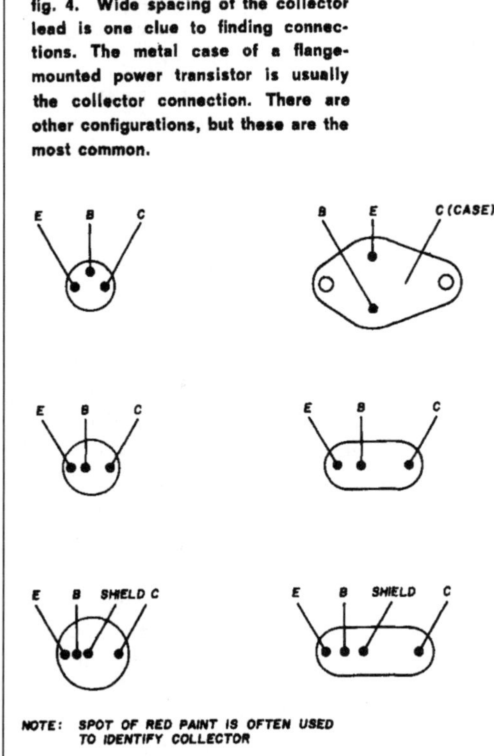

fig. 4. Wide spacing of the collector lead is one clue to finding connections. The metal case of a flange-mounted power transistor is usually the collector connection. There are other configurations, but these are the most common.

supply voltage. If it doesn't, the base circuit is not controlling collector current as it should. With zero bias, which is what you have when you short the base to the emitter, very little collector current should flow—only as much as is permitted by leakage in the transistor. With almost no collector current flowing, the collector voltage should be almost the same as the voltage at the source end of the load resistor (no drop across the resistor).

Finally, suppose you've decided the transistor is faulty and you want a final double-check. You can get that with your ohmmeter. A transistor tester is handy, but your ohmmeter is adequate if you know how to use it. The secret of checking transistors with an ohmmeter lies in measuring the backward and forward resistances of the junctions in the transistor. What you should find is shown in **figs. 3A and 3B**. A PNP transistor is shown at **A** and an NPN transistor at **B.** As indicated by the diagrams, the ratios of forward-to-backward resistances are more important than any specific values. The readings in one direction must be much higher than in the other. For small-signal transistors (they are physically small, too), the ratio should be 500:1 or better. In power transistors (the large ones with metal flanges), the ratio can be as low as 100:1, and occasionally even lower. When you're in doubt, check the readings of a suspected transistor against a new one of a similar type.

The mechanics of the test are simple. Connect the ground lead of your ohmmeter to the base; then with the other ohmmeter lead, check the reading from base to emitter and then from base to collector. Reverse the leads by touching the probe to the base and use the ground lead to check first the emitter and then the collector. If you use a small chart like those in **fig. 3,** which you can draw on any scrap of paper, you can jot down the readings and then compare them. This test method applies to either NPN or PNP, although the highs and lows are exactly opposite. Nevertheless, **ratio** is what is important; in either type of transistor, the ratios should be high.

To wind up this month's column on "quickie" transistor troubleshooting, I've included some base diagrams of several common transistor types in **fig. 4.** You'll find them easy to memorize, but you can also sketch them on a piece of cardboard and post them on the wall of your shack or on the back of your workbench.

Rarely will you find a transistor that varies from those shown in the diagrams. If you happen to have a piece of equipment with one that does vary, the manufacturer's service data that comes with the unit will show the proper lead configuration. Be very sure you are using the right configuration, particularly when you make the base-emitter "jumper" test. Remember, shorting the base to the collector, even accidentally, can ruin the transistor.

It is obviously impossible to cover all transistor troubleshooting possibilities in one short column. However, if you have specific questions about troubleshooting transistor equipment, drop me a line. Or, if you use some special technique in tracking down trouble in your own transistor gear, tell me about that. I'll use some of the more interesting and helpful ideas in future columns.

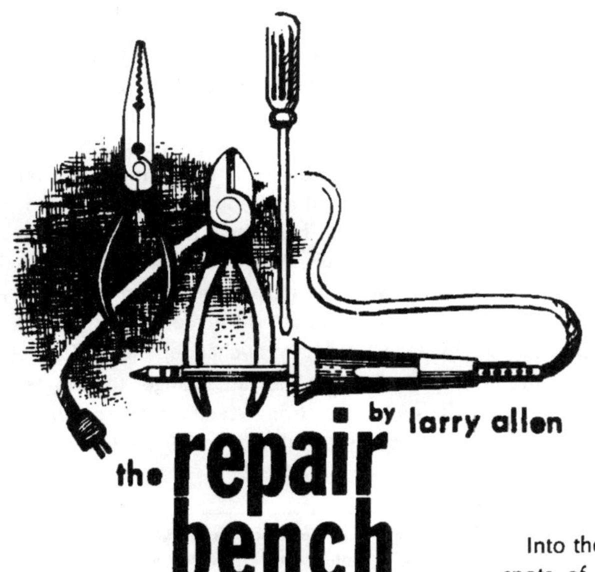

by larry allen

the repair bench

troubleshooting around fet's

Several articles about field-effect transistors have appeared in **ham radio,** and commercial equipment is beginning to show up with FET's in some circuits. Sooner or later, you'll build or buy a piece of new equipment using this special breed of transistor, and eventually you'll have to troubleshoot it. So, you ought to know a few of the idiosyncracies FET's display.

They're different from ordinary transistors. In some respects, they may remind you of tubes. But they **are** transistors and if you have one go bad in a receiver or something, you troubleshoot it much the same as any other transistor. Before I mention FET troubles and how to spot them, it seems like a good idea to recap how they differ from ordinary transistors.

two main types

Basically, a FET (rhyme it with "jet") starts with a **channel** of semiconductor material through which current can flow. Take a look at **fig. 1A.** The bar is N-type semiconductor material. Current flows in the direction of the arrows, carried through the N-material by electrons. The end of the channel the current enters is the **source;** the end from which it leaves is the **drain.**

Into the bar of N-material are diffused two spots of P-material. Connected together as in **fig. 1B,** they form a **gate.** The gate material makes a junction with the channel material, and the whole device is called a junction field-effect transistor, or just **JFET.**

Current through the channel in **fig. 1A** depends on the resistivity of the material and the voltage of the battery. In **fig. 1B,** however, the current is controlled by the negative voltage applied to the gate. Battery 2 is the gate-bias supply and reverse-biases the PN junction: negative voltage is applied to this P-material, and positive voltage would be applied to N-material. Reverse bias on the gate elements tends to **pinch off** the flow of current through the channel from source to drain. The higher the negative gate-bias, the less the current from source to drain. Naturally, if you override part of the battery-2 voltage with a signal, the signal makes corresponding fluctuations in channel current. (This is the way a grid controls current in a vacuum tube.)

The JFET is only one type of field-effect transistor. The other type is called **MOSFET,** an acronym for metal-oxide-semiconductor field-effect transistor. In a MOSFET (**fig. 2A**), the gate doesn't form a junction . . . at least not like in a JFET. The gate is a metal electrode, and between it and the channel is an insulating coating of oxide. The channel itself (an N-channel is the one shown) is surrounded by opposite-type material, called the **base**—or sometimes, substrate.

In a JFET, there may be slight leakage

across the gate-to-channel junction; there's none at all in a MOSFET, because the gate is entirely insulated by the oxide layer. Another name for the MOSFET is IGFET, for insulated-gate FET. The JFET has a high input impedance because a reverse-biased PN junction is the input element between gate and source. The MOSFET input exhibits almost infinite impedance since the gate cannot draw any current at all.

The MOSFET channel current is pinched off by a reverse gate bias. This action in the channel is called **depletion,** since the bias depletes all current carriers in the vicinity of

Another type of MOSFET works only with forward bias. (In that respect, it resembles a bipolar transistor.) The structure of this MOSFET is shown in **fig. 2B.** The source and drain are separate, and the channel doesn't exist at all until a strong forward bias is applied. With no bias, there is no current from source to drain. As forward gate bias is applied, the substrate or base material changes character. The base is P-material, as it would be in an ordinary N-channel MOSFET; the section beneath the gate changes to N-material under the strong electric field set up by the gate. That section becomes the N-channel

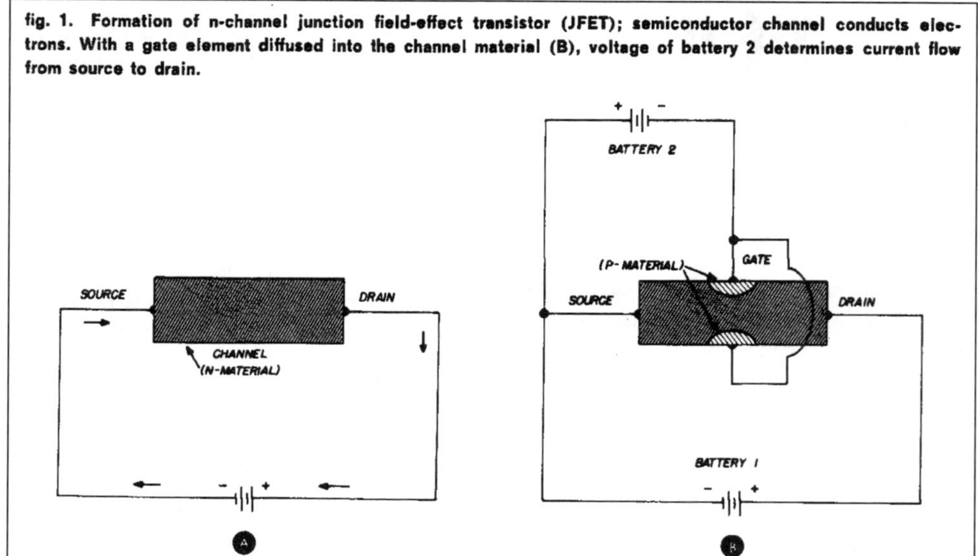

fig. 1. Formation of n-channel junction field-effect transistor (JFET); semiconductor channel conducts electrons. With a gate element diffused into the channel material (B), voltage of battery 2 determines current flow from source to drain.

the gate elements. A depletion MOSFET operates very much like a JFET. Its channel current is highest when there is no bias. Increasing the reverse bias cuts down the current in the channel. Enough bias drives the channel into cutoff—or, as it is called in a FET, into pinch-off.

The gate bias in a JFET cannot be permitted to get near the forward-bias mode. If it did, positive half-cycles of an input signal would make the junction draw current, and distortion would result. A depletion MOSFET doesn't draw gate current under these conditions, and can accept some forward bias. Nevertheless, it works best with some reverse bias.

for this MOSFET. Because gate bias therefore enhances current flow instead of depleting it, the unit is called an **enhancement** MOSFET.

In operation, then, you find the JFET can be operated only with depletion bias and the MOSFET that can be designed for depletion or enhancement operation. Most depletion MOSFET's can work slightly into the enhancement mode, too. A circuit for this kind of MOSFET is shown in **fig. 2C.** Drain voltage is applied through a load resistor. Gate bias, a reverse bias in this depletion MOSFET, is developed across a resistor in the source lead (the way a cathode resistor develops bias for a vacuum tube).

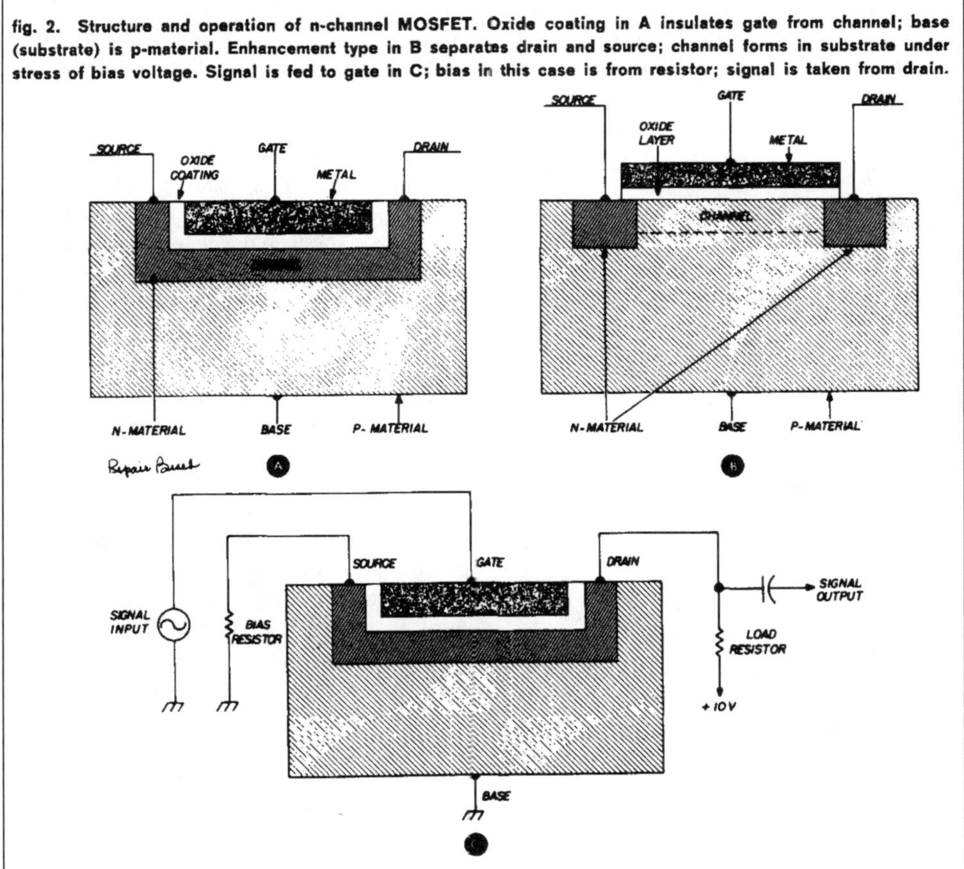

fig. 2. Structure and operation of n-channel MOSFET. Oxide coating in A insulates gate from channel; base (substrate) is p-material. Enhancement type in B separates drain and source; channel forms in substrate under stress of bias voltage. Signal is fed to gate in C; bias in this case is from resistor; signal is taken from drain.

bipolar—unipolar

Terminology sometimes gets in the way of understanding. Special words fog up the very things they're coined to simplify. Conventional transistors, the ordinary junction kind that you find in profusion nowadays, are lately called **bipolar.** There's a reason.

Current flows in semiconductors by means of two kinds of carriers: holes and electrons. In N-material, the chief carriers (called majority carriers) are electrons; in P-material, the majority carriers are holes. In all ordinary junction transistors, the minority carriers contribute considerably to operation. Therefore, the transistors are said to be bipolar.

In FET's, the flow of current is not across junctions, but through the channel—all of which is (or becomes) one type of semiconductor material. Carriers, therefore, are of only one character. In N-channel FET's, the carriers are electrons; in P-channel types, the carriers are holes. FET's are therefore said to be **unipolar.**

If you understand the operation of FET's, as I've already described it, the two terms have little meaning one way or the other. Nevertheless, you should know—if only for reference—that bipolar transistors are the ordinary kind and that unipolar transistors are a family to which FET's belong.

what happens to fet's?

Since FET's are so new, not much data has been gathered on their breakdown habits. Yet, enough is known to make it pretty simple to troubleshoot circuits that use them, and also to test them outside the circuit.

The most common fault in a FET is a short between gate and channel. In a JFET, forward bias for too long a time might damage the junction, or the reverse-voltage breakdown rating might be exceeded with the same result. Either way, the gate can no longer control channel current, and too much gate current flows. In a typical JFET, gate current should be only a few nanoamps; in a MOSFET, there should be none. Gate breakdown in a MOSFET is caused by over-

fig. 3. Simple audio amplifier using a p-channel JFET.

voltage in either direction, forward or reverse. The result is gate current, of which there would otherwise be none.

Excessive gate current in a JFET, whatever the cause, instead of shorting the junction, may burn it open. The effect is to prevent control of channel current although no gate overcurrent is noticeable. A MOSFET gate may open up, with a similar result, although excess gate current won't be the cause.

An open channel can develop in either kind of FET. It is usually caused by too much voltage between source and drain. The overcurrent that results just burns the channel open, usually at the narrow part near the gate. Afterward, no current can flow from source to drain. Depending on where the open is, there may or may not be slight gate current in a JFET that has an open channel; of course, there is never any gate current in a MOSFET, anyway.

The chief thing to guard against, then, is obviously overvoltage. If you're experimenting with FET's, pay attention to the voltage ratings. Next most detrimental is overcurrent. Even slight overcurrent may eventually overheat a JFET junction, resulting in permanent damage.

FET's are delicate in some respects. Even static that builds up on your body can blow the gate junction or gate insulation of a small-signal FET. Because of this, new FET's are "zot-proofed" by the simple expedient of twisting their leads together. You can keep them safe while you handle or install them by sliding a ring of solder or soft wire around all three or four leads. Bend the leads slightly outward to hold the ring in place. After you have the FET installed, clip or melt off the zot-ring.

finding faulty fet's

There are three ways to track down a FET that has given up. One is by signal injection or signal tracing—methods I discussed in the April and May 1968 issues. You simply pin down the faulty stage and then test it to see if the FET in that stage is at fault.

A second way is by checking voltages in the circuit with a high-impedance voltmeter. (A vom won't do; it loads down the circuit too much.) Once you know how a FET fails, you can figure out the dc voltage changes that result. Thus, measuring circuit voltages is another means of troubleshooting FET stages.

The third way is by removing the FET's from their circuits and testing them, one at a time. If you choose this method, take care to zot-proof the FET before you take it out of the printed board or socket. Also, use heat-sink pliers to unsolder and resolder; FET's are heat-sensitive, too.

You can see one ordinary FET circuit in **fig. 3.** It's an audio amplifier, using a P-

fig. 4. MOSFET supply circuits. Depletion device in A unit gets its bias from a source resistor. Enhancement type in B is biased by resistive voltage divider network fed from regular dc supply.

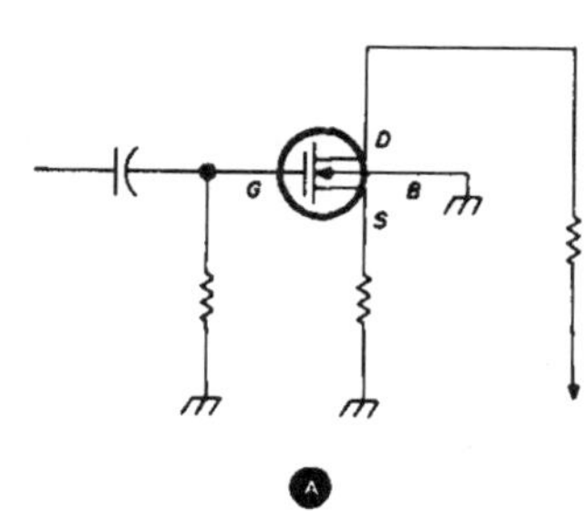

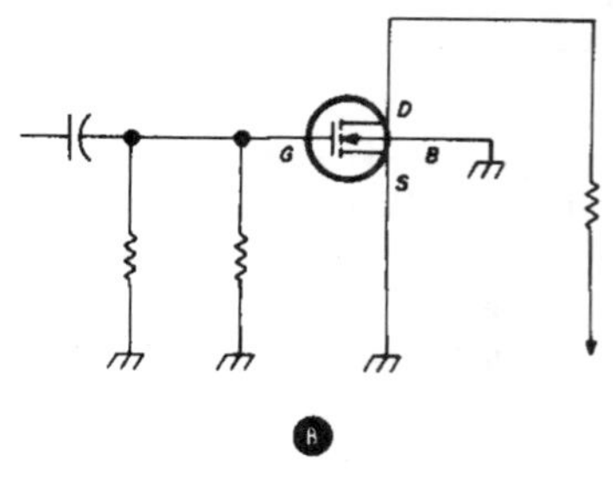

channel JFET (you can tell it's P-channel because the arrow points outward). Gate bias is developed in the source circuit, across the 1.5k resistor.

Suppose you are measuring voltages in this circuit and the drain voltage is −48 volts on your meter. Obviously, the FET isn't dropping supply voltage across the 100k resistor. You'd also find almost no voltage at the source terminal.

Suppose there is a very slight negative voltage at the gate terminal. You know that either the input coupling capacitor is leaky or the FET has gate-to-channel leakage. If the leakage is in the capacitor, and persists, the FET could be damaged. If in the FET, replace it. To find out which, disconnect the capacitor; if the voltage returns to zero, the capacitor was at fault.

Suppose the drain voltage is low . . . say, only −5 volts. The 100k resistor could have increased in value, but the FET is more likely drawing too much current. If so, the source-terminal voltage should be high. With the bias thus high, the gate should theoretically keep the FET from drawing much more than normal current; it might be open and having no effect. Finally, with drain voltage low, you might find source bias removed entirely or partially—probably the result of a shorted bypass capacitor or 1.5k resistor.

All those symptoms you could reason out for yourself, knowing the nature of FET breakdowns as you now do. Faults in MOSFET circuits present slightly different symptoms, but only because of the difference in structure. A simple MOSFET amplifier stage is shown in **fig. 4A.**

If you measured too much voltage at the drain of this N-channel MOSFET, it would likely be because the channel is open or the gate is overbiased. If it happened to be an enhancement-type MOSFET (**fig. 4B**), even zero bias might cause the lack of current through the channel; you recall that an enhancement MOSFET needs forward bias to cause curent flow. Zero bias in either kind could be due to an open gate element. Only an enhancement MOSFET would need the bias arrangement in **fig. 4B;** either resistor might upset bias. A gate-to-channel short near the source end of the channel would eliminate bias, as would a gate-to-base short. A gate-to-channel short nearer the drain might cause excess current through the drain-supply resistor and thus lower the voltage measurable at the drain terminal.

fig. 5. Resistance readings you can expect with different types of n-channel FET's. Readings are the same for p-channel types except that the "diode" readings are reversed.

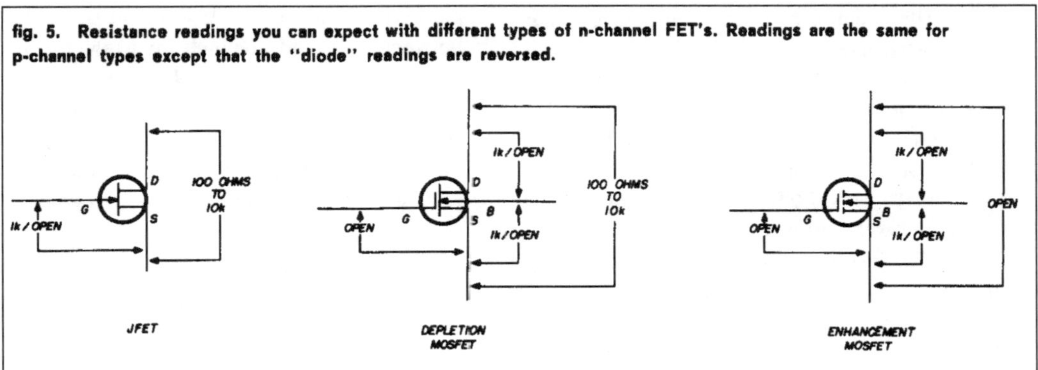

These are far from the only possibilities. Any incorrect dc voltage in a FET circuit can be reasoned out. You may find it quicker, though, merely to determine that the trouble is not in an associated component. Then you can remove the FET from the circuit for testing.

There aren't many FET-checkers around yet, but your trusty ohmmeter is an excellent tester provided its battery voltage isn't too high. The 1.5-volt type is best.

Diagrams that will help you check different FET types are given in **fig. 5.** The readings you should get are indicated clearly. Each type has it own peculiarities.

The source-to-drain readings of the JFET and the regular MOSFET are similar. You can connect your ohmmeter leads in either polarity; the normal reading lies between 100 ohms and 10k, depending on the particular FET. Source-to-drain resistance of an enhancement-type MOSFET is something else; because of the way the source and drain junctions are applied to the channel, they act as back-to-back diodes, and therefore you read an open between the two.

Both MOSFET's show open circuits from the gate to any other element, since the gate is insulated. The JFET, from gate to source, reads like a diode: about 1k in one direction and open in the other. A low backward reading indicates leakage. Open both ways means the gate is open. The JFET shown is an N-channel; the only difference in a P-channel JFET is in the polarity of the gate-to-source readings—they are just reversed.

The drain-to-base and the source-to-base resistance of both MOSFET's are similar in that they measure like diodes: about 1k in one direction and open in the other. Which direction is which depends on whether the MOSFET is N- or P-channel. All types act as back-to-back diodes, though. If the negative ohmmeter lead is on the base of a particular MOSFET and the source checks about 1k, so should the drain. If it checks open in that direction, it should check open to the drain also.

little oddities

There you have the basics of checking most FET's. Not that those shown are the only types—they're not. All those I've talked about this month are symmetrical types. That is, their gates are about half-way between source and drain. In most of them, drain and source are interchangeable. You can connect either end to function as either source or drain.

Nonsymmetrical types have the gate located elsewhere on the channel, near one end or the other. Ordinarily, the gate is placed near the source. Any faults that occur between gate and channel are likely to tie gate and source more closely together. The difference in result, as far as circuit voltages are concerned, may be noticeable. However, the resistance tests shown for symmetrical FET's are equally applicable to nonsymmetrical types.

trouble shooting by resistance measurement

In a recent **repair bench** column* I said you should only work on transmitters with power off. I suggested checking resistances to find the trouble. Since then, readers without repair experience have asked how to go at this kind of trouble shooting. No one has argued the advice, but many don't see the approach.

The best excuse for using resistance measurement as a fault-hunting method is safety. You can check out most of a transmitter without any lethal voltages applied. Sure, for a few tests the transmitter or receiver has to be "live." But so many tests can be handled more safely with power off, there's just no excuse to ignore these advantages.

A few doubters may contend that the method isn't complete. Certain parts in high-voltage circuits break down under stress, yet test okay by resistance measurements, they say. That's true. But you soon learn to spot those troubles right away—usually by their smoke. Also, certain shorts and arcs in transmitters occur only after high voltage and drive are applied. The way to track them—safely—is by a process of elimination, part by part . . . but that's a story I'll save for a later column.

Resistance trouble shooting can be done with a simple volt-ohm-milliameter, or vom. That's a major advantage; other techniques demand fancier instruments. A vom is simple to operate, portable, and usually inexpensive. The main criterion is that the ohmmeter readings be accurate.

A vom has an advantage over a vacuum-tube voltmeter, even when the vtvm is more accurate. Since a vtvm must be plugged into the power line, it

Adjusting the mechanical-zeroing screw.

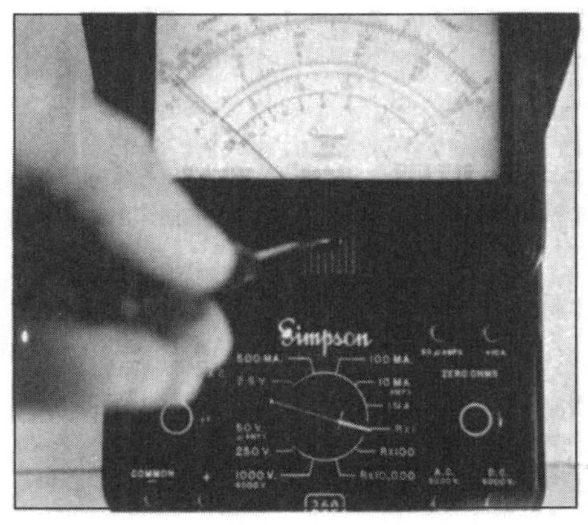

* *repair bench*, August 1968, page 52.

can introduce problems in certain equipment. With the vtvm common lead connected to some non-ground point in a transmitter or receiver, you may get some faulty readings. A vom is usually battery-powered; there's no power-line connection. You can get better readings without worrying about "unseen" ground paths caused by a test connection.

The needle should rest somewhere near midscale for best accuracy during tests.

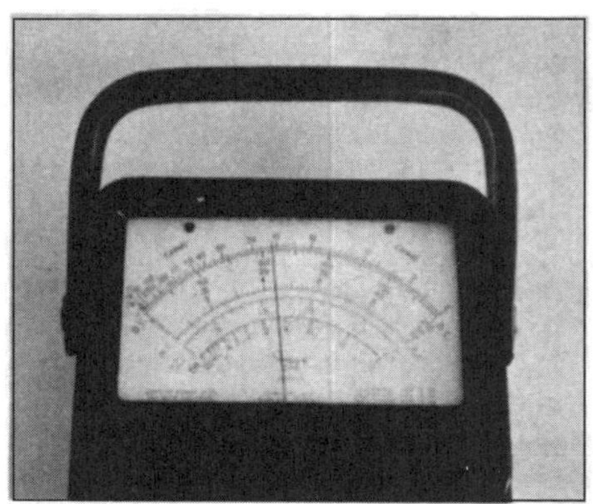

normally. That takes care of calibration.

A second aid to accurate resistance measurements concerns the range switch and the multipliers you use with the scale readings. After you connect the test leads across the resistance you're measuring, rotate the range switch to a position that makes the pointer rest somewhere in the middle section of the scale. That's where

The needle should move up scale when the test leads are clipped together.

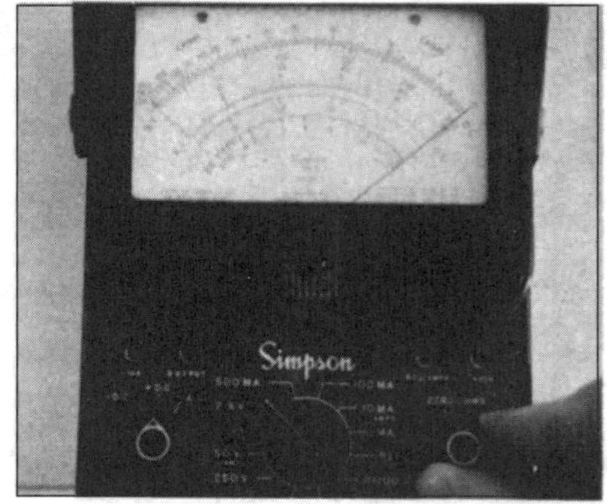

reading the ohms scales

Dependable ohmmeter measurements are useless if you don't read them accurately. You can't even go on to the next step—interpreting their meaning—unless you get the numbers right. Three aids will help you read your ohmmeter correctly.

First of all, **calibrate** the instrument. Place the vom upright or lying down, in whatever position you'll use it. (Position will affect vom readings.) With the test leads plugged in but not shorted together, make sure the meter pointer rests exactly over the left edge of the scale. You may have to adjust the mechanical-zero screw that is just below the meter face. Tap the meter glass to jar loose any pointer friction.

Next, clip the two test leads together, making zero ohms between the tips. The pointer should move up to full scale. Zero ohms is at full scale since the ohms-scale numbers go right-to-left. Again, tap the meter glass to be sure the pointer rests

readings are the most accurate because the numbers there are easy to read. You can see that they are, if you examine the ohms scale in the photos; it's the top row of numbers.

For a third aid to accurate readings,

Pay close attention to the multipliers on the range switch.

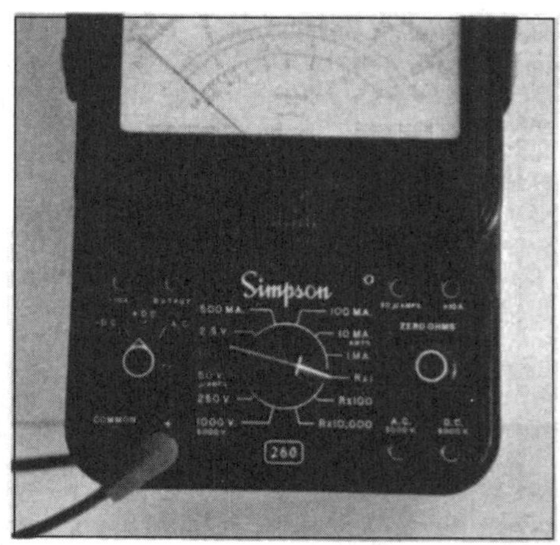

make sure you multiply the scale reading properly. Look at the multipliers on the range switch. Pay close attention to which one you set the switch at. Then multiply whatever number the pointer stops at by the multiplier from the switch. Do it carefully. It's awfully easy to drop a zero or add one and end up with a wrong reading. Use scratch paper to be sure. Whenever the pointer stops toward the left end of the scale, you'll have zeros in the scale reading as well as in the multiplier; be sure you get those right, too.

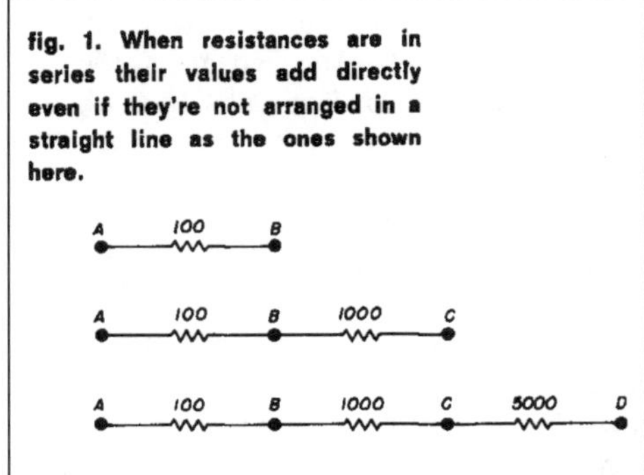

fig. 1. When resistances are in series their values add directly even if they're not arranged in a straight line as the ones shown here.

simple series paths

What do you do with those accurate resistance readings once you have them? Analyze them, of course! But that brings up the next problem: How?

What you do first is learn to recognize the different kinds of resistive paths. There are plenty of them. Every ham receiver and transmitter has dozens of resistors. Besides them, many, many other components show readings on an ohmmeter. For example, the heaters of tubes; the windings of transformers; the leakage of electrolytic filter capacitors; the forward and backward effects of diodes; and, of course, transistors, which produce some readings that can really fool you. Resistive paths exist all around the chassis.

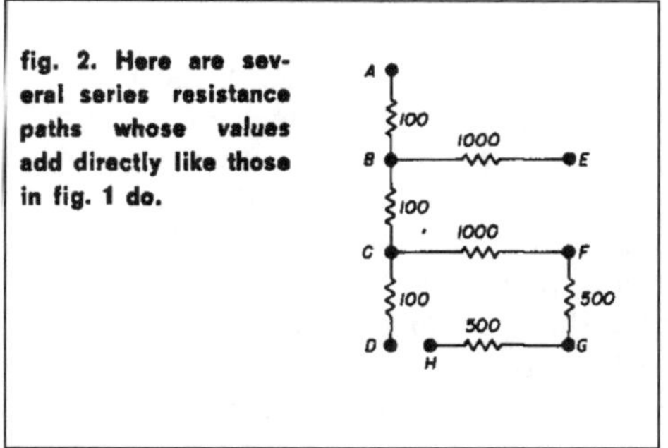

fig. 2. Here are several series resistance paths whose values add directly like those in fig. 1 do.

The trick of resistance trouble shooting is finding paths that aren't where they should be or paths that have too much or too little resistance. To do it, you'll have to know how to spot the paths on diagrams and figure out where and what they should be in the chassis.

The simplest, both to spot and to analyze, are series paths. **Fig. 1** shows a bunch of different kinds. The A-B path shown first can't very well be mistaken; it's 100 ohms, between points A and B.

In the second series path shown in **fig. 1,** the resistance between A and B again is 100 ohms. In series with the A-B path, however, is a path from B to C. Again, the over-all circuit is easy to figure out; the A-C path is the sum of the A-B path plus the B-C path. Resistance in the A-C path totals 1100 ohms.

The third series path is A-D. This one, too, is pretty easy, even if it does have three series resistances. Add all three of them, and you have a total A-D resistance of 6100 ohms.

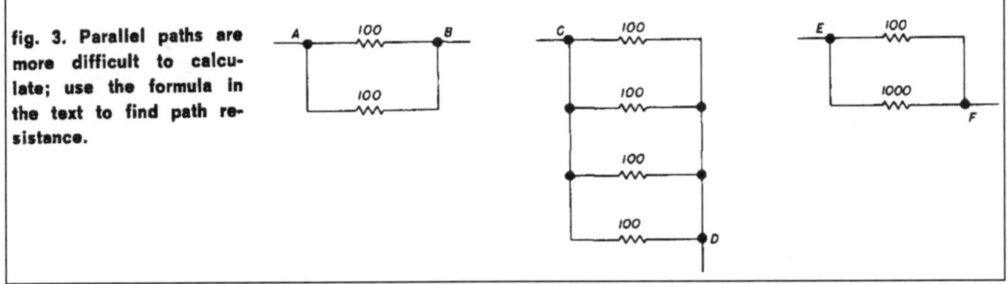

fig. 3. Parallel paths are more difficult to calculate; use the formula in the text to find path resistance.

It's important that you recognize another relationship from **fig. 1.** Connecting the B-C and C-D paths had no effect on the A-B path; in all three cases, an ohmmeter connected to A and B measures 100 ohms. You can add a dozen more paths *in all directions*, but A-B remains 100 ohms—as long as all the paths are added **in series.**

Another thing. You can measure path B-C independently, and path C-D independently. The other series paths don't interfere with measuring either one. Also, you can measure path B-D, if you want to, getting a direct 6000-ohm reading between points B and D.

Take a look at **fig. 2.** Paths don't have to be drawn in a straight line to be in series. Nor do they have to be connected directly to each other. An ohmmeter connected to A and D measures the A-D path directly; the resistance is 300 ohms. None of the other resistances is of any consequence to that particular path, because they are not in series with it.

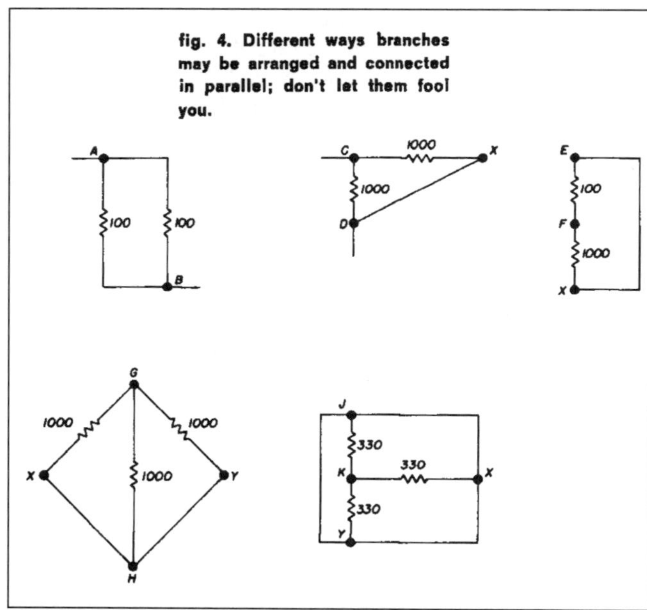

fig. 4. Different ways branches may be arranged and connected in parallel; don't let them fool you.

What about path A-H? Only the resistance **between** A and H have a bearing on the resistance of that path. An ohmmeter connected to A and H measures only the resistances between them. The resistances of path B-E and path C-D are ignored because they're not in series with path A-H. Path A-H measures 2200 ohms. Trace its path—from A to B, to C, to F, to G, to H.

Trace the path from E to G. It goes from E to B, to C, to F, to G. An ohmmeter connected between E and G measures only that path, and indicates 2600 ohms.

The D-H path is 2100 ohms, the B-F path is 1100 ohms, and the F-A path is 1200 ohms. Because these resistances are in series, they're easy to measure and easy to analyze.

Suppose you're trouble shooting a circuit like **fig. 2** and measure 2700 ohms between B and G. The schematic shows series resistances totaling only 1600 ohms in the B-G path. At least one resistance has changed value, and you have to *figure out which one.*

One way is to measure each resistance individually. There may be several in an actual circuit, and that can take a lot of time. A better way is to work your way toward one end or the other. Leave one ohmmeter lead on B and move the other one from G to F. If the reading drops to 1100 ohms, that means the B-F path is okay, so the F-G path must be at fault. Suppose, though, the reading only drops to 2200 ohms. That means F-G is okay; the 500-ohm change tells you so. The trouble is therefore in the B-F path.

Move the test lead from F to C. If the reading drops to 100 ohms, B-C is okay. You can then move both leads to measure path C-F, just to be sure. If it measures 2100 ohms, you've found the trouble; it should measure only 1000 ohms.

Those are the principles. As long as paths are in series, tracking down trouble is a cinch. Plenty of resistive (and continuity) paths in ham equipment are series types. With them, you'll have an easy time. However, the other kind can be a real pain.

resistive paths in parallel

The simplest kind of parallel path is shown in **fig. 3.** Parallel paths are hard to describe. As far as an ohmmeter is

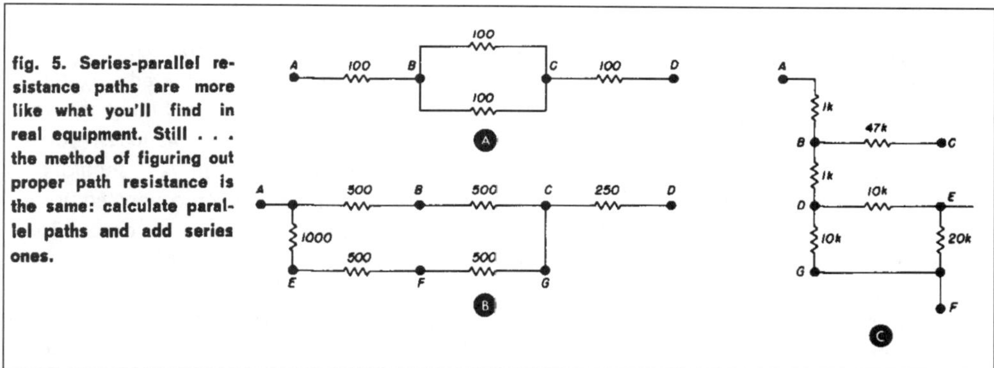

fig. 5. Series-parallel resistance paths are more like what you'll find in real equipment. Still . . . the method of figuring out proper path resistance is the same: calculate parallel paths and add series ones.

concerned, the path between points A and B is merely path A-B. As you can see from the diagram, however, it isn't that simple. There are two paths. One is from A to B . . . and, yet, so is the other! You'll find several paths like this in electronic equipment, and very often they aren't at all obvious on the schematic diagram.

Resistances in parallel add inversely (another word for upside-down). Put two equal-value resistors in parallel and the resistance of their path is only half the value of either one alone. Take path A-B in **fig. 3,** for instance. The two 100-ohm ressistors in parallel make a path measuring only 50 ohms; that's what an ohmmeter reads. The C-D path, with four 100-ohm resistances in parallel, measures only 25 ohms with an ohmmeter.

From this you can figure that adding a parallel path always lowers the ohmmeter reading between two points. Parallel path resistance is, in fact, less than the lowest value of resistance in any one branch of the parallel path. The E-F path in **fig. 3** shows a situation where one branch has a much lower resistance than the other. Calculated by the formula for parallel resistances,* path E-F has a resistance of 91 ohms. That's what an ohmmeter measures between points E and F.

Consider what it means, then, if you measure path E-F and find a reading of 100 ohms instead of 91. You can assume the 1000-ohm resistance has been removed, is defective (open), or has changed value to some extremely high resistance. On the other hand, if the reading is 1000 ohms, the 100-ohm resistance is at fault. Get the idea?

$$\frac{1}{R_T} = \frac{1}{R_1} + \frac{1}{R_2} + \frac{1}{R_3} + \cdots \frac{1}{R_n}$$

odd parallel connections

The trouble with parallel paths in typical ham gear is that they are seldom as obvious as those in **fig. 3.** For example, they could be arranged in all sorts of odd shapes, like some of those in **fig. 4.** Strange as they may appear, these diagrams are all of parallel resistance paths.

A-B is easy to recognize; it's like the one from **fig. 3,** only turned up on its side. Path C-D looks at first glance like a 1000-ohm path. Follow path C-X, though, and you'll find that X and D are the same point insofar as the circuit is concerned. The C-D path is therefore a parallel one involving both resistances. The true resistance is 500 ohms between C and D; that's what an ohmmeter reads.

Consider the E-F-X path. The two resistors appear in series; but they're not, because the jumper connection makes E and X electrically the same point. So, any current from E flows through both resistances to reach F. The true path is E-F (or F-E if you prefer). An ohmmeter at E and F measures the combined resistance of the two—combined in parallel, not in series. The path from E to F measures 91 ohms.

Path G-H is fairly easy to see, now that you're looking. The path is through three parallel resistances. Points X and Y are mere connections, and are the same (electrically) as point H. Combining the three resistances by the parallel-resistance formula, you can figure out the net resistance of path G-H: 333 ohms. An ohmmeter shows it to be around 330 ohms.

Whether it looks that way or not, path J-K is a parallel path through three branches. J, X, and Y are tied together by several jumpers, and can all be considered point J. The net J-K resistance is 110 ohms.

The point of studying odd resistance hookups like those in **fig. 4** is that you won't overlook a parallel connection just because it isn't obvious. In fact, many you'll encounter in actual equipment are even less obvious. Some are through components not drawn with resistor or wire symbols. In every case, keep in mind that a **parallel** path, whether visible on the schematic or not, always **lowers** the ohmmeter reading to some value **less than** the lowest value in any branch of the parallel path. If you forget that fact, resistance trouble shooting can be confusing; remember it, and you can use the technique to pin down some very elusive faults.

series-parallel combinations

A combination of paths is what you'll find most in equipment circuits. Some parts form series resistance paths, and some form parallel ones. You'll find them in all kinds of arrangements. A few possibilities are sketched in **fig. 5,** just to give you an idea.

In diagram **A,** paths A-B and C-D are in series with two-branch parallel path B-C. To know what resistance to expect between A and D, you first calculate the resistance of path B-C; that's 50 ohms. With that settled, it's easy to figure the rest: all paths are in series. A-B is 100 ohms, B-C is 50 ohms, and C-D is 100 ohms; the total is 250 ohms.

Fig. 5B is a little different. The paths are complicated. The path from A to C is made of two parallel branches, both of which have several series resistances. To find out what each branch should measure, solve one at a time. Path A-G totals 2000 ohms. Path A-C totals 1000 ohms. Points G and C are jumpered together, so are the same point. Combining the resistances of both parallel branches, you'll find the resistance of path A-C should be 667 ohms; an ohmmeter connected to A and C measures 660 or 670 ohms. The whole path, of course, extends from A to D. Combining path A-C with path C-D is simple addition, since they're in series. Path A-D should measure slightly over 900 ohms.

Fig. 5C shows what you might find in a real circuit. You can figure out what your ohmmeter should measure from any convenient point to any other point. Suppose F (and G) is ground; that's usually a good place to clip one test lead. Path F-E should measure about 10,000 ohms (10k), because the G-D-E path is in parallel with it. Path F-D should measure 7500 (30k in parallel with 10k). The path from F to B includes F-D in series with B-D and should measure 8500 ohms. The F-C path includes F-B as well as B-C and thus totals 55.5k (47k plus 8.5k). From F to A totals 9500 ohms or 9.5k; path B-C doesn't enter into the F-A path at all.

next month

You can see from all these diagrams that trouble shooting by resistance measurement is based entirely on figuring out from the schematic diagram just what a resistance path should be and then measuring it with the ohmmeter. If the resistance isn't what you expect, you isolate the one that's wrong. There, you'll find the trouble.

Principles alone aren't enough to make you feel at home with the resistance-measurement technique. In the next column, I'll show you some real trouble-chasing with this method. I'll show you how to track down parallel-series resistive paths, and how to use resistance tests to pinpoint specific parts troubles. By the time you're through with this subject next month, you'll know how to find faults in any kind of ham equipment with no power applied at all—the safest way you can troubleshoot.

the repair bench

by larry allen

high-power troubleshooting —keeping alive

On nights when the bands are open in all the right directions, and you're hearing reports that you're getting into Deep Zamba-Zamba-Land with a five-nine signal, and the XYL comes by the shack with fresh coffee and tells you she thinks you **should** buy that new beam, life is really worth living. Why take a chance of spoiling it? Operating high-power has responsibilities as well as pleasures. The key responsibility is preventing it from doing any harm—to you, to your family, and to visitors.

If you housekeep the shack like you should, and designed and built it well to start with, the last two mentioned are well taken care of. Danger arises when you—lord and master of the domain—get careless during troubleshooting. Let's face it. When you haul a 2-kW linear up on the bench and open it up, you're dipping into a powerful piece of machinery.

what the dangers are

When testing a high-power transmitter, or its power supply, you have three dangerous voltages to contend with. One is the primary supply. It may be "only" 115 volts of house ac, but it can be the most dangerous voltage in the equipment. Another is the high-voltage dc that supplies the plate of the power amp. Not only is it several thousand volts, but the supply has a powerful current capability—and it's current that does the killing. Third, and just as important, is the rf voltage developed in the final tank and along the antenna feed line. Rf is usually confined to a well shielded cage; but, remember, we're talking about when the unit is on the repair bench —you may take the lid off for some reason.

Each of these three dangerous voltages behaves differently. The primary power, usually 115 or 230 volts ac, has an awesome ability to push current through your body. Once contacted, it can contract your muscles so tightly you can't turn loose, haul your heart up motionless, and hold your lungs powerless to breathe. You can die of asphyxiation as easily as from heart fatigue.

An acquaintaince of mine once survived a tangle with a shorted electric drill on a TV tower. He was bound up so tightly by the current, he couldn't release the trigger; he couldn't even yell. He did manage to grunt, and someone on the ground yanked the plug. Just plain lucky, he was. At that, he blacked out and hung upside-down for a while on the tower, and spent some time in the hospital overcoming shock. Rest assured, you had better respect that common old "everyday" line voltage.

Just as deadly is the high-voltage dc inside the power supply or transmitter. A fellow I had only met once was killed by a 3,000-volt plate supply. He was troubleshooting the transmitter modulator. The trouble was an overload, and he had to cheat the interlocks **and** hold in the overload relay. He took all the normal precautions for working on live gear; he was a trained broadcast engineer. But, he lost his balance. One hand hit the transmitter cabinet and the elbow of the other arm landed on the 3,000-volt terminal of a coupling transformer. The horrible jolt to his body, heart, and lungs was too much. Revival was impossible.

I'm not trying to scare you, although if that's what it takes to teach you respect for transmitter voltages, then you'd better be scared. Safety in high-power equipment is a serious subject—you bet your life.

The rf voltage in a high-power transmitter has characteristics very different from line ac or power-supply dc. You don't even have to touch the wire or terminal carrying high-energy rf; it can jump out and burn you badly if you even get close.

If you're in good health, the rf voltage from a ham transmitter might do no more than give you one of the nastiest burns you can imagine. But high-power rf has been known to kill, and you shouldn't take the chance. Stay away from it. That isn't always easy, for reasons I'll explain, but you should learn how.

general precautions

The most obvious step to protect yourself from high voltages is to keep the equipment **off** while it's on the bench.

"That sounds nuts," you say. "How can you troubleshoot it?"

Well, in most cases you can. Practice using **resistance measurements** to guide you to the trouble. You can do that with the unit turned off. There aren't really so many parts and dc paths in the average high-power transmitter or linear. It might turn out to be faster than going to all the trouble of wiring up the high-power unit for operation on your bench.

If you're convinced you have to fire up the equipment on the bench, first make sure there are no bare wires trailing anywhere. Use exactly the neat installation-type wiring you'd use if you were installing it permanently. The transmitting antenna lead must not be exposed. To "haywire" a test setup is courting danger. The primary power wiring must be through a cable and plug just the same as in a permanent installation; alligator-clipped connections won't do. Be especially careful of the wire carrying the high voltage from the power supply to the transmitter. It **must** be one solid piece, and should be the highest-quality ignition wire you can find.

Try to do your troubleshooting without cheating the interlocks. They're included to automatically disconnect primary power from the high-voltage supply when the lid or door is opened. Sure, you may have to open up the unit to reach test points. But it's dangerous to cheat interlocks. Again, the old re-

fig. 1. Protective bleeder resistors in a typical high-voltage power supply. R1 is an added safeguard, and the switch is an interlock feature to discharge the power supply if the cage is opened for service.

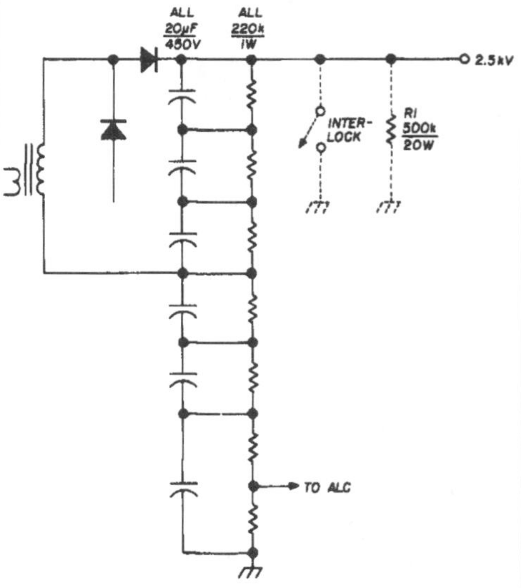

sistance-measuring technique may save your life. When you have the interlocks cheated, and power is applied to all circuits, the open equipment is at its most dangerous. The primary ac, the dc high voltage, and the powerful rf energy are all right there waiting to zap you the instant you get careless.

Even with the set turned off, you're still not safe. The better equipment has built-in safeguards against some of the voltage hazards that lurk in a dead transmitter, but they can't always be trusted. (Suspicion is a useful companion where high voltages are concerned.)

For example, filter capacitors in the high-voltage section can hold a body-jolting charge for days and weeks after a set is turned off. To prevent this, bleeder resistors are almost always included. **Fig. 1** shows a high-voltage doubler circuit, with bleeders. What happens if one of them opens? The charge won't drain off. Safer transmitters include another resistor as bleeder for the whole supply (R1 in **fig. 1**). The safest units further include a switch that shorts out the high-voltage dc output whenever the protective shields are removed or the housing lid is opened. This is in case the primary interlocks are cheated, purposely or accidentally. The discharge switch also protects in case bleeders aren't doing their job.

Still, the watchword is: **don't trust them.** All these devices are fine, unless for some unusual reason they aren't working. It only takes once to ruin you. So . . . use a grounding stick, ALWAYS. This is something you can make. A long bakelite or fiberglass rod is best (**fig. 2**). Screw a small hook—stainless steel or aluminum—into one end. Attach a 2-foot piece of wire to the hook, making very sure of good electrical connection. To the other end of the wire, which can be heavily-insulated ignition wire, solder a good, heavy-duty clip.

When you open up a transmitter or power supply, clip the grounding wire to a **bare** chassis spot (paint insulates). Hold the rod at its free end. Then touch the hook to all bare wires or terminals that you have any slight suspicion may have carried dc voltage. Finally, hang the hook over the main high-voltage terminal—usually at the power-supply output. Leave it there until you are finished troubleshooting. That gives you protection just in case you or someone else accidentally breaks an interlock.

Another precaution, often overlooked on the repair bench, is grounding. A ground from the chassis to a cold-water pipe or to an 8-foot copper ground rod is as important on the bench (maybe more so) as at the operating console. Without it, the chassis may become electrically hot with respect to ground, and offer a dangerous situation. The chassis might also, when you fire up the transmitter, take on some of the rf energy; that, too, is dangerous. Make a point of grounding the chassis carefully.

the live ones

Sooner or later you're going to insist on troubleshooting a transmitter **live.** It may even be necessary. You'll be tampering with dangerous stuff, though, and should act accordingly. There are some protectice measures you can take.

First of all, READ THE INSTRUCTION BOOKLET. If you won't or can't, you have no business working on dangerous high-powered equipment. The manufacturer of your transmitter or power supply is well aware of the danger points, and will likely have printed cautions in the manual. Study them. Compare the schematic diagram with what you see in the chassis. If you have **any** doubts about where a test point is, trace it down with the unit turned **off** and with all the earlier precautions. Then you won't have to probe around later hunting for it. A live transmitter is no place to "learn" where the test points are.

Next, rehearse each test you're going to make. Sound silly? Nothing's silly that might save your neck. A written list of what you're going to do and what you expect to find will prevent that "hunting around" that breeds carelessness. A dress-rehearsal run-through, with the set turned off, will give you the confidence to go straight ahead with your plan when the power is on. You can't afford many mistakes—maybe none.

Once you start, the old ploy about keeping one hand in your pocket is still a good idea. It's based on the theory that you won't complete a circuit to the chassis and therefore won't get bitten by voltage. That theory doesn't always work, because you can contact two points of voltage with one arm or hand. However, it can help keep the current from passing through your rib cage, and might very well save your life even if you get a nasty jolt.

Standing on a rubber mat is another good idea, especially if your floor is concrete. Again, it may not prevent your getting shocked, but it could keep the effects to a minimum.

fig. 2. Grounding stick for discharging high-voltage circuits. Heavy insulation on wire keeps it from contacting any circuit but the intended one.

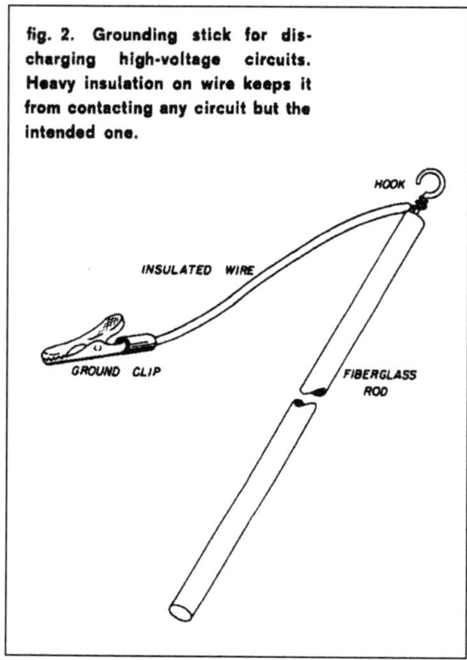

In tuning a high-power transmitter, you may have to change taps on a coil. To make each change, shut down the transmitter, hook the grounding stick to the coil (other end of the wire clipped to chassis, of course); make the change; remove the stick; then fire up the transmitter again. This takes a little extra time, but not enough to matter when your health is at stake.

When messing around with tuning, keep away from that rf. It can reach out in some of the most unexpected ways. I saw a guy "testing for rf" (he said) by bringing his screwdriver blade near different parts of the plate coil. He judged by the little rf arc he could draw. He laid the screwdriver out of reach, so he tried it with a pencil. The burns he got on three fingers and a thumb took three weeks to heal. He forgot the pencil-lead ran all the way up through the pencil. Good thing it was only a 250-watt transmitter. The only safe place for your hands in regard to a radiating transmitter is away from anything that even looks like a coil.

You have to watch how you handle tools, too. Not only can you damage your transmitter, you can damage yourself. Metal tools in a powerful rf field will overheat quickly, even though many inches away from direct contact. Keep them out of live transmitters. If you have an adjustment to make inside a radiating transmitter (not a very good design, and not likely in commercial units), use a plastic tool; plenty are available.

the bench itself

This is another factor worth consideration if you expect to work on a high-power transmitter. The bench should be wood; metal just isn't a good idea. Receptacles to fit various power plugs are important; you don't want to get tangled up in any jury-rigged power connections.

The all-important ground connection mentioned earlier is most handy if it terminates right there at the bench; that makes it quick to connect to. The wire lead to the pipe or the ground rod should be as short as possible and of the heaviest wire you can round up.

A well-protected antenna connection is a good idea, if you can't manage a dummy load. Coaxial cable is always best, but not feasible at all frequencies or with all transmitters. Just make the terminations at the rear of the bench space, so you don't accidentally touch them while the transmitter is radiating.

Finally, the bench should have an accessible switchbox that shuts off everything. You never know just when you might want to close down the whole operation suddenly, or need someone else to. Make it easy. One precaution: whoever can turn it off for you could also turn it on when you least expected or needed it. Add extra safety by having a hidden switch in series with the main one; when you want the bench to stay off, open them both.

the repair bench

by larry allen

analyzing wrong dc voltages

There isn't really a "best" way to troubleshoot electronic equipment. The most successful radio and electronic technicians combine several methods. I've already told you about some of them in past columns, but this month's method is probably the most popular; at least it's one you have to understand if you expect to repair every trouble in all sets. It's called **voltage troubleshooting.**

Just because this method is the best known doesn't mean it's always best. Most experienced guys save this one for last. It's a quick way to find the faulty part in most circuits, once you know which circuit to look in.

You can find which circuit this way too, but other ways are usually quicker. I'll use this month's column to tell you how to pin down faulty parts within the circuits; that's the best way to use voltage troubleshooting.

how the method works

In theory, it's simple. You start by measuring each dc voltage in the suspected circuit. Then compare yours with correct voltages. Most instruction books and schematics nowadays include voltages—at least at each tube or transistor terminal. When your circuit diagram doesn't show the working voltages, use past experience and your knowledge of how the circuit works.

When you find a voltage that isn't right, figure out what could cause it. If you know Ohm's law for voltage, current, and resistance, it's not too hard to decide what's causing the voltage to change. In fact, it's really cut-and-dried.

Here's an example. In **fig. 1** are two versions of a simple series resistance circuit connected across a dc voltage. The first, **fig. 1A,** shows normal dc voltages (which I'll call operating voltages). The other, **fig. 1B,** shows voltages you might measure in this circuit when you start troubleshooting. (Voltages are always measured with respect to ground.)

There's 150 volts at the R2-R3 junction in **fig. 1A** because current flows through R3 and R4, causing (says Ohm's law) a drop of 100 volts. There's only 10 volts at the

fig. 1. Simple series resistance circuits show how voltages divide according to the various resistance ratios.

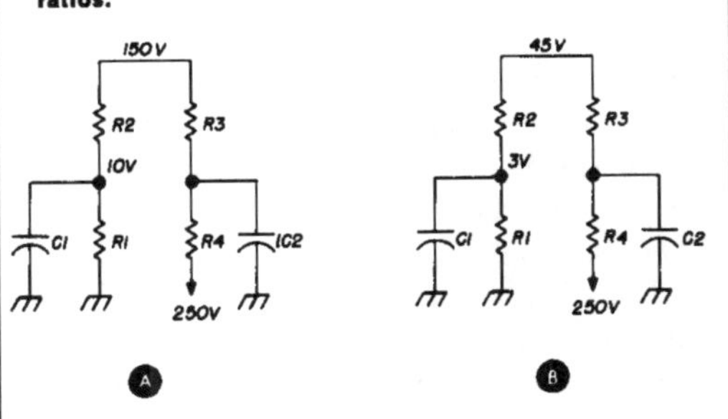

R1-R2 junction because another 140 volts is dropped across R2. The same amount of current flows in all four resistances, since they're in series, so the voltages divide up exactly according to the resistance ratios.

You analyze voltage to determine why the two voltages in **fig. 1B** have changed from normal. What part has gone bad? For some reason, the resistance ratios must have changed. Your first step, then, is to figure out what the new ratios are.

If you consider the 45 volts, you can quickly see two possibilities. The first is that R1 or R2 may have lowered in resistance. In that case, the ratio of their resistance to that of R3-R4 lowers, and less of the total voltage appears across them. On the other hand, the combined resistance of R3-R4 may have increased. That causes more voltage to drop across them and lowers the voltage reaching the top of R2.

So . . . which is it? The clue lies in the fact that **both** the R1-R2 voltage **and** the R2-R3 voltage have kept the same ratio between them, even though both are lower than normal. The 3 is to 45 as the 10 is to 150; the ratio is still 1 to 15. The trouble is more likely in R3 or R4; the value of one of them must have increased. To find out which one, you could disconnect them and measure with your ohmmeter. However, further voltage measurements will show one of them still retains its ratio to R1-R2; the other is the culprit.

This simple reasoning is the basis for analyzing all dc voltages in series circuits. You study the ratio of resistances, compare the ratio of voltages, and then figure out what's causing the foulup.

voltage dividers and ohm's law

You should already know about series and parallel circuits.* For voltage troubleshooting, you must know how voltages spread themselves around both kinds of circuits.

Start with the simple series circuit in **fig. 2A.** The resistors divide the voltage in proportion to their values. The ratio is 2:2:1. The R1-R2 junction is 80 volts below the supply voltage, and the R2-R3 junction is 80 volts further below supply voltage. Looking at the divider string from the standpoint of the voltmeter: R1 drops 80 volts, R2 drops 80 volts, and R3 drops 40 volts. The 2:2:1 ratio is maintained.

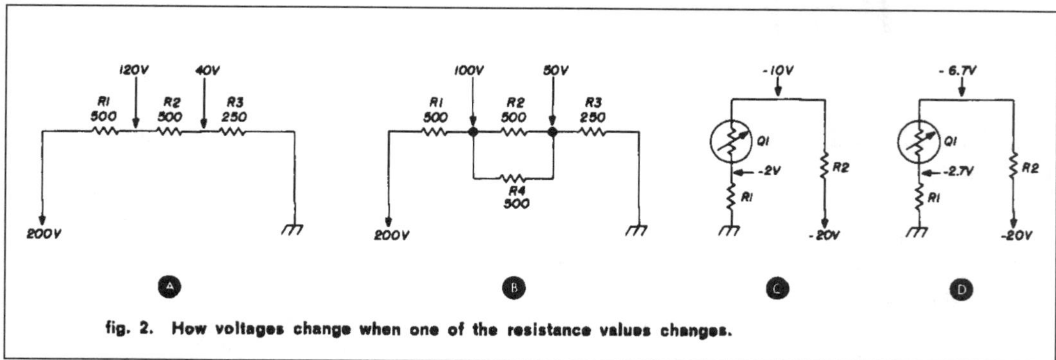

fig. 2. How voltages change when one of the resistance values changes.

Look what happens when you alter a resistance value, as in **fig. 2B.** Some additional resistance, in the form of R4, has been added in parallel with R2. It lowers the effective resistance; R2 and R4 are both 500 ohms, so their parallel value is 250 ohms.

That goofs up the resistance ratio of the divider chain. It has become 2:1:1, and the voltage divides accordingly. R1 drops 100 volts, R2 (with R4) drops 50 volts, and R3 drops 50 volts. So you see, among the voltages, the 2:1:1 ratio holds.

Now, imagine that **fig. 2C** is a supply circuit in a solid-state receiver. The variable resistance labeled Q1 is actually the tran-

* For a rundown on series and parallel circuits, check the **repair bench** in the November, 1968 and January, 1969 issues of **ham radio.**

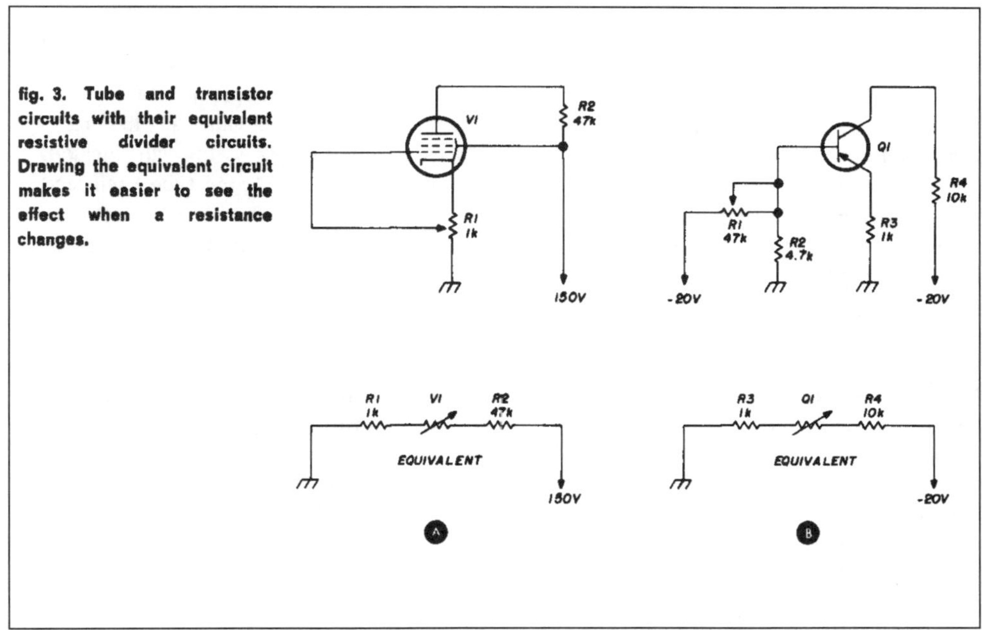

fig. 3. Tube and transistor circuits with their equivalent resistive divider circuits. Drawing the equivalent circuit makes it easier to see the effect when a resistance changes.

sistor, but imagine it here as a variable part of the divider chain. Voltages are shown at each point in the circuit, so you can figure out the voltage (and therefore resistance) ratio. Across R1 is dropped 2 volts, across Q1 is dropped 8 volts, and across R2 is dropped 10 volts. The ratio is 1:4:5.

Suppose something goes wrong with the circuit. You measure with your voltmeter and find the voltages changed to those in **fig. 2D.** Now you have a new ratio of voltages. Across R1 is 2.7 volts, across Q1 is 4 volts, and across R2 is 13.3 volts. If you write out the new ratio, however, it's easy to see where the voltage changed. The new ratio is 2:3:10. The old ratio was 1:4:5, which is the same as 2:8:10; only the center factor has changed. That means the resistance of Q1 has changed, and that's what upset the voltages.

tubes and transistors as dividers

Thinking in ratios is easy as long as the resistances are simple. Experience will teach you to estimate voltage ratios close enough to give you a clue to which resistance has changed.

Tube and transistor circuits are different. Changing the bias on either one alters its dc operation. **Fig. 3** suggests how to consider a tube or transistor as a variable resistance in the dc supply circuit. In **fig. 3A,** the tube bias can be changed by moving the slider on cathode-bias resistor R1. The equivalent circuit shows the tube's dc plate resistance as variable; it actually changes as bias is varied.

Simplicity is lost because R1 is also in series with the dc supply circuit. When the resistance of V1 changes, current changes all through the series circuit. A change in current through R1 affects bias, which changes the plate resistance of V1, which changes the bias, which changes plate resistance, and so on. The interdependency makes it tough to treat tube circuits as simple ratio dividers.

This problem also applies to transistor circuits; to see how, check **fig. 3B.** Bias for the transistor base is controlled by voltage divider R1-R2; since R1 is a potentiometer, bias is variable. As bias is altered, more or less current can flow in the emitter-collector circuit. The effect is to alter the emitter-collector resistance, represented by resistance Q1 in the equivalent circuit.

To see how complex these relationships can get, take a look at **fig. 4**—which is from a ham rig. Equivalent diagrams are shown at the right. The one at **fig. 4B** is for the collector circuit; it's the supply path through which dc must flow. **Fig. 4C** is the dc supply path for the base circuit; it includes the biasing voltage divider and the path through the base-emitter junction (shown as Q1B). Notice that R3 is in this path, too, so a current change in either circuit changes voltage across that resistor.

Finally, notice the combined paths in **fig. 4D.** Figuring out ratios in this circuit would be hard even if the resistances were simple. But they're complicated by the fact that the value of Q1C is controlled by Q1B. It's just too much to be simple.

a step at a time

However—and this is important—even though you can't use voltage ratios directly in tube and transistor circuits, you can apply the principles. Concentrate on one supply path at a time. At each step, ignore the effects of other circuits. When you decide what happened in one circuit, then study whether another could possibly be causing the voltage upset.

The transmitter tube stage in **fig. 5** is an easy example. Suppose you've traced trouble to this stage and have decided to measure voltages to help you find the faulty part. Your voltmeter tells you the plate voltage is low—only 180 volts instead of the 235 called for by the schematic. All the other voltages are about normal.

The plate-supply circuit (all the resistances plate current flows through) includes R5, one winding of T1, the tube itself, and R3. One at a time, consider the effect each part **might** have on voltage at the plate. Redrawing the equivalent plate-supply circuit, as in **fig. 5B,** can help. Here's the way your thinking might go.

If R3 increases in value, the dc voltage at the plate increases instead of decreases. If R3 decreases, the voltage lowers—but only a little, since R3 is such a small portion of the over-all circuit. Conclusion: forget about R3 for the moment.

If the V1 resistance increases for some reason, the dc plate voltage goes up. However, if the resistance of V1 decreases, the plate voltage goes down—exactly the symptom you have. Conclusion: the trouble might be V1. But don't rush; check other possibilities, too.

The winding of T1 can be eliminated as a suspect. If it opens, there's no voltage at the plate of V1. If it shorts, it changes the

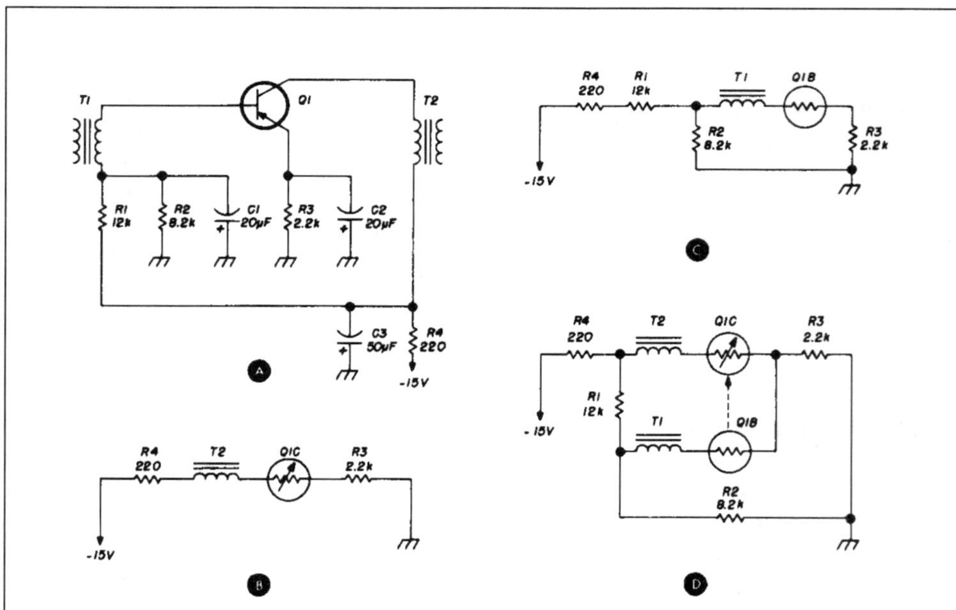

fig. 4. Transistor audio amplifier; the equivalent circuits in B, C and D show how difficult it is to use the "ratio analysis" method of troubleshooting a stage like this.

voltage too little to notice. If the value of R5 increases, the voltage at the plate of V1 lowers. So, there's another suspect! If R5 decreases in value, voltage goes up instead of down.

Now you have two suspects: R5 may be too high in value, and the dc plate resistance of V1 may be too low. Which is it? To find out, you measure other voltages in the stage.

For example, what about screen voltage? It's okay. If the dc plate resistance of V1 is low, it also lowers the voltage at the screen. But it isn't, so the tube and its bias must be okay. (You can make sure bias is okay by measuring it.) That leaves R5 as the likely culprit. You can just measure R5 with your ohmmeter to be sure.

Suppose you do find the screen voltage low, too. What then? The cause is probably the tube plate resistance, but what's making it low? It could be a faulty tube or incorrect bias. Either one could allow too much current through the tube, and thus lower the plate resistance. So—measure the bias. If it's okay, the tube must be bad. If the bias is wrong, go into that circuit to find out why.

The secret to voltage troubleshooting is to take it step-by-step, through each portion of each dc supply circuit. Decide how each part could cause the symptom you observe. If you trace the trouble to the tube or transistor, find out if it's bad or if some other operating voltage is making it work wrong.

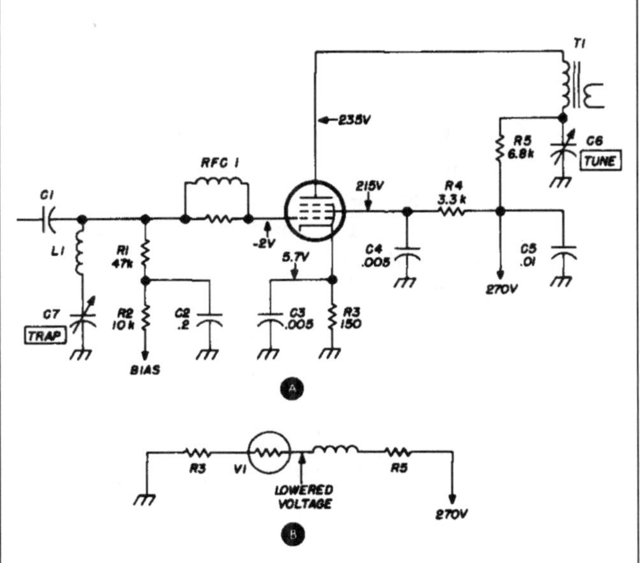

fig. 5. Driver circuit, with normal voltages listed at tube pins; B show equivalent plate circuit for troubleshooting.

voltages in transistor stages

The emitter-collector resistance of a transistor is set by bias current in its base circuit. Just as plate voltage of a tube may change if bias isn't right, the collector voltage of a transistor can, too.

The transistor stage in **fig. 6** is a mixer from a communications receiver. The transistor is a pnp type, forward-biased (base more negative than emitter). Dc operating voltages are listed alongside the emitter, base, and collector leads.

Suppose you trace trouble to this stage and want to use voltages to pin down the faulty part. The voltages you measure are listed in **fig. 6B.** Collector voltage is high, suggesting there's either too little current through the transistor or R4 is too low in value. If the trouble is in R4, however, the voltage at the emitter must also be high —and it isn't.

The best clue is the positive voltage on the base. It is reverse-biasing the base-emitter junction, cutting off collector current through the transistor. You have to trace the cause of the positive voltage. As it turns out, the other side of C1 goes to a positive voltage in the preceding stage; C1 is leaky, and is coupling the wrong voltage to the base of this stage.

Suppose, instead, you measure the voltages listed in **fig. 6C.** Collector voltage is very low, meaning there's excess current through the transistor or R4 is too high in value. Base voltage is a little low, but it hasn't changed nearly as much as emitter voltage has. Study the circuit. You'll quickly decide that C2 has shorted. With 1.9 volts of forward bias (instead of the normal 0.5), the transistor conducts heavily, lowering collector voltage drastically. The base voltage is slightly lower because of the effect of heavy current through R4; the base also gets its voltage from R4, through R2.

There are countless transistor-supply circuits you can analyze this way. Always pick out the voltage that is most wrong and find the cause of that first. Then check possible causes of other wrong voltages. The part that's common to all symptoms is usually the culprit. Each symptom leads to another, and ultimately to the bad part.

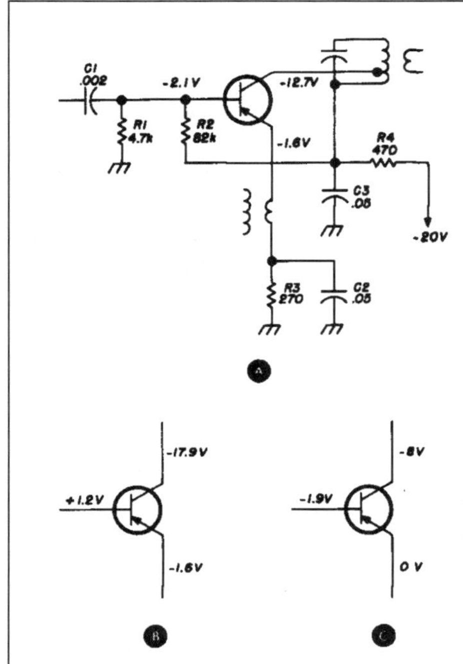

fig. 6. Transistor receiver mixer circuit; B and C show voltage changes in different troubleshooting situations.

testing individual components

This is the end of the line in troubleshooting—testing individual parts to make sure they're really bad. Resistors you can test most easily with your ohmmeter. Capacitors and other parts can be evaluated by voltage tests, combined with simple logic.

For example, C1 in **fig. 6** should couple absolutely no dc voltage from the preceding stage. If the preceding voltage is positive and the capacitor is leaky, the result is what you get in **fig. 6B.** If the preceding voltage is negative, it could drive the base of the transistor more negative and wind up reducing collector voltage. Leakage through C1 might also reduce voltage on the preceding stage—another helpful clue.

There is a reasonably sure way of testing a capacitor for leakage. If one end is connected to a dc voltage, disconnect the other end. With a vtvm, measure there for dc voltage. If voltage is leaking through, the voltmeter will detect it.

Don't be misled by electrolytic capacitors. They have a certain amount of natural leakage. Best way to check them is with an ohmmeter; leakage resistance should be no lower than 30k–50k.

You can tell if a coil or transformer winding is open, provided the winding carries dc. Just check at both ends with your voltmeter. An open winding blocks voltage.

You can also check for leakage between windings. Make sure one winding is connected to a voltage. With the other winding completely disconnected from any circuit, touch the voltmeter probe to one of its wires. If you measure any dc voltage, the transformer is leaky.

Any short-circuit can be pinned down quickly. Just find where the voltage has dropped completely to zero. This technique is most helpful in supply circuits where a filter, a bypass, or a high-power tube has shorted.

coming up

Awhile back, I promised to devote a column to using an oscilloscope at the repair bench. Apparently more amateurs own scopes than I suspected, but not many know how to use them or for what. Of course, everyone (just about) knows how to hook a scope up as a modulation monitor for a-m. But there are dozens of uses for this versatile instrument.

In my next column, I'll tell you exactly what a scope is and how to go about using it most easily. I'll give as many uses for it as I have room for. Then, after that, I'll explain a bunch of troubleshooting shortcuts with the scope. If you've got one, you should be using it on the repair bench. It's doggone handy.

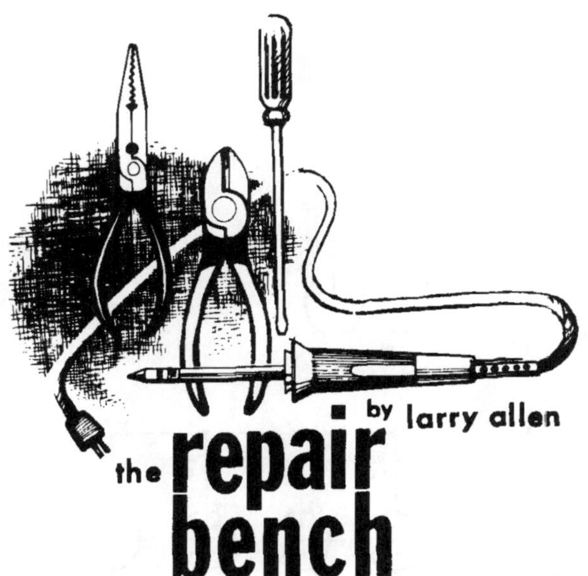

the repair bench

by larry allen

more ohmmeter troubleshooting

Last month, I explained how to use your ohmmeter for safe troubleshooting. Before I ran out of space, I reviewed series and parallel paths and also told you how to get accurate readings on your ohmmeter.

I hope you learned to recognize the "odd" series-parallel paths I described. When you're tracking down a fault in some receiver or transmitter, how easy it is for you depends on how well you spot the resistance paths that aren't always obvious. This month, I'll start off by telling where you might find some of those paths.

hidden resistance paths

There are two small parts most likely to cause false ohmmeter readings: electrolytic capacitors and transistors. Electrolytics are part of all power supplies and are used for decoupling along dc supply lines. In either use, they often get in the way of ohmmeter tests. Transistors foul up ohmmeter readings because of their "diode" nature. To make matters worse in transistor equipment, many coupling capacitors are electrolytic. They mix their leakage currents with the transistor leakage currents, and you can drive yourself nuts sorting out the true paths.

Fortunately, these two paths follow known behavior patterns. If you know the patterns, you can figure out if you're being misled. Here are some of them.

1. In **power-supply paths.** The leakage resistance of most electrolytic capacitors is nearly 50k. Learn to allow for it when you're checking resistances along B-plus lines. The diagram in **fig. 1** shows power-supply paths in one receiver schematic (simplified, of course).

Suppose you clipped the common lead of your ohmmeter to ground and the test lead to the plate of V4. Since there is no apparent path to ground, you might expect to read an open circuit, or infinity. Instead, you read around 600k. As you can guess, the ohmmeter is reading the resistance of R13 **plus** the leakage resistance of the power-supply capacitors.

2. Allow for **diode action.** Suppose the reading from the plate of V4 to ground is almost exactly 470k. What does that mean? Maybe a short in one of the electrolytics? To check, you move the test lead to the junction of R3-C4. A very low resistance there seems to mean there's a shorted electrolytic.

But before you jump to a false conclusion, reverse the test leads: put the common lead on the R3-C4 junction and the other (red) lead to ground. The low reading disappears!

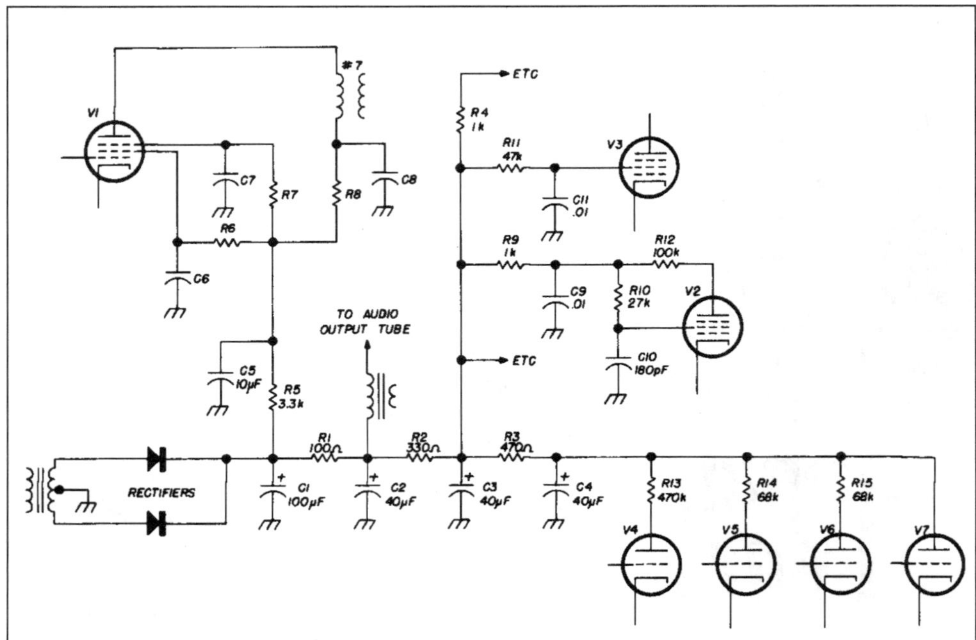

fig. 1. Power-supply from a popular receiver. It's simplified, to make tracing easy. If you must troubleshoot such a set, you can draw a simplified diagram like this to help you "see" the circuit branches.

It's confusing, isn't it?

What happens is that the voltage of the ohmmeter battery causes the rectifier diodes to conduct and give the low reading. Whenever the negative side of the ohmmeter battery (sometimes, but not always, the red test lead) is touched to the cathode end of the rectifier, the ohmmeter reads the diode's forward resistance (which is low). An ohmmeter thus connected from R3-C4 to ground might read through R1-R2-R3 and the diode. Connected to the plate of V4, the same ohmmeter might read 470k almost exactly, since the 1k (approximate) resistance of the R1-R2-R3-diode path is tiny by comparison. In either case, reversing the test leads eliminates the false path.

3. Consider **transistor leakage.** A form of diode action takes place in transistors. That is, the ohmmeter battery can forward-bias a transistor junction and make it draw current. The ohmmeter then reads the junction resistance as well as whatever circuit-path resistance you're measuring. There's an example in **fig. 2.**

The pnp transistor is wired up as a mixer. Suppose you connect your ohmmeter across R1 to measure it. Instead of the expected 4.7k, you read about 1k and figure R1 is bad. However, if you reverse the ohmmeter leads, you measure 4.7k. How come?

The base-emitter junction of a pnp transistor is forward-biased when the base is negative and the emitter is positive. If you connect the test leads the wrong way, the ohmmeter reading includes the base-emitter resistance (100 ohms or so). It is in series with R3, and both are in parallel with R1. The result: a wrong measurement.

How about when you measure R2? There should be no diode action between base and collector. What would you think, then, if you tried measuring R2 and got a 40k reading? Another hidden path is probably the cause. **Fig. 2B** shows the possible cause: leakage through the power-supply electrolytic capacitor (which may not even be drawn at that location on the schematic). The path through R4, the capacitor's leakage, and R1 lowers the resistance read by an ohmmeter connected across R2. You can thus get an erroneous reading, even if the transistor is normal.

A significant clue, though it might fool you, is the fact that reversing the ohmmeter leads in this case doesn't change the reading. The trouble is therefore **not** diode action.

4. Watch out for **paths drawn elsewhere.** The path just described is a common example. There are others harder to spot. One is shown in **fig. 3.**

The situation in **fig. 3A** is apt to be overlooked in tube circuits. The trouble is that schematics show only the common power-supply connections. If there happens to be a ground path somewhere else on the schematic, you can get a low ohmmeter reading you can't account for.

In **fig. 3A,** for example, if you measure from the plate of VI to ground, the voltage-divider network in the screen circuit of V9—which may be drawn at the other end of the diagram—causes an incorrect reading. (It should be infinity.) Even if you disconnect the power-supply capacitor, the too-low reading persists. The equivalent circuit is redrawn in **fig. 3B** to show how the circuit looks to the ohmmeter. (R9 has no effect, since it does not go to ground.)

You'll run into this voltage-divider situation more often in transistor sets than in tube equipment. Base-bias and emitter-bias dividers are standard practice. They cause trouble only when they are in parallel with each other but are widely separated on the schematic.

5. Something else to watch for in transistor circuits is shown in **fig. 4.** This one can fool you for several reasons. First, there is the diode action of the base-emitter junction. Second, there are parallel paths created by the power-supply electrolytic (not shown, as usual). Third, the parallel path depends on where you connect the ohmmeter.

The four rearrangements of the circuit show what the ohmmeter "sees" as it is connected between ground and each of the four points: A, B, C, and D. Anytime you have trouble figuring out paths, redraw them the way I have these.

From A, the ohmmeter should measure either about 850 ohms or about 4.6k. The diode is the reason for the alternative. One way, the ohmmeter makes the diode conduct and measures its forward resistance plus R4; the other way, it reads only the resistance of R1, since the diode can't conduct. The resistances of R1, R2, and R3 are comparatively so high, they have little effect even though they are in parallel.

From B, the ohmmeter should measure about 35k. The diode shows little effect, because R1 and R4 are relatively small com-

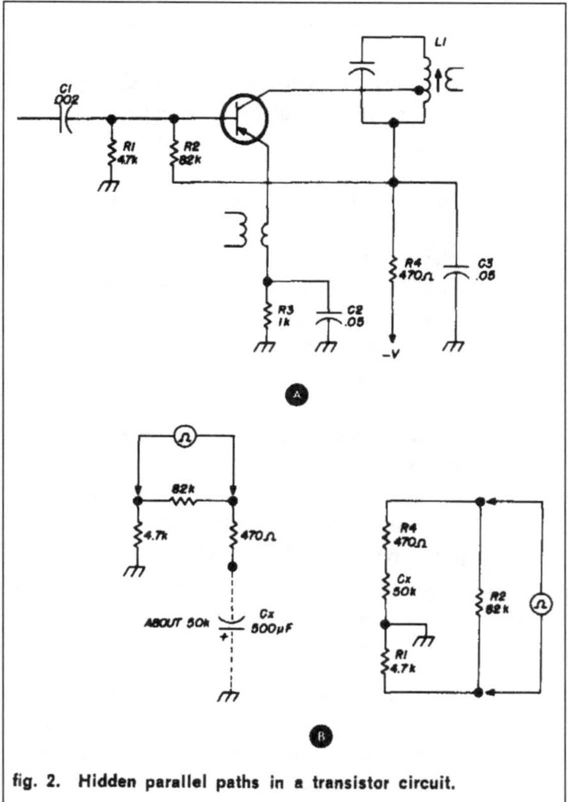

fig. 2. Hidden parallel paths in a transistor circuit.

pared to R2. The ohmmeter therefore reads about the same either way. The 68k–72k is in parallel with the (approximate) 77k.

From C, there are several considerations. First of all, you don't have to worry about the effects of R2, R3, and the power-supply electrolytic. They are so much larger than R1 that the 4.7k is the only resistance that can affect that leg of the circuit. The diode action of the base-emitter junction is important, though. Connected one way, the ohmmeter

fig. 3. Sometimes a parallel path is hidden by the way the schematic is drawn.

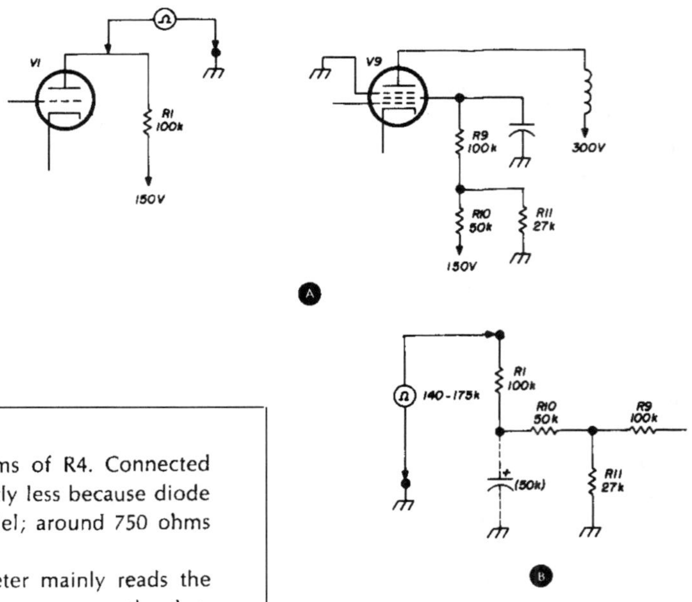

reads only the 820 ohms of R4. Connected the other, it reads slightly less because diode action puts R1 in parallel; around 750 ohms seems about right.

From D, the ohmmeter mainly reads the value of leakage in the power-supply electrolytic—about 30k. The much higher values of R2 and R3 isolate any action of the base-emitter junction, or the effects of R4 or R1. The condition of the electrolytic determines exactly what the ohmmeter reads.

knowing the cause of wrong readings

These general descriptions of where to look for "hidden" parallel paths also suggest ways to deal with them. When your ohmmeter readings are different from what the schematic leads you to expect, here are some quick steps to run down the cause.

1. If a reading is low, reverse the ohmmeter leads. If the reading increases, there's diode action in the circuit, which may be caused by a diode, a power-supply rectifier, or a transistor. In unusual cases, **slight** differences may be traced to electrolytic capacitors that don't "form" under reverse polarity.

2. If an ohmmeter reading starts low and builds up slowly, an electrolytic capacitor is causing it. This is the normal charging action of the electrolytic. Reversing the ohmmeter leads gives a lower reading, which builds back up again to the higher one. The reading is anywhere from 25k to 50k, depending on the condition of the electrolyte in the capacitor. The charging action is what you watch for.

3. If you're still looking for the reason for a low reading, see if you can find any other parallel paths drawn elsewhere on the schematic diagram. If you don't find any, you are probably on the trail of the fault.

what to do about parallel paths

There are times when it's faster and easier to eliminate the effects of parallel paths or to get around their effects some way. Once you spot a wrong reading, track down the guilty parallel component. You can't always figure it out from Ohm's law, either.

One of my **first** suggestions is: Don't make resistance measurements with respect to ground. Don't clip the common lead of your ohmmeter to ground. Make each measurement between specific points **in** the circuit.

As an example, think about a circuit like the one in **fig. 1.** You can clip the common ohmmeter lead at the junction of R2-R3, and measure the branches through R9, R11, R4, etc., directly. That way, you eliminate any effects from power-supply electrolytics. To measure in the branch that starts through R5, you can clip the common lead at the junction of R5-R1 and eliminate any diode-action problem with the rectifiers.

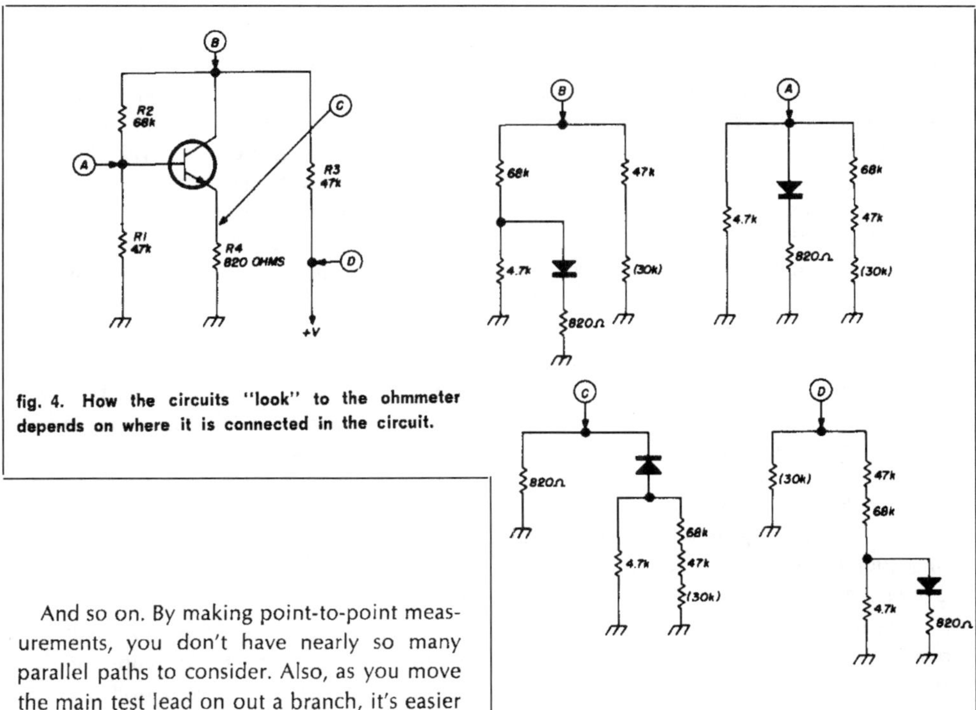

fig. 4. How the circuits "look" to the ohmmeter depends on where it is connected in the circuit.

And so on. By making point-to-point measurements, you don't have nearly so many parallel paths to consider. Also, as you move the main test lead on out a branch, it's easier to figure out the effects of added circuit resistances. Analysis is much simpler.

My **second** suggestion is to disconnect unwanted parallel paths when they get in your way. This happens seldom, if you plan your point-to-point measurements. But when it does happen, it's usually simple to unsolder a parallel component or branch.

On printed-circuit boards, you can often unsolder just one lead. Or, use a single-edge razor blade to slit across the foil and disconnect a circuit branch; that's easily restored afterward by a solder "bridge" across the tiny slit.

My **third** suggestion is to pull a tube or transistor out of its socket if you think it's fouling up your attempts to interpret resistance readings. Transistors that are soldered into their printed-circuit boards aren't that simple, yet can be handled without much trouble. Just unsolder the base connection (or slit the foil as already described).

In a circuit that contains a FET (field-effect transistor), unsolder the gate connection **and** the source connection. You can get low-resistance readings through some FET channels, even when the gate is disconnected.

My **fourth** suggestion for dealing with parallel paths is a last resort. When you can't analyze a circuit any other way, disconnect and test each part in the circuit—one at a time. It takes time, but if you've done your preliminary testing right, there aren't too many parts to test. At least you'll have narrowed the trouble down to one small section of the circuit.

in dangerous circuits . . . the safe way

I brought up resistance measurements in the first place to explain how to troubleshoot transmitters without exposing yourself to dangerous high voltages. Hunting a fault in the "live" plate circuit of an ssb linear can be a real pain in the neck. You need a special meter to measure high voltage, and usually you can't even reach test points without cheating a bunch of interlocks. There may be an overcurrent relay, too, or an undercurrent relay. In other words, the practical way to troubleshoot is with the transmitter or linear completely disconnected from power, and opened up. For that, you need ohmmeter measurements.

The plate circuit of a commercial linear

amplifier is shown in **fig. 5.** Mainly, you check for continuity. However, you can disconnect the power-supply lead and make measurements between ground and points in the circuit.

If an interlock switch isn't working, your measurements show it. The normally open (n.o.) switch is there to short out high voltage across the capacitors when you open the transmitter compartment. At the same time, the normally closed (n.c.) interlock switch removes applied voltage. Start your tests with power disconnected and the interlocks pushed in the way they would be with the compartment closed.

If everything's normal, you should get a low resistance reading from the plate cap of either tube to the clip that connected to the power-supply terminal. Check the opening and closing of the overcurrent relay with your finger. An open choke, tuning tank coil, relay contact, relay coil, n.c. interlock, or metering resistor (R1) spoils continuity, but each one is easy to track down. Just clip your common ohmmeter lead to the power-supply clip, and work your way to the tube platecaps with the other test lead.

A shorted or leaky by-pass capacitor (C1, C2, C4) shows up when you connect your ohmmeter, set for its highest megohms range, between the plate-cap and ground. Be sure the PA current meter is not connected; it could be an unseen parallel path. Disconnect the power-supply clip, too, for the same reason. Set the ohmmeter to its lowest range to test the n.o. interlock switch; it should read zero ohms to ground when the cover is off the transmitter high-voltage compartment and infinity with the cover in place.

In high-voltage circuits, especially the ones that carry high-energy rf, certain troubles only show up when the circuit is energized. For example, capacitor C2 or C1 might break down under high-voltage stress, yet test okay by the ohmmeter. The only sure test is to remove it from the circuit and substitute another one. This is costly, though, when the capacitor is an expensive one, such as a bathtub type. For these unusual cases, there are some special tests I'll talk about in a future column. You can make these tests **without** working inside the "live" transmitter or linear amp.

individual parts tests

As a wrapup for this two-month coverage of ohmmeter tests, here are some parts tests. Usually, these tests are made with the part disconnected from its circuit. When other tests and resistance readings point to a particular part, the final test is of the part itself. The following tests are quick and dependable.

Test **resistors** and potentiometers by measuring their resistance. For potentiometers, also check how smoothly the resistance changes. Connect your ohmmeter between one end lug and the center lug, and turn the control shaft slowly; any unusual or sudden resistance change means you need a new control.

Test ordinary **capacitors** by measuring with the highest range on your ohmmeter. There should be absolutely no leakage (infinite resistance). Touch the two capacitor leads together for a moment before applying the ohmmeter. Capacitors above 0.01-μF show a slight charging "kick" of the ohmmeter pointer. Those that don't are probably open. A shorted capacitor reads zero ohms between the two leads.

Test **electrolytic capacitors** by applying the ohmmeter leads in first one direction and then the other. Settle for the connection that gives the highest reading. It should exceed 30k, but not go beyond 60k or so. If it's too low, the capacitor is leaky. If it's too high, the capacitor is becoming ineffective. Notice how long it takes to "build up" to the highest ohmmeter reading. If it takes more than 3 or 4 seconds, the capacitor needs replacing.

There are three tests for **coils and transformers.** Test them first for continuity of windings. The schematic diagram may list the dc resistance of each winding; if some turns are shorted, the reading will be much too low. If a winding is open, the ohmmeter shows nothing.

Test coils and transformers second for leakage between windings. Use the highest ohmmeter range. Even several megohms—an awfully high resistance—is no good between transformer windings; they should be totally isolated (infinite resistance).

Test coils and transformers third for leakage from windings to ground. With the transformer mounted as usual but with all its leads disconnected, clip one ohmmeter lead to

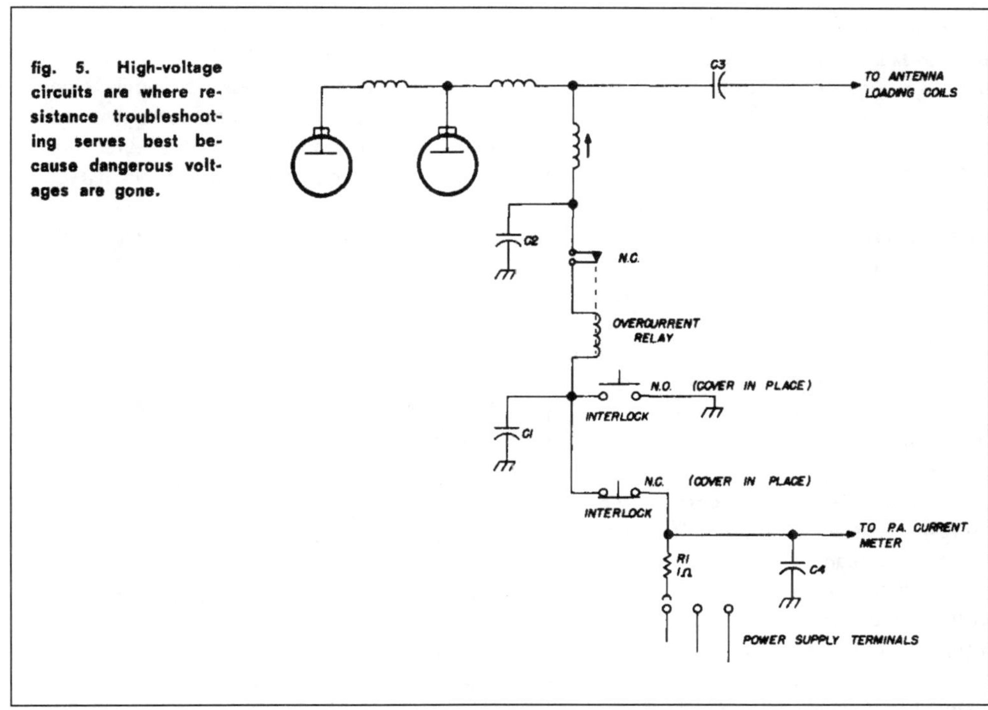

fig. 5. High-voltage circuits are where resistance troubleshooting serves best because dangerous voltages are gone.

ground and touch the other to each winding lead. There should be infinite resistance, since each winding should be totally isolated from the frame of the coil or transformer. If you think maybe the coil frame isn't grounded, clip one ohmmeter lead to the frame and touch the other to each winding. There should still be no reading. A reading other than infinity indicates leakage.

Test a **relay** by checking its coil the same way you test a transformer winding. Check the coil for continuity; if you know the dc resistance, measure it. If the ohmmeter reads infinity, the relay coil is open. Then check the relay contacts, with the Rx1 ohmmeter range. Normally closed contacts should measure zero ohms when the relay is at rest; at the same time, normally open contacts should measure infinity. Push the armature in with your finger. The normally open contacts should measure zero ohms that way, and the normally closed contacts should measure open (infinite resistance).

Test **speakers and headsets** by measuring them with the Rx1 ohmmeter range. **Listen** as you make the ohmmeter connection; the voltage of the ohmmeter battery usually makes speakers and headsets "pop." The ohmmeter pointer should show very low resistance for speakers and a few hundred ohms for headsets.

Test **microphones** in one of three ways, depending on type. For **carbon** microphones, connect the ohmmeter across the element leads and whistle into the mike. The reading should vary with each whistle. If it stays steady, the granules are packed and the element needs replacing. If the ohmmeter reads infinity, the mike cord is probably open, or the element is making poor contact.

For **dynamic** microphones, the exact resistance depends on the impedance of the mike. However, there should be some sort of reading, even if an input transformer is used inside the mike. Also, if you listen, there is usually a "pop" in a dynamic mike when you connect or disconnect the ohmmeter.

For **ceramic** microphones, the easiest test is to connect the ohmmeter and listen for a slight "click." The ohmmeter shows a resistance reading only if the mike is defective; a good ceramic or crystal mike reads infinity resistance.

Finally, connect the ohmmeter across the **push-to-talk** leads, and operate the button on the mike; a zero-ohms reading when the button is held means the p-t-t contacts and wires are okay.

Test **diodes** by measuring them first one way and then the other. A diode should measure low resistance in one direction and high resistance in the other; if the two readings are closer together than 100-to-1, the diode is probably bad. Power-supply rectifiers usually measure 100–200 ohms in the low-resistance direction; other types measure much lower.

A conventional **bipolar transistor** should show high resistance between emitter and collector, no matter which way the ohmmeter is connected. Between the base and emitter, however, resistance should be low in one direction and high in the other—just as in a double. If the base-emitter resistance is high in both directions, the transistor is open; if it's low in both directions, it's shorted. If resistance between emitter and collector is lower than several thousand ohms in either direction, the transistor is probably useless.

A **field-effect transistor** normally shows low resistance between drain and source—usually no matter which way the ohmmeter is connected. It should show extremely high resistance in one direction between gate and drain, and fairly high in the other direction. A MOSFET, sometimes known as an IGFET, should show *infinite* resistance between its gate and any other element, no matter which way the ohmmeter is connected.

next month

Everyone has his pet way to troubleshoot. In past **repair bench** columns, I've shown you several ways. You've read about signal tracing, signal injection, and now resistance measurement.

The best commercial technicians use several methods in various combinations that fit particular servicing problems. In my next column, I'll talk about one that is probably the most common: voltage testing. When a part goes bad, it usually upsets the dc operating voltages in its circuit. Those voltage changes, if you know how to interpret them, are a valuable clue to what part is bad. Next month, you'll find out how.

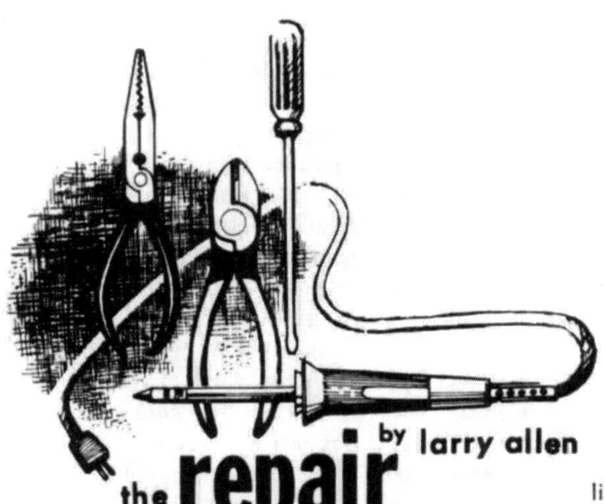

by larry allen

the repair bench

troubleshooting amateur gear with an oscilloscope

Well, guess I'd better begin where I left off. Before my spring sabbatical, I promised to tell you about oscilloscopes. And so I will.

A scope lets you **look** at a radio or audio signal. That's why it's so valuable as a tool for troubleshooting. It measures signal voltages, and at the same time shows the shape of the waveform. Thus, if distortion in an amplifier has fouled up a signal so it doesn't come through plain enough to understand, a scope can show you which stage makes it distorted. In those transistor-switching power supplies, you can look at the switching waveform and even track down troubles **about** to happen. You're probably familiar with the scope as a modulation monitor. And, if you are interested in amateur television, you simply have to know how to use a scope.

Most hams don't bother with a scope much because they don't really know how. Hooking one up as a modulation monitor is simple compared to twisting the knobs so you can view some of the strange-looking waveforms a serious ham runs into.

The scopes best suited for ham troubleshooting are the same kind used by radio and tv repair technicians. They should be wideband. You can buy them ready-made, like the one in **fig. 1.** Or, there are scope kits; **fig. 2** is one of those, in a compact size. The most common service-type models are recurrent-sweep scopes, called that to distinguish them from a more elaborate type called a triggered-sweep scope.

The dials and controls on the front of various brands and models differ only slightly. They're arranged in all sorts of positions, but the labels are always similar. Once you know one scope, you can quickly get acquainted with any of them.

getting the scope fired up

There's a group of knobs, usually beside or just below the screen, that turn the scope on, light up the trace, and get it set up to show a display. They are the **intensity, focus,**

fig. 1. One scope that's popular with tv service technicians. Other well-known brands are listed in the box on page 57.

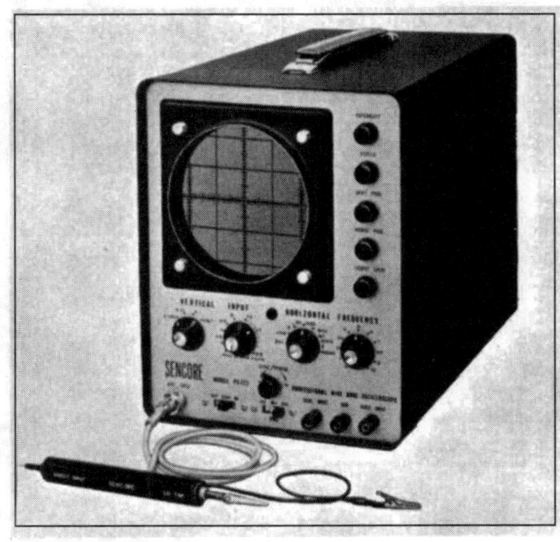

and **position** controls, and they're shown at the right in **fig. 3.**

The power switch may be on the **intensity** control. You turn it on, and turn the control all the way up. If the power switch is separate, turn the intensity up anyway. After a minute or so, either a dot or a line should show up on the screen.

Adjust the **focus** to make the dot or line as fine as possible. The top two photos of **fig. 3** show how a scope dot looks when it's out of focus and then when it's focused. The bottom two show out-of-focus and in-focus line traces.

fig. 2. Compact, 3-inch scope you can build from a kit. Most kit-built scopes cost much less than comparable factory-built models.

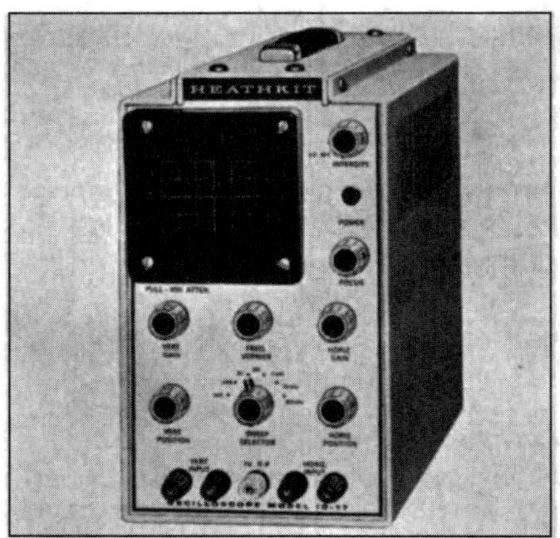

If the dot or line doesn't show up after $1^1/_2$ minutes, the two **position** knobs may be set wrong. Move them slowly from one end to the other, first separately and then simultaneously. **Vertical position** moves the display up and down on the screen, and **horizontal position** moves it from side to side. When the scope first comes on, if a **position** knob is near one end of its rotation, the dot or line may be out of sight. Set the knobs to center the line or dot as well as you can.

To make a dot into a line, which you'll have to do to display waveforms, find the group of controls shown in **fig. 4.** The important ones for this are **horizontal frequency** and **horizontal gain.** On some scopes, the main frequency dial is labeled **sweep.** It controls the sawtooth generator that scans the beam (which makes the dot) rapidly from side to side. Turn the **horizontal frequency** switch to a position that makes the sawtooth generator sweep the beam back and forth 20 or 30 times a second. The switch in **fig. 4** is set between the 10 and 100 marks. That's close enough for now; later, with a waveform on the screen, you can refine the frequency of the scope's internal sweep generator with the lower left control.

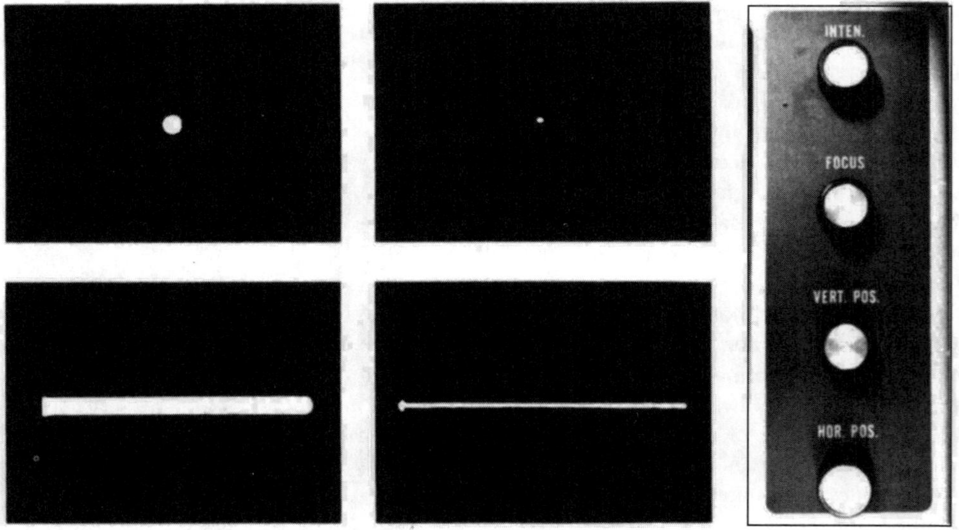

fig. 3. Scope dot or line when it's out-of-focus (large and fuzzy) and when it's in-focus (sharp and fine).

fig. 4. Frequency-control knobs on a kit-type servicing scope. Ignore phase knob at lower right.

There may still not be a line, or it may be very short. Turn up the **horizontal gain** until the line extends most of the way across the scope screen. The scope is ready now for you to display a waveform and learn how some of the other controls work.

looking at an ac waveform

One handy waveform is the 60-Hz voltage from the power line. In fact, so much radiation from it is picked up by the capacitance of your body, you can use that as a test signal. Just grab the tip of the probe with your fingers. Leave the ground clip dangling.

You'll have to set the **vertical input** controls (**fig. 5**) so the scope is sensitive enough to show the rather weak 60-Hz signal picked up by body capacitance. On the scope from which the **fig. 5** photo was taken, the main input knob is set to X1, the most sensitive position. As you can see from the photo, the **vertical gain** doesn't need to be turned up much. (It may on some scopes, especially if

fig. 5. Vertical input knobs affect size of display on screen; on some scopes, they (along with graticule lines on the screen) help with voltage measurements.

fig. 6. Two top photos, and bottom left, show how the ac waveform may look before you adjust line frequency. Bottom right shows solidly locked three-full-cycle display.

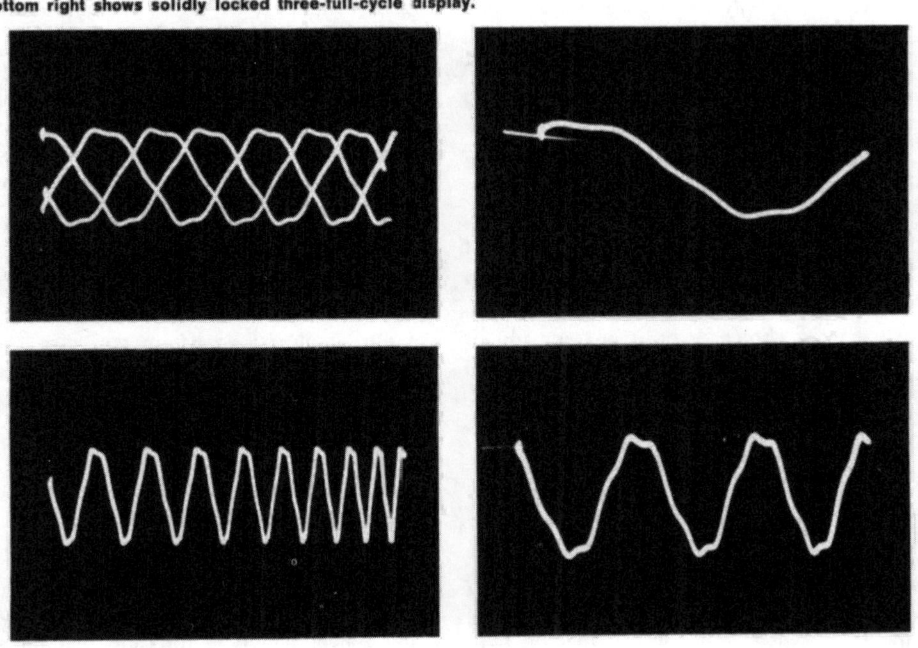

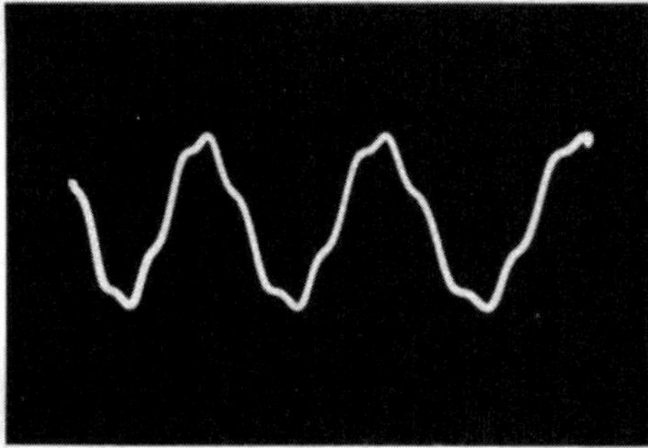

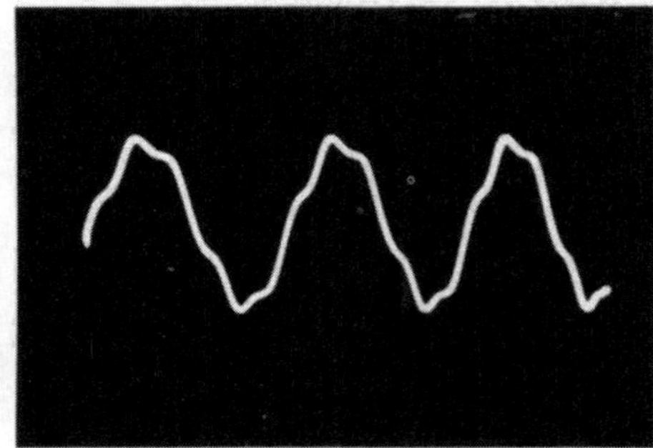

fig. 7. Ac waveforms locked, by sync switch, on negative slope (left) and positive slope (right). Irregular shape of sine ac is due to fluorescent-lamp radiation.

60-Hz radiation is mild around your shack or bench.)

The waveform display you see on the screen, if you set the **horizontal frequency** switch as I mentioned earlier, looks something like the top photos or the bottom left one in **fig. 6.** Best viewing is with two or three cycles of a waveform on the trace. You have to adjust the fine frequency control (labeled **frequency vernier** on the scope in **fig. 4**). Just turn it carefully back and forth till you have only three waveforms on the screen—as in the bottom right photo in **fig. 6.**

The waveform may not snap into place quite that easily. It may appear to be running one way or the other, and you may not be able to stop it with the fine frequency control. Holding it in position solidly is the responsibility of a knob or switch labeled **sync.** In fact, you can adjust where the left side of the trace starts, merely by how you set **sync.**

In scopes that have only a **sync** switch, you have a choice of positive-going (+) or negative-going (−) internal sync, external sync, or sync from the scope's power transformer (line). You'll almost always use internal sync, and whether you choose + or − depends on the general shape of the waveform.

With the waveform at top left in **fig. 7,** the signal display is shown synchronized on the negative-going portion of the test waveform; the **sync** switch is at **−int.** In the top right photo, the waveform is synchronized on the positive-going part of the signal; the **sync** switch is set at **+int.**

A scope with a control instead of a switch is synchronized much the same. The knob is usually marked with a center zero, and can be turned either way. One direction locks the signal on the negative slope, and the other locks it on the positive slope. If your scope has the control, notice once you get the display stopped how you can work the control back and forth slightly and shift the point on the slope where the trace starts. That isn't too important with an ordinary sine-wave display, but it is when you work with oddly shaped waveforms, like in amateur tv.

looking at odd waveshapes

Sine waves are the most common type you'll look at when you use the scope on your repair bench. They're what you find in the power supply before the rectifier and filters get at the voltage. And you often inject sine waves and then trace them to check out audio amplifiers, clippers and other speech stages.

But sine waves are far from being the only waveshape you'll see. The only problem you can have viewing the odd-shaped ones is making them stand still on the screen. But there are a couple of simple tricks for that. If you know what's important about a waveform, that's enough to let you sync it in tightly on the scope—every time.

As one example, look at the rounded sawtooth at the top in **fig. 8.** It's taken at the output of a rectifier, across the input filter. The important thing to notice is the steepest slope. That's always easiest to synchronize on. In this one, the steepest slope is upward, which means it goes in the positive direction. So, you set the **sync** switch to the + position. And, as you can see in the photo, the trace locks (and starts) on the upward slope.

You have to learn to recognize the dominant steep slope and which way it goes. Often the waveform can only be slowed down a little with the **frequency** control, and you have to see the waveshape "on the run." Once you're sure which way the steep slope goes, you can set the sync for the polarity that will lock easiest. Then you can go back to the fine frequency control and slow the waveform down from running. When you get the scope sweep close enough to an exact multiple of the frequency of the signal you're viewing, the waveform locks in.

An example of locking in an odd-shaped waveform appears at the lower left in **fig. 8.** This is the video waveform from a flying-spot scanner such as you might build for ham tv. This display shows three cycles of video, which means the scope's sweep must be set to one-third the horizontal scanning rate of the tv set.

The horizontal sweep for commercial TV is 15,750 Hz. The scope, then, must be set to about 5,250 Hz. That way, the scope display has time to show three cycles during each scan of the scope beam. The **horizontal frequency** switch in **fig. 4** must be set between the **1000** and **10kc** (10 kHz) marks. Then the **frequency vernier** is turned slowly back and forth until you catch a glimpse of the shape of the waveform.

You'll see that its dominant slope is downward-going, meaning the waveshape is main-

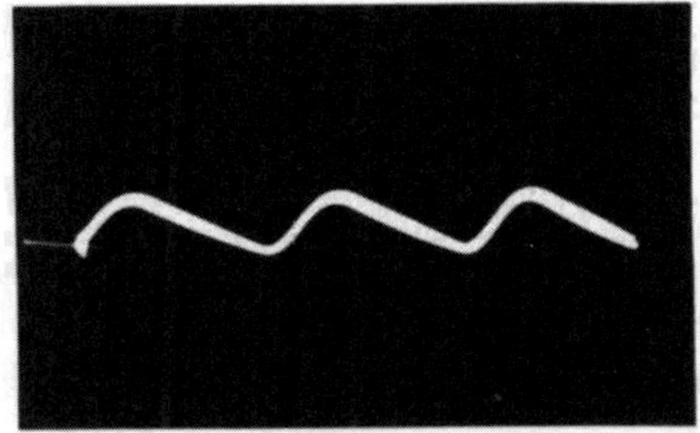

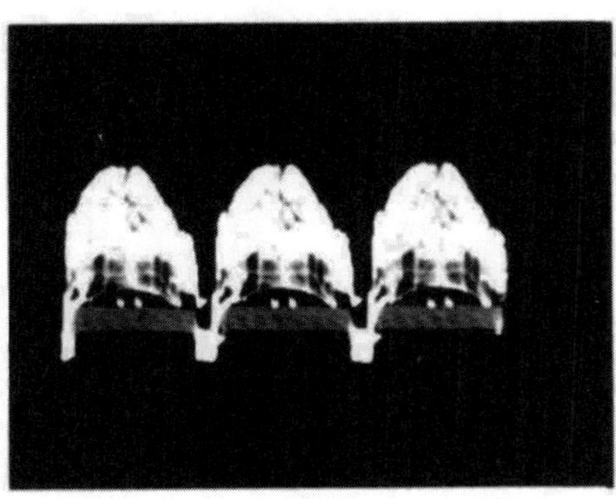

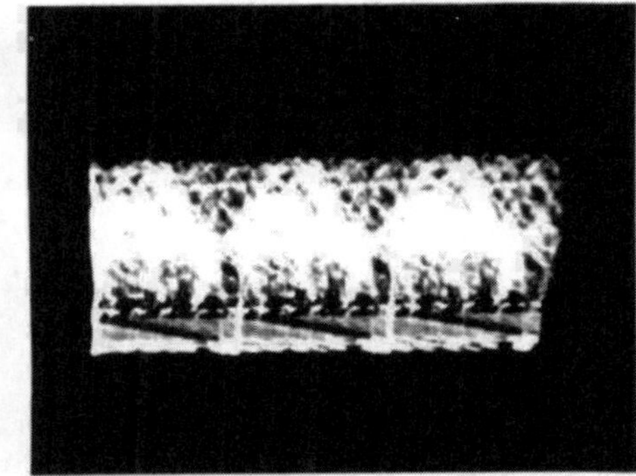

fig. 8. To lock in unusual waveshapes, you need to know their approximate frequency and which direction the dominant slope goes. All of these were taken in a home-tv flying-spot scanner.

ly negative. So, to lock it in solid, the **sync** switch is set on the **−int** mark.

With the sync switch inadvertently set wrong, the waveform can't be settled down much better than in the lower right picture in **fig. 8.** This photo was taken with the switch at **+int.** You can see only a vague outline of the waveshape, if you look close.

using a scope for troubleshooting

Until you've used an oscilloscope for tracing faults in ham units, you can't appreciate how helpful it can be. And it's a mistake to try working on ham tv without one. Now that you know how to stabilize the waveforms on the screen, you can put a scope to work on your repair bench.

Next month, I'll tell you more about it. There are dozens of different tests you can make with it, and lots of ways to save time. I'll show you the waveforms you can expect to find in different kinds of equipment, how to measure them, and how to tell if they're not what they should be. And I'll tell you how to set the scope to look at each of them.

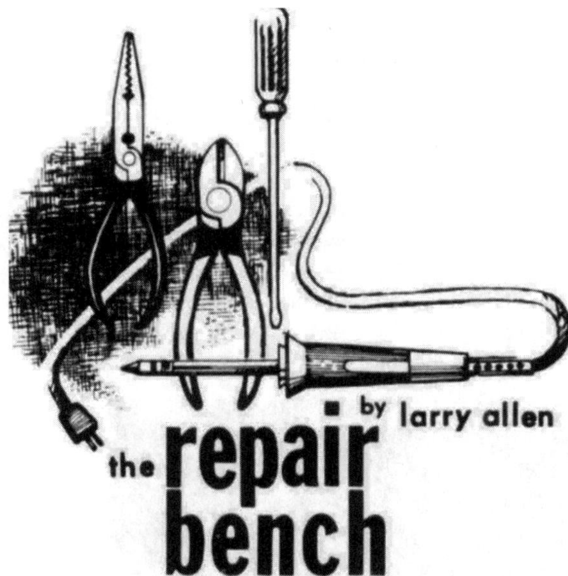

by larry allen

the repair bench

putting a scope to work in ham gear

Here I am to continue with scopes. I told you last month how to fire up the scope and get displays locked into position on the screen. This month I'll give you some specific uses for it on the ham repair bench.

A scope, remember, lets you **see** the exact shape of an ac waveform. You can identify sine waves, square waves, tv sync pulses or any unusual waveshape. More than that, and often more important, you can measure peak-to-peak values of waveform voltages, whether they're nice round sine waves or crazy, misshapen oddballs.

measurement methods vary

Probably the chief difference in various brands of service-type scopes lies in how you measure voltages with them. Yet, they're all fairly easy. There are three main ways. Each uses the scope's variable **vertical input** control, but in different ways. The **vertical input** switch is always a decade-type multiplier. And the marked graticule on the face of the scope is important to voltage measuring.

In one method, you first feed in a signal of known value and adjust the **vertical input** control to make the trace coincide with that voltage marking on the scope graticule. That calibrates the input amplifier. You can thereafter read the voltage value more or less directly on the scope graticule.

The second way of measuring peak-to-peak voltage is almost identical, except that the calibrating voltage is inside the scope. You just turn a switch to **calibrate,** and adjust the **vertical input** knob till the waveform display fits between two "calibrate" lines on the graticule. Then, you can read any signal-voltage value from where it fits on the graticule. The switch and the waveform are shown in **fig. 1.**

You've probably figured out that you can't mess with the **vertical input** knob once it's set. If the display is too small, you turn the input multiplier switch to a more sensitive position. If the display is too large (offscreen), switch to a less sensitive multiplier.

The third common method of measurement lets you read the voltage value directly from the **vertical input** knobs. **Fig. 2** illustrates the controls. You adjust the **vertical input** controls to make the waveform display exactly 2 inches high on the graticule (which is marked in inches). Then, you read the number pointed to by the variable knob and multiply it by the multiplier setting. That tells you the peak-to-peak voltage of the input waveform, no matter what its shape.

The waveforms that follow are photographed from a B&K model 1450 servicing scope. Its lighted graticule, shown in **fig. 3,** has two numbered scales: 0-2 and 0-6. The multiplier switch determines which scale

fig. 1. Before measurements, the vertical input amplifier must be calibrated to a standard voltage. Switching to vertical applies internal signal to deflection amplifier in this scope. Display is adjusted to fit the full-scale graticule mark.

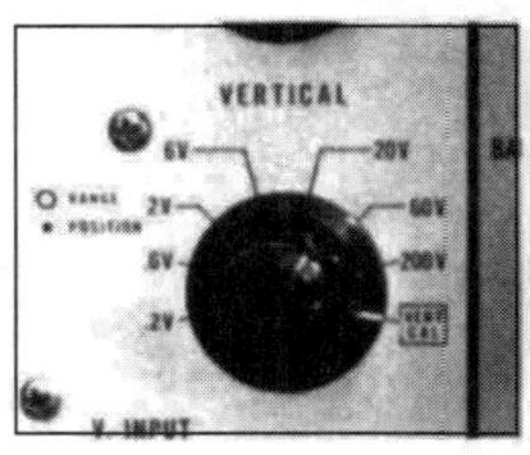

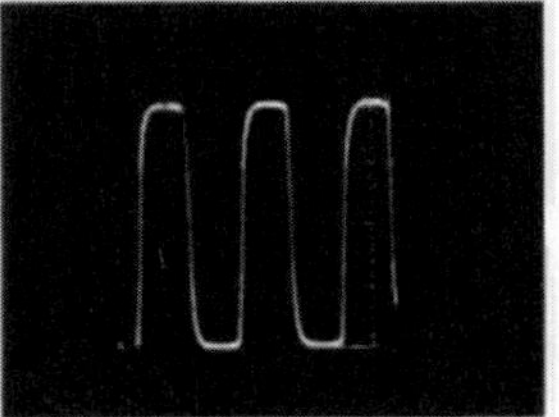

lights up. The correct full-scale voltage is listed on each waveform photo so you can read peak-to-peak amplitude accurately. As you go along, reading voltages on a scope like this becomes second nature.

tracing input power

A simple use for the scope, and a good one to practice with, is checking input ac voltage to a receiver or transmitter. It's the same thing you'd ordinarily do with your ac voltmeter.

The diagram in **fig. 4** shows the test points in a popular ssb transmitter. The photo below the input wiring shows the waveshape. Ac line voltage is a sine wave.

The 120 volts your voltmeter reads is a root-mean-square (rms) value, and the scope shows peak-to-peak (p-p) values. Line voltage is 335 volts peak-to-peak. The highest full-scale voltage reading on this scope is 200 volts, but its probe has a 10X switch on it. That attenuates any signal voltage by a factor of 10.

The multiplier switch for the photo in **fig. 4** is set at 60. A voltage of 600 will therefore reach all the way to the top of the 6 scale. The waveform you see goes about a subdivision and a half above the 3 mark.

The **sync** is set for **−int,** but it could just as easily be set for **line;** either would lock the line-voltage waveform in tightly.

The **sweep** control must be set for 10–50 Hz, and the **fine frequency** adjustment is turned so four cycles appear on the display. Adjust positioning controls so the bottom of the waveform is on the base line of the graticule.

Use an isolation transformer between the

fig. 2. Waveform peak-to-peak amplitudes can be read directly from the knobs of this scope. The display is made exactly two inches high on the graticule by turning the knobs; then voltage is read.

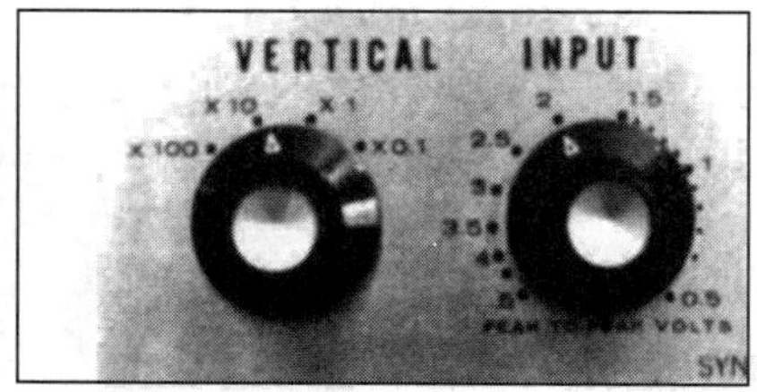

plug of any chassis you're testing and the power line. Otherwise, you can blow fuses with a direct short between the scope ground lead and one side of the power line. (Not to mention the big chance of a nasty—even deadly—shock.)

Clip the ground lead of the scope to the ground side of the power cord. Then, move the test probe from point 1 to point 3 as shown. You should get the same waveform at every point. If you get a waveform when you probe the grounded end of the trans-

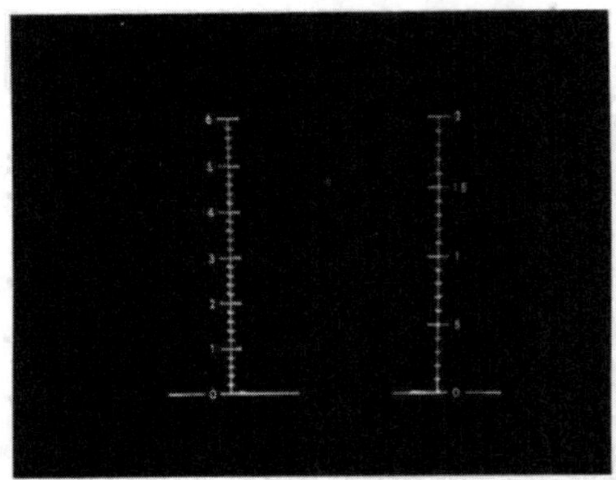

fig. 3. You read p-p voltages on scales with this scope, much like you do with a voltmeter. Either scale is used with decimal multiples chosen by the vertical input switch. Scales light one at a time.

former primary, however, it's a sign the ground is open. You can test the primary winding, too, by opening its ground connection and probing at the ground end; if you get a waveform, the primary is okay (**fig. 4** shows this test).

You can test the secondaries, too. Just probe points 4, 7, 9, 12, and 13. Remember that the **Vac** readings listed on a schematic are rms values. To know what p-p reading you should expect, multiply rms voltage by 2.8. Probing at point 10 also tests the doubler filter that couples ac voltage to the first rectifier.

power-supply filter capacitors

A scope is a fine tool for testing power-supply filters. It lets you actually see how much ripple is left in the dc output of the supply.

Here's a good rule-of-thumb: Ripple should

not exceed 1% of the dc voltage across the filter, **with the equipment drawing full load.** Thus, in a 500-volt dc supply, output ripple under load should be no more than 5 volts. If it is, the filter capacitors aren't doing their job.

For a regulated supply, the ripple percentage should be even lower. The regulator circuit or its capacitors are faulty if the ripple at the output exceeds 0.1% of the dc value. Ripple voltages are always measured peak-to-peak.

The waveform photo in **fig. 5** is taken across the filter of a 400-volt dc power supply. The full scale is 20 volts, so the ripple waveform is 4 volts p-p. The scope **sweep** is set for 10–50, and **fine frequency** is set to display four cycles. **Sync** is **–int.**

sine waves in amplifiers

Here is where a scope begins to shine. It can show up distortion at the same time it's measuring gain in an amplifier. Suppose you suspect trouble in the speech amplifier section of a transmitter.

You feed about 50 mV of audio signal into the microphone input jack **(fig. 6).** Any frequency around 1000 Hz works fine. Then you move the scope probe from stage to stage, at inputs and outputs, all the way to the modulator. Each amplifier stage should multiply the signal voltage by at least 10 or 20.

A clipper, on the other hand, is supposed to prevent overmodulation. It limits how much audio signal reaches the modulator. You can check the clipper with a scope.

Connect the probe at the modulator grid or at the output of the clipper stage itself. Turn up the audio generator output, watching the scope as you do. When the audio signal gets strong enough to cause 100% modulation, the clipper starts acting. The sine wave begins flattening out as the second photo of **fig. 6** shows. From that point on, you should be able to turn the generator wide open without getting a higher waveform on the scope. The shape only gets flatter.

You can trace rf amplifiers with the scope, too. Just feed in a modulated rf signal, and use a demodulator probe with your scope.

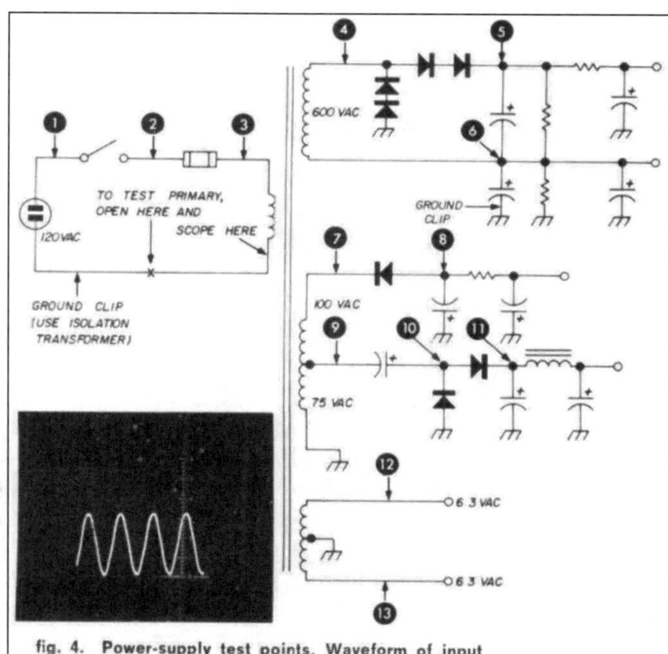

fig. 4. Power-supply test points. Waveform of input voltage. Full scale in this photo is 600 volts; peak-to-peak voltage therefore appears to be about 335 volts.

With amplitude-modulated signals, any circuit fault that cuts down the rf signal also cuts down the audio modulated on it.

You can even peak up receiver alignment with a scope fed by a demodulator probe. Just use the amplitude of the waveform on the scope as an indicator of peaking as you turn the adjustments. You can move the

fig. 5. Ripple voltage found in one fullwave power supply. Taken across input filter. Full scale of graticule is 20 volts; 4-volt p-p reading indicates filter capacitor is doing its job fairly well.

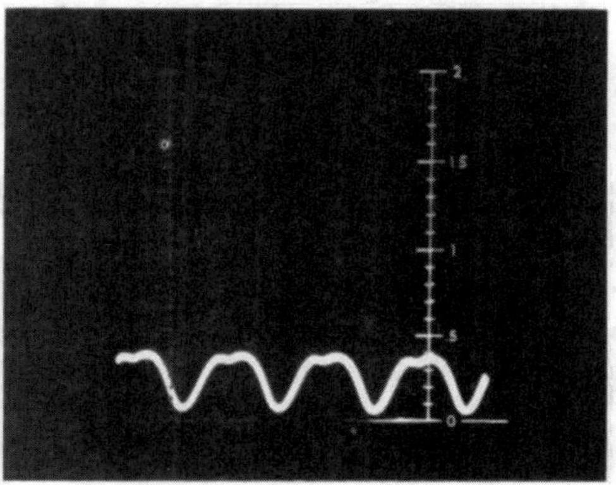

probe from stage to stage as you progress with tuning adjustments, or you can connect it following the last rf or i-f stage. If you go beyond an a-m detector, you don't need the demodulator probe. However, you can align the i-f and rf stages in an fm or ssb receiver using an a-m signal and the demodulator probe with a scope.

ham television

In recent years, hams have found it logical to stick with the commercial tv standards. Thus, they can use standard receivers with modified tuners. Also, surplus and used transmitting equipment is more available.

For examining and measuring signal voltages in television stages, a scope is indispensable. The photos in **fig. 7** are the most common waveforms.

Figs. 7A and **7B** are photos of the signal voltage that contains both the video that makes the picture and the sync pulses that keep it steady. The scope probe is connected to the output of the tv set's video detector. Full scale of the graticule when these photos were taken was 6 volts p-p; the waveform is just under 5 volts p-p.

In **7A,** the scope sweep is set for about 5 kHz; that shows three cycles of the signal at the 15,750-Hz line rate. **Fig. 7B** is the same signal, but with the scope sweep set for about 20 Hz. That displays three cycles at the 60-Hz frame rate. Peak-to-peak voltage is the same, no matter what display rate is used.

From the vertical sweep section of the tv set, the signal voltage in **fig. 7C** is taken. It is a trapezoid, displayed on the 600-volt scale. Its amplitude is 220 volts p-p. The scope sweep is set for 20 Hz.

Fig. 7D is also a trapezoid, this time taken in the horizontal sweep section of a television receiver. The scope is set to sweep at about 5 kHz, to display three cycles. The full-scale voltage range on the graticule is 200; the waveform is 120 volts p-p.

You'll discover that waveforms in any television receiver are viewed at either of just two scope-sweep rates. The 5-kHz sweep rate is for all waveforms that bear any relation to the horizontal line rate of the tv receiver. The 20-Hz sweep rate is for waveforms that relate to the vertical frame rate.

Fig. 7E is a special waveform found in the automatic frequency control circuit of the tv-set horizontal oscillator. The scope sweeps at the line rate. The graticule is at 200 volts full scale; the waveform is therefore 100 volts p-p in amplitude.

Another special circuit in TV receivers is the automatic gain control. In most modern sets, it is a keyed type. That is, a pulse is applied to it from the horizontal sweep transformer (called the flyback). The photo in **fig. 7F** is a sample of the keying signal voltage. The scope is set to display horizontal-rate signals. Peak-to-peak amplitude of the keying signal is 175 volts.

next month

These last two installments have given you a sample of how you can use a scope on the ham repair bench. A little practice makes

fig. 6. Sine-wave audio signal gets flattened in clipper when too much signal is fed in. That's normal, because clipper should keep signal at modulator from exceeding voltage for 100 percent modulation.

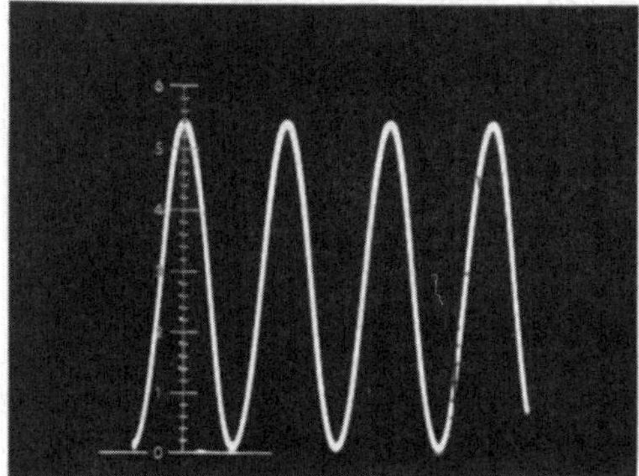

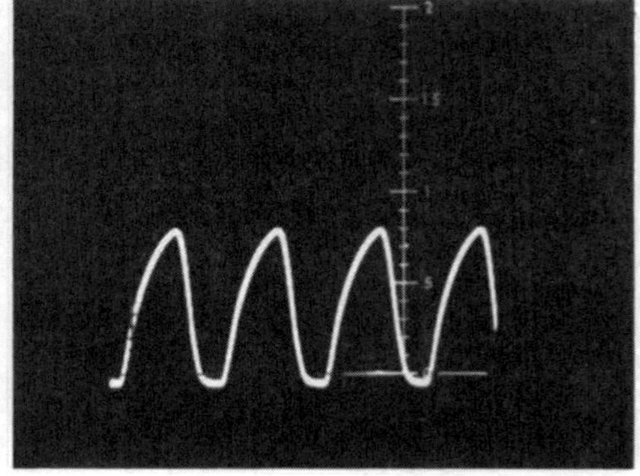

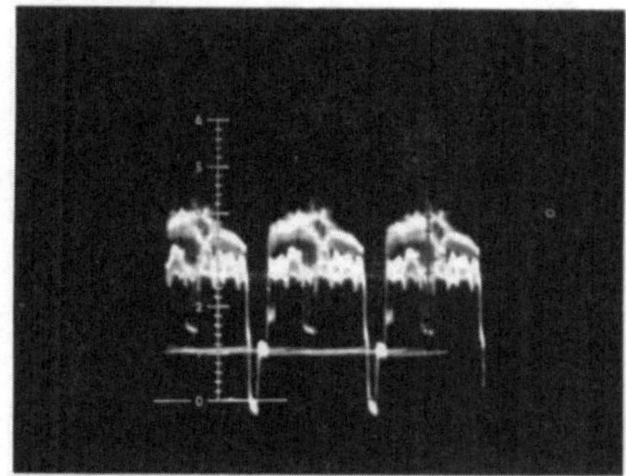

A. video waveform at line rate

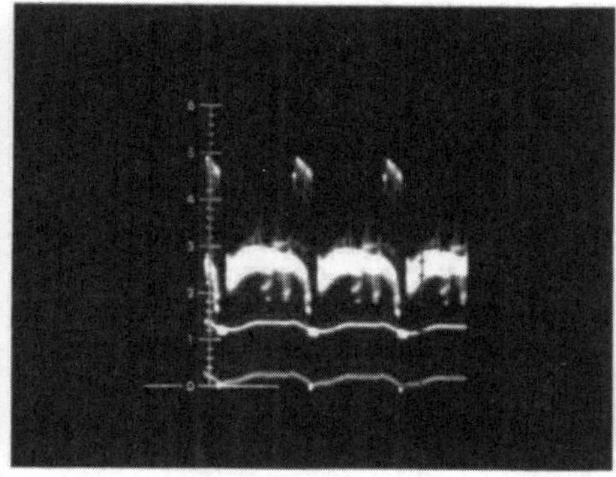

B. video waveform at frame rate

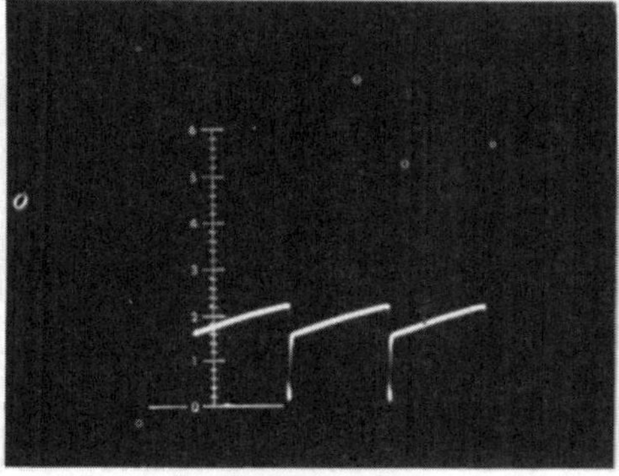

C. vertical oscillator waveform

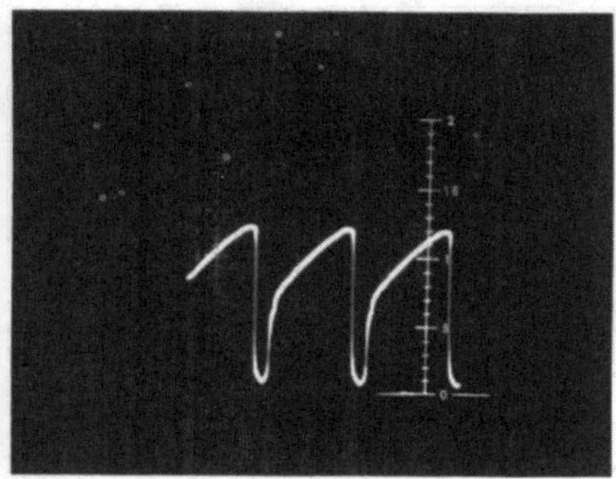

D. horizontal drive signal

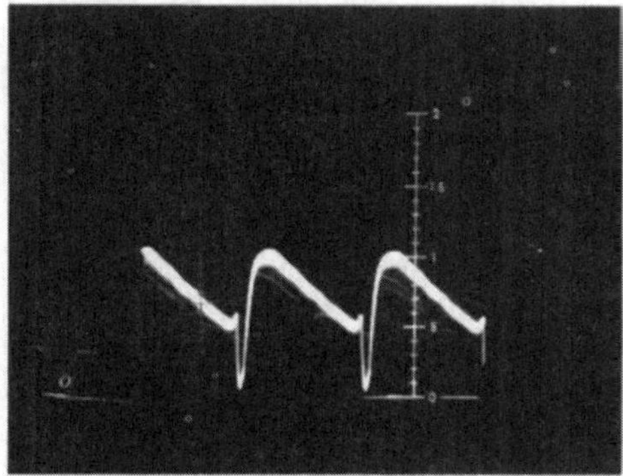

E. afc comparison signals

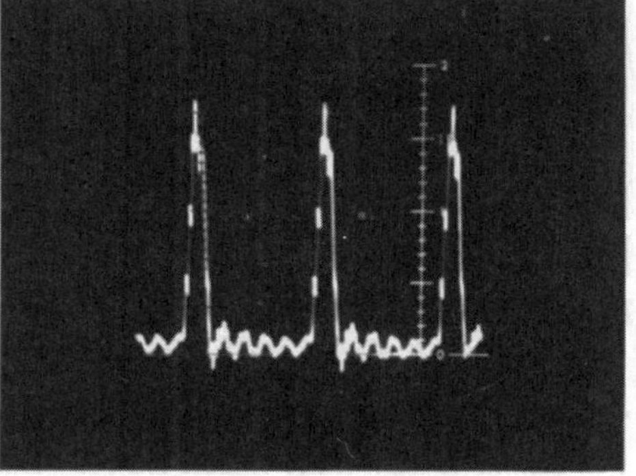

F. agc keying pulse

fig. 7. Typical waveforms from television receiver that can be used for amateur tv reception. Video waveform, displayed at line rate (A), video waveform, displayed at frame rate (B), vertical oscillator waveform, at frame rate (C), drive signal to horizontal output tube, displayed at line rate (D), waveform of comparison signals in afc stage, shown at line rate (E) and keying pulse for agc stage, at horizontal rate (F).

you as familiar with your scope as you are with your voltmeter. And you'll find the scope is far more versatile.

Next issue, I shift to a new topic: single-sideband. Tuning up an ssb transmitter has some things in common with any other ham transmitter. But, there are important differences. Those are the subject of **repair bench** next month.

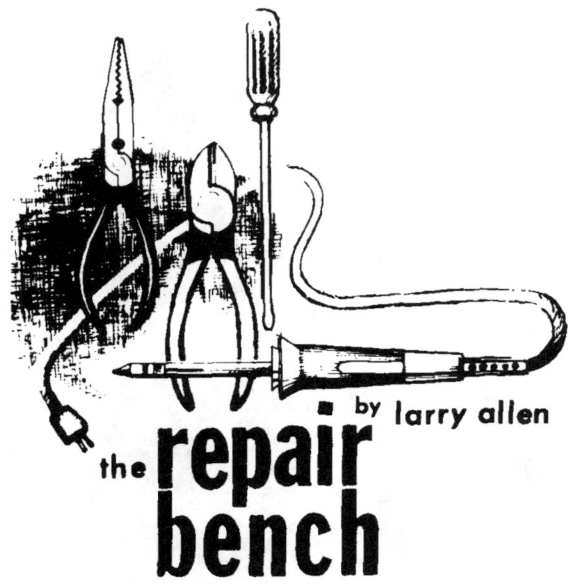

the repair bench

by larry allen

finding faults in rf and i-f amplifiers

What do you do if you can only pick up nearby or strong stations? What if they come in weak and perhaps noisy? First thing you should do is suspect the rf amplifiers in your receiver.

This isn't unusual, particularly with transistor front ends. If the manufacturer (or you, if the receiver is homebrew) designed the rf amp to use the sensitive but delicate field-effect transistor, a gate punctured by static surge isn't at all uncommon.

The problem lies in recognizing a bad front end. If the mixer and i-f stages are naturally quiet, and transistor stages often are, you may not know an i-f failure from an rf one. Or the fault may be in the automatic gain control (agc) system. Only careful testing will tell you for sure.

amplification vs noise

Familiarity with your receiver is the best assurance of knowing when there really is trouble. You should get to know how much natural receiver noise to expect when there's no station.

Then, when suddenly you can't pick up stations you know should be there, make a listening test. Tune the receiver dial to an empty spot. Turn rf and af gain up.

Is receiver noise (the background thermal hiss) up to snuff? If so, the i-f stage must be amplifying. Also, the mixer stage is probably okay; much front-end thermal noise normally originates there.

Yet, some of today's field-effect transistor (fet) front ends are too quiet for this kind of analysis. It's normal to hear almost no receiver hiss. You have no choice then but to rely on other testing methods. You try to determine rf-stage sensitivity. Or, you can just test the stage by regular dc-measurement methods.

what's in a rf stage

Most of what I say about troubleshooting rf stages can be applied to i-f stages, too. There's little difference.

In most a-m receivers, rf stages tune over several different bands. The i-f stages are fixed-tuned. But it's common to tune ssb receivers nowadays by synthesis—the same as the transmitter. In that case, the receiver rf stages are fixed-tuned. You troubleshoot them the same as i-f stages.

A typical old-fashioned tube-type rf stage is drawn in **fig. 1**. Only one deck of the bandswitch is shown; the input band coils are omitted for simplicity.

Most hams can figure out a way to track down trouble if they are sure what a stage is supposed to do. I don't mean in just a general way, but specifically—each part of the stage. Take the stage in **fig. 1** for an example.

First and foremost are the *input* and *output* circuits. T1 and C2 are the input

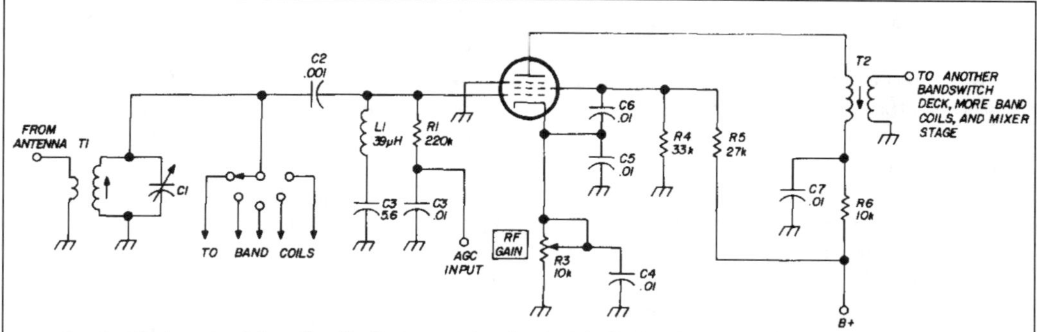

fig. 1. Rf stage contains circuits that must be checked individually if you don't use test procedures that check overall stage performance.

coupling components. R1 is the input load, decoupled by C3. Ordinary signal tracing or injection is the way to make sure these components work.

The output circuit is T2, decoupled by C7. Those decoupling components are an important part of the input and output circuits, so don't forget them when you analyze stage operation.

There are also tuned circuits to contend with, although part of them are not shown in **fig. 1**. T1 and C1 make a tuned circuit. But they are in parallel with band coils you can't see. The band coils, with the inductance of T1, set the band the stage is to tune across and C1 tunes the specific frequency. If there's a fault in one of these coils, that band won't tune properly. If the fault is in T1 or C1, *none* of the bands tune as they should.

Next, consider the *supply* circuits. They carry dc voltage to the tube elements.

The plate supply is through R6 and the primary of T2. Capacitor C7 is important in the plate dc supply circuit only because of the possibility it might short. You therefore must consider it part of the plate supply circuit when you're diagnosing.

Resistors R4 and R5 are the chief components of the screen supply circuit. C6 is a potential part of it—if the capacitor happens to become leaky or shorted. Leakage in C6 would put voltage intended for the screen onto the cathode. Consider that possibility when you analyze the screen or cathode dc circuit.

The cathode dc circuit, which can be considered dc supply even though it's just a ground return, is through R2 and R3. C4 or C5 can become part of this circuit if either happens to short. Otherwise, they don't affect cathode dc voltage.

Notice that R3 is variable. It's the *rf gain* control for the receiver. (In the receiver from which this example is taken, R3 also is part of the i-f stages. That connection is omitted here for simplicity.) Changing the value of R3 between R2 and ground varies cathode bias on the tube. The pentode is a sharp-cutoff type; its gain depends sharply on its bias. Thus, by changing bias, R3 controls rf amplification.

Voltage at the grid is controlled from the agc stage. The agc control voltage is fed through R1. Decoupling capacitor C3 is part of the grid-supply circuit only if it shorts or gets leaky. C2 and C3 might become part of that circuit if either went bad.

There's another resonant circuit. It isn't tunable. It may also be called a trap circuit, because that's what it's there for. L1 and C3 form it, and it's resonant to 9 MHz, the i-f of this receiver. It traps out any stray 9-MHz signals, preventing them from being amplified and running through the mixer to upset the i-f stages.

That about sums up the circuits in this rf stage. Input, output, tuning, bypass or decoupling, and trap: those are the signal-carrying circuits. Plate supply, screen supply, cathode return, and grid bias: those are the dc-carrying circuits. You have to consider each when you set about troubleshooting an rf stage like this one.

transistor rf stages

The transistor input stage from one receiver is drawn in **fig. 2**. You can probably identify the circuits. They may look a bit different from those in **fig. 1** because they're in a transistor stage.

The input circuit comprises L1, C1, C2, and both R1 and R2 as load resistors. Decoupling for the two resistors isn't in the diagram, but it's understood. The agc line has a bypass capacitor not shown; so does the 12-volt supply.

L1 and C1 are fixed-tuned, although adjustable with a tuning tool. They form a broadband tuned circuit. (This particular rf stage is part of a one-band transceiver.) The taps on L1 are for impedance matching.

The output circuit comprises L2, C4, R5, and C5. L2-C4 are a tuned tank, with R5 as a band-broadening load across it. L2 is adjustable for band-peaking. C5 couples amplified rf energy to the mixer stage (through another tuned circuit, omitted for simplicity).

The only other signal circuit is emitter bypass capacitor C3. If it opens, substantial degeneration can occur, but it only makes the stage weak—it doesn't make it dead.

There are only three dc supply circuits. That's because a transistor has only three elements to receive voltage.

The transistor is pnp. Therefore, normal forward-bias operation puts the emitter positive, the base less positive (same as more negative than emitter), and the collector far less positive (same as far negative from the emitter).

The emitter gets voltage directly from a positive 12-volt supply line through R4. C3 is the decoupling capacitor, and is a concern to the dc circuit only if it shorts or gets leaky.

Collector goes to ground through L2. The winding has no appreciable dc resistance, so for dc the collector is grounded. That puts it far negative with respect to emitter.

The base has the only complicated supply network. The main dc comes from the 12-volt line through R1. However, a connection through R2 lets the actual voltage—and therefore bias on the transistor—be varied by the agc line and by the setting of *rf gain* control R3. The transistor operating characteristic is such that bias controls amplification. Thus the *rf gain* control sets optimum gain of the stage, and agc varies it to accommodate signal strength.

Of course, C2 is part of the base dc circuit only if it comes defective. A faulty decoupling capacitor in the agc line could also affect base bias. You need remember these capacitors only if you're troubleshooting and find the base voltage is wrong.

external influences

One other thing you can't forget when

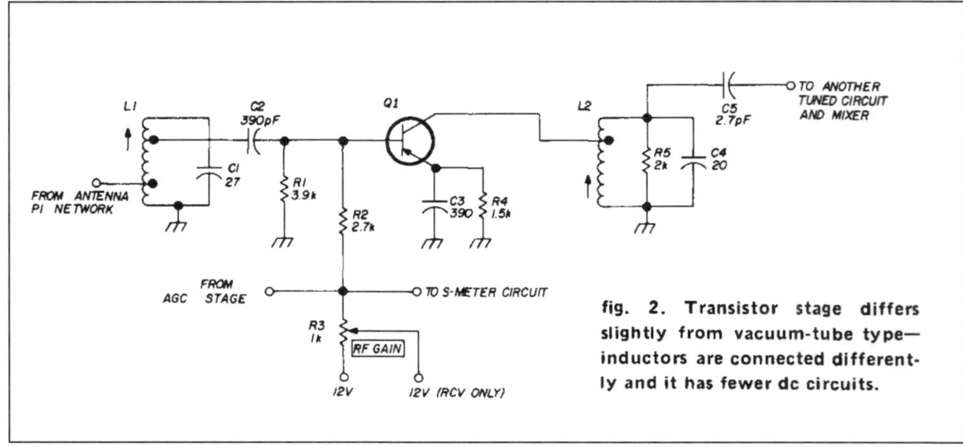

fig. 2. Transistor stage differs slightly from vacuum-tube type—inductors are connected differently and it has fewer dc circuits.

you're troubleshooting rf and i-f stages. A trouble in the stage may be caused somewhere else.

A blocked i-f or rf stage is common. Sometimes, that's traced to an agc stage overdoing its bias thing. Or, the i-f or rf amp may block on strong signals—a sign of overload. That, too, may be traceable to a faulty agc stage—this time not doing enough.

The receiver block diagram is sometimes helpful in spotting stages that affect rf or i-f. Some don't show that much detail and you have to rely on your ability to read the schematic. **Fig. 3** is a partial block diagram of the set from which **fig. 2** is taken. Its detail is enough to be helpful.

You might find the voltage upset in an i-f stage, yet the agc stage works normally. Suspect the s-meter hookup. If any part of that circuit shorts to ground, it could foul up bias on rf or i-f stages. A short inside the s-meter might make the *rf gain* control work wrong.

In other words, examine the schematic or block diagram before you go tearing into any rf or i-f stage. If external stages affect the rf stage, check them out or isolate them from the rf stage some way.

testing rf amplification

One way to see if an i-f or rf stage is working right is to measure its gain. In modern transistor receivers you can expect to find a voltage-gain factor of 20 or more. A tube stage usually gives even higher gain.

You can make this measurement fairly easily if the output of your rf generator is calibrated. First, clamp the agc line with whatever dc voltage produces normal idling (no-signal) bias on the rf or i-f stage you're testing. That bias is usually written on the schematic or on the voltage chart.

Clip your vtvm to the a-m detector output, or through an rf probe to the output of the last i-f amp. Feed the generator signal to the *input* of the rf or i-f stage being measured. Note the dc meter reading. Set the generator output level for some meter reading that's easy to remember. Make a note of the rf output level of the generator.

Move the generator signal to the output of the stage being tested. Turn up the generator level until the meter reads the same as before. Divide the new generator output-level reading by the earlier one. The result is the voltage gain of the stage.

As an example, suppose 0.7 microvolt of signal drives the meter to 1 volt dc when the generator is connected to the stage input. When you connect the generator to the stage output, you have to turn the generator up to 14 microvolts to get

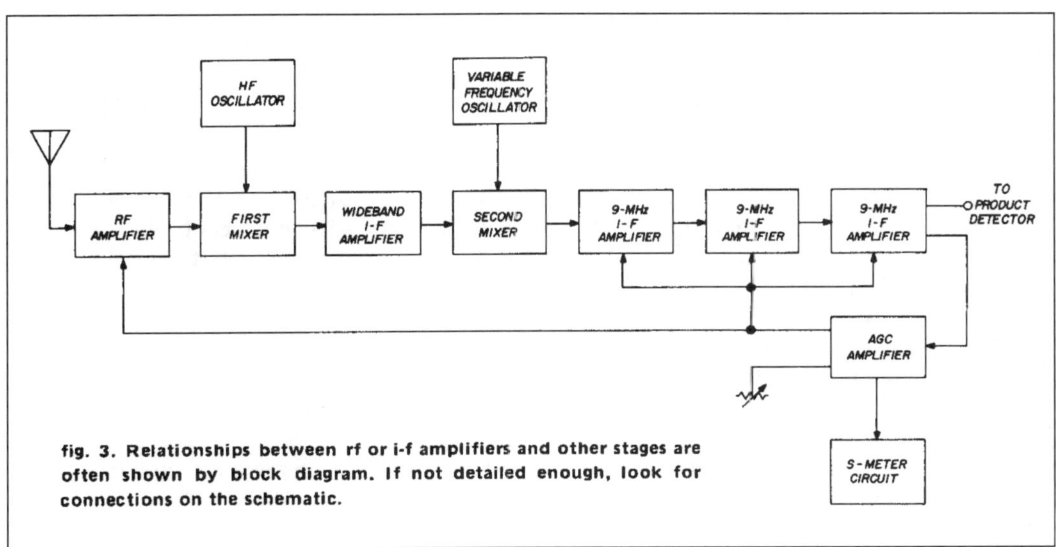

fig. 3. Relationships between rf or i-f amplifiers and other stages are often shown by block diagram. If not detailed enough, look for connections on the schematic.

fig. 4. Circuit for a voltage-doubling probe you can use with a vtvm to measure relative rf signal levels in rf and i-f stages. Entire probe should be shielded to prevent hand capacitance from upsetting the reading.

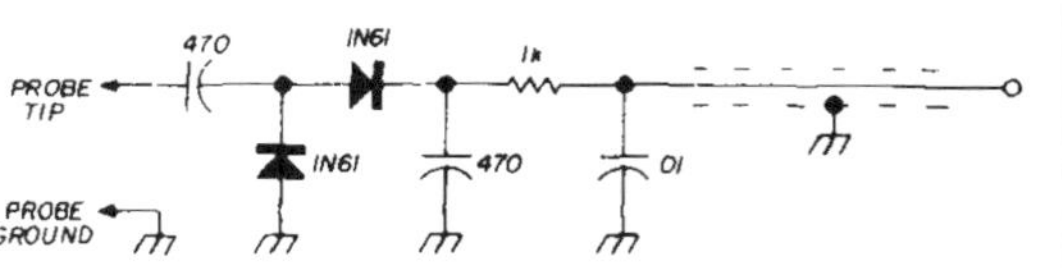

that 1-volt dc reading on the meter. Dividing 14 by 0.7 gives 20. That's the voltage gain of the stage.

Unfortunately, only the more costly signal generators have calibrated output. You may have to use a less accurate way. It's only relative, and your best bet for using it is to make measurements while your receiver is working normally and record them for reference.

You'll need a doubler-type rf probe for your vtvm. A suitable circuit is sketched in **fig. 4**. It's more sensitive than the ordinary single-diode probe. The vtvm should be a very sensitive one, with lowest full-scale reading 1.5 volts or less.

Again, clamp the agc. Turn the *rf gain* control wide open. Keep the generator unmodulated.

Feed the rf signal to the antenna input jack. Tune the generator to the center of the band if the stage is fixed-tuned; if the stage is tunable to one frequency, set the generator precisely to that frequency.

Connect the vtvm probe first to the base of the rf transistor. Set the meter on its lowest range. Turn up the generator signal until you get a perceptible reading on the meter. Then set the generator output for some very small but definite voltage indication—say 0.01 volt. Don't change the generator setting.

Move the vtvm probe to the base of the mixer. The reading should be much higher now—say nearly 0.2 volt dc. That represents a gain of about 20 if you're using the doubler probe.

Obviously, these figures are approximate. Voltage gain for a tube is roughly the same. For tubes or transistors, however, the surest system is to make a record of normal amplification while your receiver is new. Then use the same measurement method when you test on the repair bench.

dc troubleshooting

You should already know the methods of tracking down the cause of incorrect dc voltages on an rf stage. You might want to make dc tests before you go to the trouble of putting the rf probe on your vtvm.

But usually you'll find that kind of fault is obvious—as when the stage is completely dead. It's for subtle weakness or abnormal overloading you need to clamp the agc and test stage gain. Then you can hunt down the small voltage problem—or bad transistor or tube—that's causing the trouble.

checking the stage another way

You can also use a form of signal injection. Connect the vtvm, without a probe, to the a-m detector of the receiver. If it's handier, keep the probe on the vtvm and connect it to the i-f input of the product detector (or at the output of the last i-f amp). Clamp the agc as before.

Connect the rf signal generator, tuned to the rf frequency as already described, to the input of the mixer stage. Turn up the generator output just enough to cause a reading on the meter. Make a note of the reading.

Move the generator back to the input of the rf stage. Note the increase in the reading. The dc-voltage increase should be about the same as the one already described. Without the doubler probe, a 10-times increase means about 20 gain in the rf stage. With a doubler probe instead of the set's a-m detector, a 20-times increase means roughly 20 gain in the rf stage.

Now go a step further. Be sure you've got the generator frequency set precisely. Try tuning the coils in the tuned circuits. If the meter reading doesn't vary, the coils may be at fault. Before you replace them, though, make sure the capacitor that decouples each coil is not open.

Finally, if tuning is erratic and you can't seem to make head nor tails of how the coils tune, check the bypass capacitors on the supply lines and on the agc line. One of them may be open.

looking to the future

Remember that the ways of checking rf stages outlined here can be used just as well with i-f stages.

There's still another way of troubleshooting rf and i-f stages: with sweep alignment. Unfortunately, no inexpensive sweep generator available today goes down far enough in frequency.

In a future column I'm going to show you how to make your regular sweep generator go down far enough to sweep 60-kHz, 455-kHz, and other i-f amps. You can do it without modifying the instrument you have.

First, though, there's a new troubleshooting system that has come to my attention. It's called *1-2-3-4 Servicing* by its originator, Forest H. Belt. In the next issue of repair bench, I'll tell you what *1-2-3-4 Servicing* is all about. That'll prepare you to understand what goes on when you use the sweep-alignment method of rf and i-f troubleshooting.

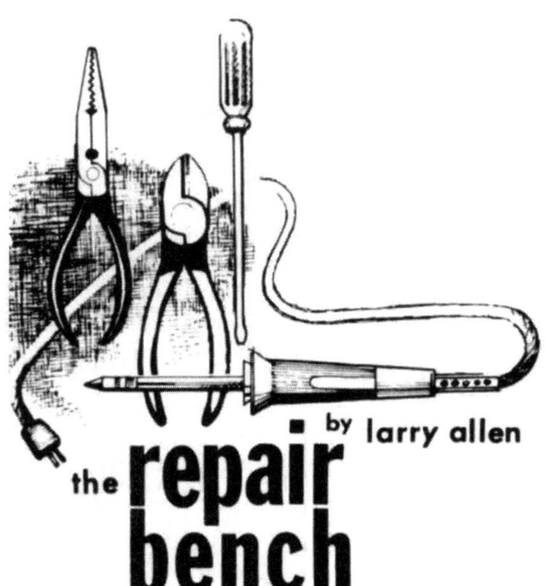

by larry allen

the repair bench

how to use a sweep generator

Television repair technicians – the good ones – learn to use a sweep generator early in their careers. It's a must. There's no other way to align a TV set, because its i-f stages are stagger-tuned. That is, they're aligned to various frequencies. It takes a sweep generator to show how they're tuned and the effect they have on the wideband television signal.

But I've discovered an awful lot of hams don't understand a sweep generator. If *you* don't, read on. I'm going to explain it.

You can use a sweep generator to help you design interstage coupling transformers and wind them with just the bandwidth you want. Or, if you tinker with ham TV, you'll need it to keep your receiver aligned. It's also a good troubleshooting tool when your ssb receiver (or any receiver, for that matter) has rf or i-f trouble. You can even watch the effects of agc (or avc) on bandwidth.

the instrument

Sweep generators for general use are scarce these days. Many new models are for television sweep alignment, and put only TV i-f and channel frequencies. But some general-coverage models are still available. Here's a list of those I've seen and used:

EICO 369
Heathkit IG-52
Knight-kit KG-687

The Knight-kit has a built-in marker generator; I'll explain its purpose, too, later.

You can see what they look like in **fig. 1.** These cover frequencies from about 3MHz to 220 MHz, the high end of the vhf TV band. You can often use harmonics of the dial frequencies up to nearly 900 MHz, with reduced output of course. Stability is usually critical at that high a harmonic. But the fundamentals, as you can see, encompass most ham frequencies.

What, exactly, is a sweep generator?

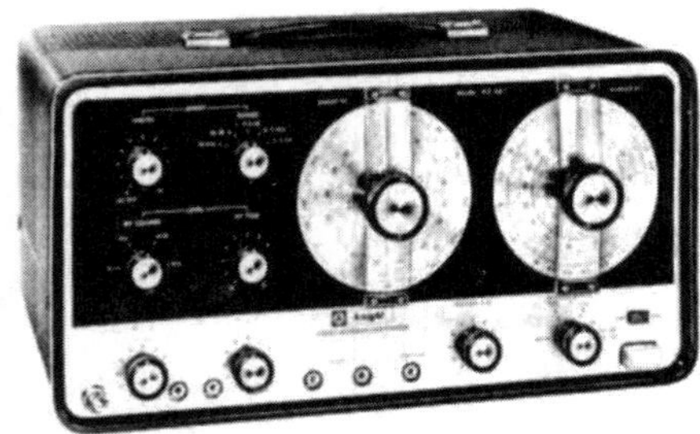

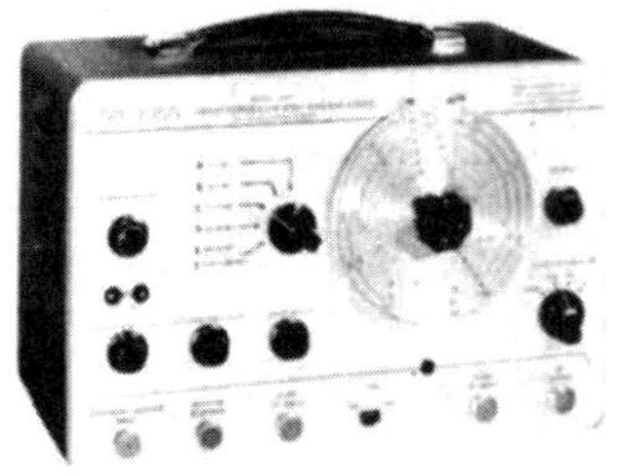

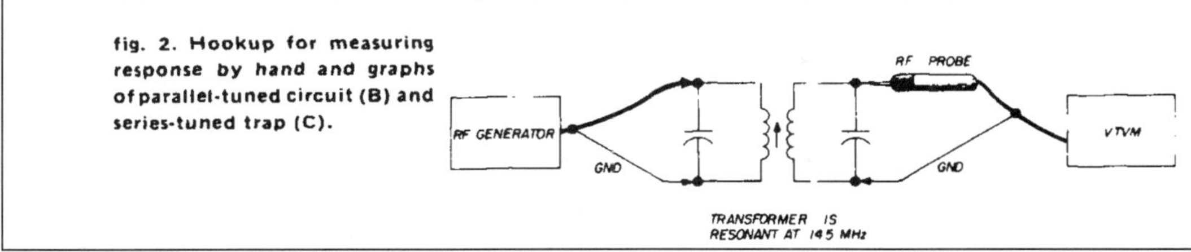

fig. 2. Hookup for measuring response by hand and graphs of parallel-tuned circuit (B) and series-tuned trap (C).

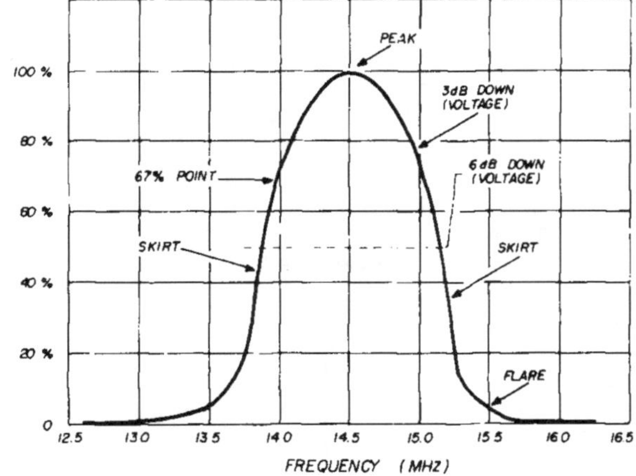

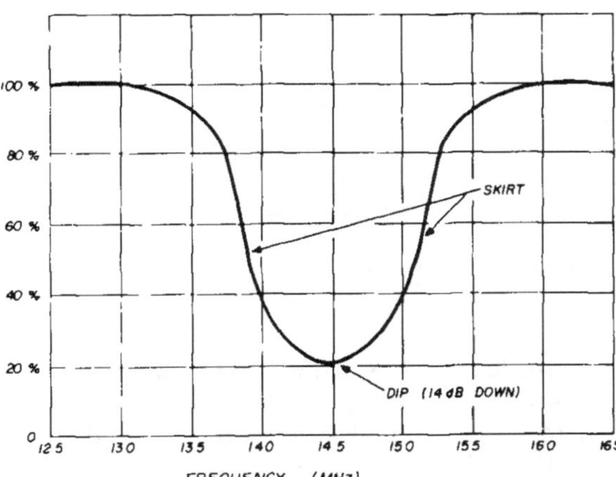

You'll understand better as I show you what it does, but here's a brief description.

An oscillator creates an rf signal at whatever frequency is set on the generator's dial. A sweep circuit in the generator is connected so it tunes the frequency of the main oscillator up and down, as if you were turning the dial above and below the frequency you first set it at. However, this sweep circuit is driven by a 60-Hz signal from the power line, and it swings the oscillator frequency up and down 60 times every second.

Suppose, for example, you set the dial for 14.5 MHz. The sweep circuit swings the oscillator above that and below it. If the generator is set to sweep 5 MHz, the oscillator is swung downward to 12 MHz, then up through 14.5 MHz to 17 MHz. It thus goes from center to one end, to the other end, and back to center, 60 times per second. The 5 MHz is called the *sweep width* and the 14.5 MHz is called the *center frequency*. The 60 Hz is called the *sweep rate;* that's the same on all service-type sweep generators.

response curve

Now for some principles of what the instrument is for. Let me review tuned circuits. They are the only reason you ever need a sweep generator.

The most important characteristic of an LC circuit is the frequency it's resonant to. Second most important is how well it rejects nearby frequencies — in other words, how sharp its resonance is. Together, these are the *response* of a tuned circuit.

You can plot the response of any tuned coil-capacitor combination. You just feed in a lot of different rf signals, one at a time, and graph how the circuit responds to each one. The hookup is shown in **fig. 2A**, and the graph of results with one tuned circuit is **fig. 2B**.

Here's how it's done. Tune the generator to whatever frequency makes the highest reading on the vtvm. Mark that on the center of the graph at the 100% line, and write the frequency directly below it along the bottom of the graph. This one happens to be peaked at 14.5 MHz.

Then start slowly downward with the tuning dial of the generator. Above the 14.25-MHz line, make a dot that represents the voltage that gets through the

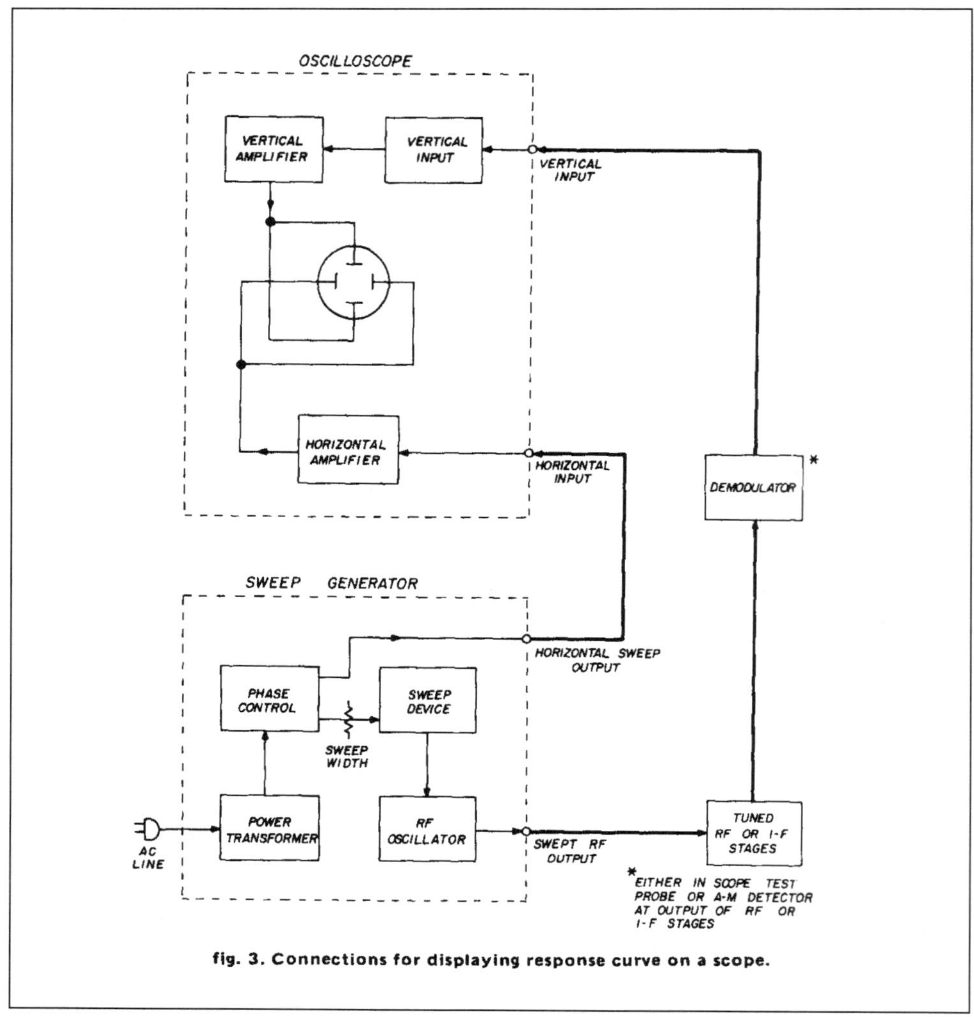

fig. 3. Connections for displaying response curve on a scope.

transformer. Show it as a percentage of what the voltage is at peak. As you tune the generator on downward, make voltage-level dots at every 0.25-MHz increment: at 14.0, 13.75, 13.5, 13.25, 13.0, 12.75, and 12.5 MHz.

Then start again at the peak and move the generator dial upward. Mark the voltages at 14.75, 15.0, 15.25, 15.5, 15.75, 16.0, 16.25, and 16.5 MHz.

These dots represent how well the tuned transformer responds to each frequency. Joining the dots together with a solid line helps you interpolate responses at frequencies in between the increments you measured. The solid line forms a curve, which is a *response curve* for that particular tuned transformer.

This particular transformer is fairly broadband. The 6-dB points (half-way down) on its curve are at about 13.8 and 15.2 MHz. Its 6-dB bandwidth is approximately 1.4 MHz. At and below 13 MHz, response is nil; and it's nil at 16 MHz and above.

If you plot the response of a tuned circuit that's connected as a trap, the curve is inverted, as in **fig. 2C.** You plot this one starting at some frequency well below the expected response of the tuned circuit. The meter registers the full signal from the rf generator. As you turn the dial toward resonance, the rf voltage reaching the meter starts dwindling – at 13 MHz. It gets less and less. Once you tune past the resonant frequency, voltage of rf reaching the meter rises again. Eventually, you tune the rf generator beyond the influence of the tuned circuit. That's beyond 16 MHz.

plotting automatically

What if you could turn the generator dial back and forth very rapidly and just as rapidly plot *all* the voltage points instead of just some? You'd develop the response curve a lot quicker. Well, you can. In fact, the whole thing can be done automatically.

You use an oscilloscope as the voltage-measuring device. And you use a sweep generator to swing the frequency up and down rapidly – 60 times per second. If the scope is synchronized to move its beam back and forth 60 times each second, the voltages measured at all frequency points by its vertical amplifier are displayed side by side. The outcome is a continuous curve of voltages right on the screen of the scope.

fig. 4. Settings of scope controls to produce response curve. Photo at bottom right shows a baseline before the demodulated signal is applied.

The sketch in **fig. 3** lets you see in detail how to make the connections. The same 60-Hz voltage that operates the sweep circuit in the generator is fed by a connecting cable to the horizontal input of the scope. Applied to the horizontal deflection plates of the scope crt, it swings the beam back and forth exactly in rhythm with the sweeping of the rf oscillator in the sweep generator.

At the same time, the swept rf signal is fed to the tuned stages of circuits whose response you want to view. A demodulator probe (or a detector with the stages) develops an output voltage for each and every frequency in the band being swept, one right after another.

Those voltages go to the vertical plates of the crt. They move the beam upward in proportion to each voltage. Since the beam is at the same time being swept from side to side, each voltage level appears one after the other. And, since it is being swept exactly in step with the sweep-generator rf signal, the voltage levels occur in the same sequence as the frequencies.

This is done over and over, 60 times a second. Your eye sees an automatically plotted curve of the response. It's the response of the whole group of tuned circuits or stages to the band of frequencies being swept by the generator.

sweeping a receiver i-f

As I said, there are dozens of ways you can use this ability to see the response of a tuned circuit or stage. So I'll show you how to set it up. The i-f of a ham receiver makes a good example, but this hookup can work for any tuned circuit or any group of tuned stages. Just feed swept rf into the input and feed the output to your scope.

First connect up the sweep generator and scope as diagramed in **fig. 3.** Horizontal output of generator to horizontal input of scope. Rf output of generator to input of tuned stages (in this example, to the input grid or base of the i-f section). Using direct probe, connect scope vertical input to output of a-m detector. (If there's no detector, use the scope's demodulator probe.)

The photos in **fig. 4** show settings of the important scope controls. Horizontal Sweep to "Ext." Vertical Input attenuator to X1 (the object is to set up the scope so a 2-volt peak-to-peak input signal makes a 2-inch vertical display). Horizontal Gain up just enough to make a base line about 3 inches wide. Positioning controls to keep base line below center of screen. (If the diode in your demodulator probe or the a-m detector is connected with its anode on the output side, position the trace above center, because the response curve will come out negative – below the base line.) The last photo of **fig. 4** shows the scope crt screen, with the base line properly set up.

You set up the generator as pictured in **fig. 5.** Rf dial at frequency near center of bandpass you expect in tuned stages. Sweep Width about twice as wide as you expect. (The i-f in this receiver has a center of 5 MHz, and a normal bandwidth of 50 kHz. A very low sweep width setting is called for – about 100 or 150 kHz. A television video i-f requires a sweep width setting of 10 or 12 MHz.) Phase control must be set after response curve is visible on scope crt. Output control set just high enough to produce 2-inch display on scope (exact setting depends on amplification in tuned stages being tested).

fig. 5. Control settings for the sweep generator.

Fig. 6 is what response curves look like. The top left photo shows it before your adjust the Phase control on the sweep generator. The top right one is a normal response curve of an operating i-f section in a ham receiver. The curve of any tuned circuit, or any group of circuits tuned to the same frequency, should look

fig. 6. Response curves.

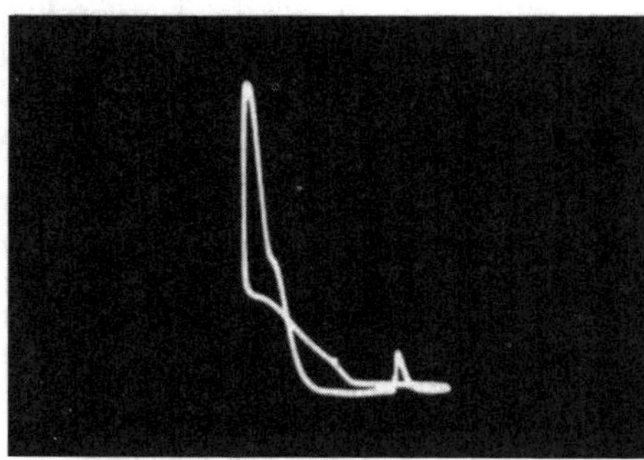

Phase control needs adjusting.

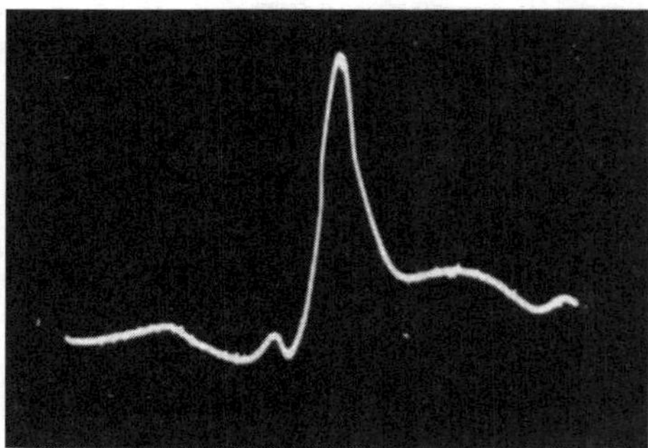

Normal narrowband curve.

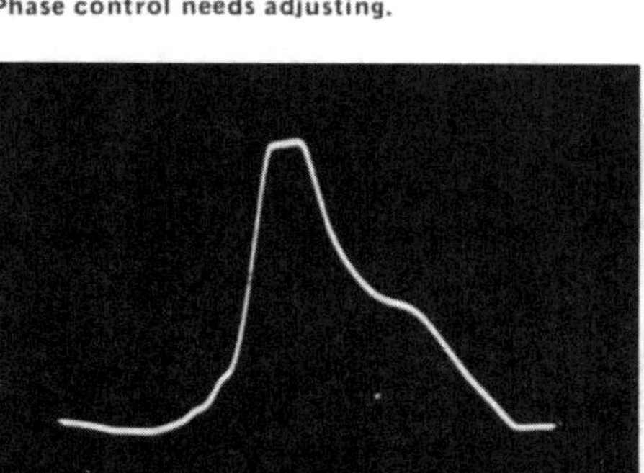

Bandwidth spread out by misadjusted slug.

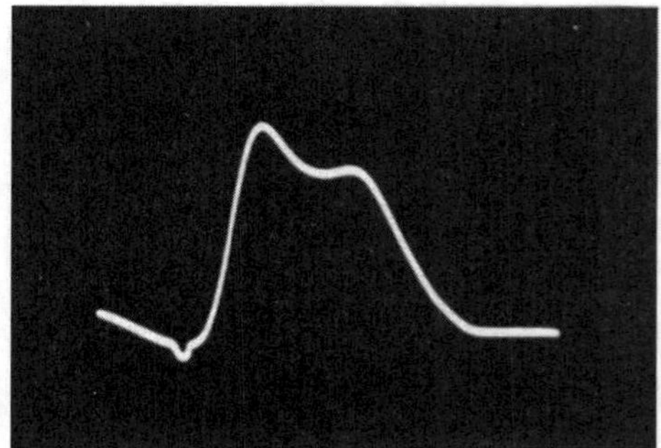

Wideband response of video i-f strip.

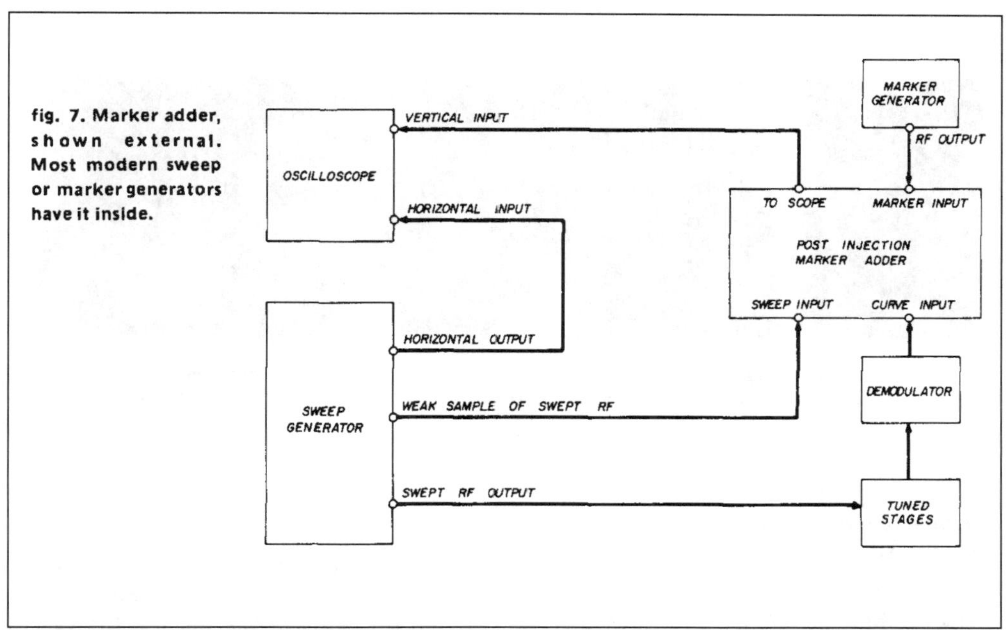

fig. 7. Marker adder, shown external. Most modern sweep or marker generators have it inside.

like this.

The bottom left photo is the response curve of the same i-f stages, but with two of the transformer adjustments misadjusted. As you can see, you can actually broaden out the response of the i-f, if that's what you want. For best QRM rejection, of course, you want the curve steep, narrow, and tall. That means, respectively, good adjacent-carrier rejection, good selectivity, and good amplification.

Incidentally, you might have to "clamp" the agc or avc line. Some receivers can't show a true response curve with the agc working. The alignment instructions will tell how much dc voltage to apply to the agc line if such a step is necessary.

wideband tuning

The bottom right photo in **fig. 6** is the response curve of a television i-f strip. Any modern TV i-f has several stages tuned to different frequencies. It's called *staggered tuning.* For example, the input to the first stage may be tuned to 42.4 MHz, the interstage circuits to 43.0 and 44.5 MHz, and the output of the third stage to 45.0 MHz. These four response curves, when the stages are cascaded as TV i-f stages are, appear to line up side by side. The result is the wideband response curve you see.

There are trap circuits in a TV i-f, too. They are at 39.75, 41.25, and 47.25 MHz. Their major purpose is to eliminate any signals at those three frequencies; they interfere with the wanted signals. On the response curve, they are responsible for how steep the skirts are. Right beside the amplifying response curves, the trap curves drop the response faster at the edges than ordinary tuned circuits could.

marking the frequency

One deficiency of a response curve displayed like this: you can't know exactly the frequency of the peak or of the points where response drops off rapidly. This is particularly important in a curve like the TV i-f response. You need to know where certain precise frequencies appear on the slope or along the top of the curve.

A *marker generator* and *marker adder* are the instruments for this purpose. At one time, they were separate instruments. Nowadays, the marker adder is part of either the sweep generator or the marker generator.

A marker generator is merely an accurate rf signal generator. Any stable rf generator can be used for marking re-

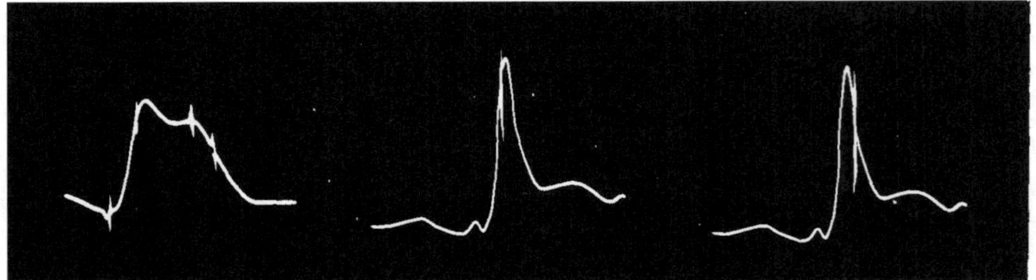

fig. 8. Frequency markers on response curves. On the left the markers are provided with a special marker generator. Markers are used in the two photos to the right to show the bandwidth of the curve.

sponse curves, if it is accurate or easily calibrated.

The best way to use markers is by what's known as *post-injection.* The marker is added to the response curve after demodulation. (The old way was to feed the marker signal right in with the sweep-generator rf; it often upset the tuned circuits and made a false curve.) The sketch in **fig. 7** shows marker adder connections. The instrument is shown separately, but it's usually part of one of the generators. In that case, some of the connections are made internally.

For television, there are multimarker instruments available, with crystal-controlled marker signals. Such a generator displays markers at several points on the response curve simultaneously. The one at the left in **fig. 8** is an example. Marker frequencies are identified, to give you an idea how they help you recognize the true bandpass of the i-f stages being tested.

The ham-receiver curve in the middle has only one marker. It's labeled, so you know where (in frequency) the down-frequency skirt of the curve is. In actual practice, you find this frequency from the dial of the marker generator. The right-hand photo is the same curve, with a marker at the start of the up-frequency skirt. The frequency spread between the two skirts is the difference between the two marker frequencies. In this case, it's 0.1 MHz. That's the bandwidth of the i-f strip from which this curve is taken.

what's to come

Armed with this information about how a sweep generator is used, you should find a lot of specific jobs for it on the ham bench. You can test and adjust filters, bandpass transformers, critical coupling between windings of transformers you wind yourself. With practice, you'll find the instrument easy to use. It tells you if your alignment job on a receiver (or the frequency conversion adjustment of an ssb transmitter) has produced the intended result. You can test any circuit that's within the rf range of the sweep generator you own.

Therein lies one drawback. None of the units available today go below a center frequency of 3 MHz. So, how do you do anything with 1.8-MHz circuits? Or the 60-, 455-, and 2.2-MHz i-f sections? If you find yourself using a sweep generator, and want to make it useful in this range of low frequencies, drop me a letter or postcard. If there's enough interest, I'll devote this department to that some month.

Next month, I write about mobile power supplies. They've come a long way since the days of the vibrator. Not every ham knows what to do with one when it's bad. Those transistor converter circuits look too complicated.

But they're not, really, Once you understand how and why they work, they seem simple. And fixing them on the repair bench is easy.

low-power dummy load and rf wattmeter

An accurate and reliable test instrument that can be built for less than five dollars

Neil Johnson, W2OLU, 74 Pine Tree Road, Tappan, New York 10983

The advantage of using a dummy load for rf tests are many. No interference is created, and station identification isn't required for prolonged test periods. If the dummy load contains a calibrated meter, you can measure rf power with good accuracy.

Most hams have dummy loads that will handle a kilowatt. While these are fine for their intended purpose, most won't provide accurate readings at power levels of 25 watts or less.

The unit described in this article is a highly accurate dummy load and rf wattmeter that can be used for testing low-power transmitters. Two versions are described: one is limited to 11 watts, and the other is good for 25 watts. Essential components cost less than two dollars. The enclosure, connector, and handle bring the total cost to less than four dollars. Sound intersting?

11-watt load

My original load used eleven 2-watt resistors in parallel. To save a buck, I used ten compostion resistors, each 560 ohms, and one 680-ohm resistor. These are rated at 2 watts, 10 percent tolerance. I chose them because they're widely available and cost less in quantities of ten units.

The resistance of this dummy load is 51.73 ohms, which provides a good match for popular coaxial cables. Power capacity is limited to 11 watts. This is because composition resistors should not be used at more than 50 percent of their power rating, otherwise their resistance will change, which will degrade the accuracy of the wattmeter.

I mention this for those who might wish to build a very low-power unit with readily available resistors. The schematic of **fig. 1** can be used, substituting the composition resistor bank for R (see photo).

25-watt load

Better resistors will allow rf power measurements up to 25 watts. The better units turned out to be another of those surplus bargains that appear from time-to-time.* I ordered two of these and connected them in parallel to obtain 50 ohms.

In an earlier project I used the 25-watt version of these resistors in a dummy load.[1] Four resistors, each 50 ohms, were connected in series-parallel. The CGW resistors can be extremely overloaded and will retain their accuracy to a remarkable degree. The 25-watt units used in the load described in reference 1 have been sold out. However, for those who may have some, the data provided in **table 1** should be helpful. I believe the overload characteristics with respect to permanent resistance change also apply to the 13-watt units.

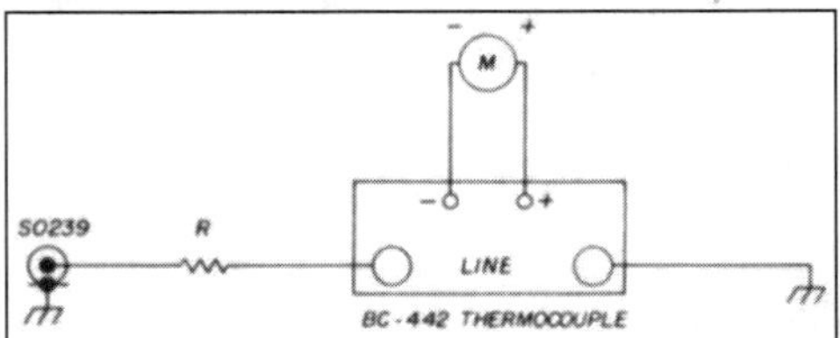

fig. 1. Schematic of the 25-watt dummy load and rf wattmeter. Resistor is two 100-ohm resistors in parallel; each resistor 13 watts, 1% non-inductive type.

thermocouple and meter

These components are another "ham special" value.† Both units came from the BC-442 antenna tuning unit. They are first-rate, rugged, accurate components originally designed for military use. The dc meter has a nonlinear scale, a relatively expensive type of meter construction not generally associated with amateur equipment. This gives an expanded scale at the low end, and rf output as low as 2 watts can be measured.

My experience with this meter and thermocouple has shown them to be highly accurate and dependable. This comes as no surprise when it's realized that these units had to withstand constant vibration in flight, hour after hour, plus the many shocks in taking off and landing at some rugged air strips.

*John Meshna, Jr., 19 Allerton Street, Lynn, Massachusetts 01904. Catalog listing: "Corning Glass Works, Tin-Oxide Film Resistors; 100 ohms, 13 watts, 1% tolerance." Price: 35 cents each.

†Fair Radio Sales, P. O. Box 1105, Lima, Ohio 45802; $1.25 for both thermocouple and meter.

table 1. Data for the CGW 25-watt resistor.

tolerance	2.5 or 10%
stability	less than 1% permanent resistance change*
operating temp	25°C
manufacturer	Corning Glass Works, Raleigh, N. C.
marking	CGW R35 25W 50 ohms

*When operated at 10 times rated power for 5 seconds.

construction

Construction details are shown in the photos. The usual procedure of keeping rf leads short and direct should be followed in wiring this simple instrument. This will ensure a vswr of 1.1 or better throughout most of the amateur high-frequency bands.

Inside the wattmeter, showing the CGW "R Series" resistors.

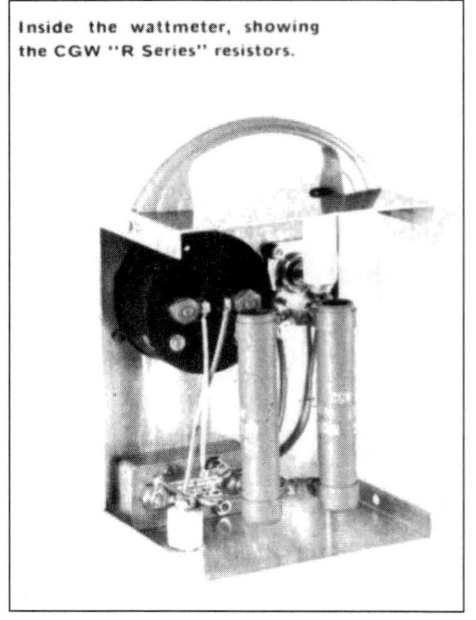

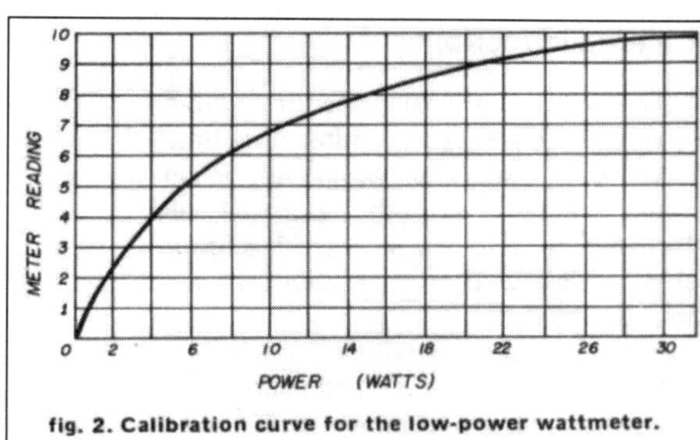

fig. 2. Calibration curve for the low-power wattmeter.

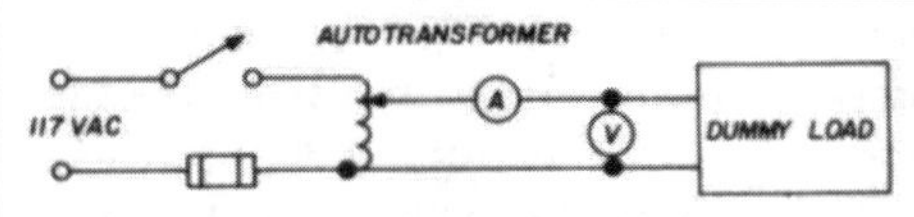

fig. 3. Circuit used for calibration. A is a Weston model 433 ac ammeter; V is a Simpson model 261 ac voltmeter (see text).

calibration

The calibration data shown in **table 2** and **fig. 2** were obtained with the setup shown in **fig. 3**. The wattmeter was calibrated at 60 Hz, using mirror-scale laboratory instruments: a Weston Model 433 ac ammeter and a Simpson Model 261 voltmeter.

The voltmeter takes power, so I checked it against the Weston 433 before starting the calibration procedure. Readings were checked going up and down the scale. Then a second check was made to ensure repeatability of the data.

Note that the meter reading at 10 **(table 2)** occurs when the resistors are operated beyond their maximum power rating. No harm is done to either meter or resistors. A temporary effect was that the resistors changed value to 50.5 ohms, and the entire dummy-load resistance increased to 51.5 ohms. The resistors returned to their original value upon removal of rf power.

The 51-ohm resistance of the unit is due to the series combination of the resistors (50 ohms) and the thermocouple, approximately 1 ohm. Note that readings below 0.23 ampere are not valid.

If you intend to use the instrument at levels above 15 watts for extended periods, some means of auxiliary cooling should be used. Removing the cover will also help.

table 2. Calibration data.

meter reading	true rf amperes	rf watts into 51 ohms
1	—	0.9
2	—	1.8
3	0.23	2.7
4	0.28	4.0
5	0.335	5.6
6	0.39	7.9
7	0.46	11.0
8	0.54	15.2
9	0.64	21.4
10	0.78	31.4

Composition resistor bank used in the 11-watt unit.

conclusion

The amount of abuse the CGW resistors will take staggers the imagination. For example, my old 200-watt load[1] withstood a kilowatt for 5 seconds maximum without resistor values changing more than 1 percent (permanent change after cooling). By the same token, the smaller resistors, which have a nominal rating of 26 watts per pair, could be run up to 260 watts.

A high-quality rf wattmeter, good for any power up to 25 watts, is not a bad investment for $4 and a pleasant hour's work.

reference

1 Neil Johnson, W2OLU, "A 200-Watt Dummy Load for $2.00", *73* May, 1967, p. 66.

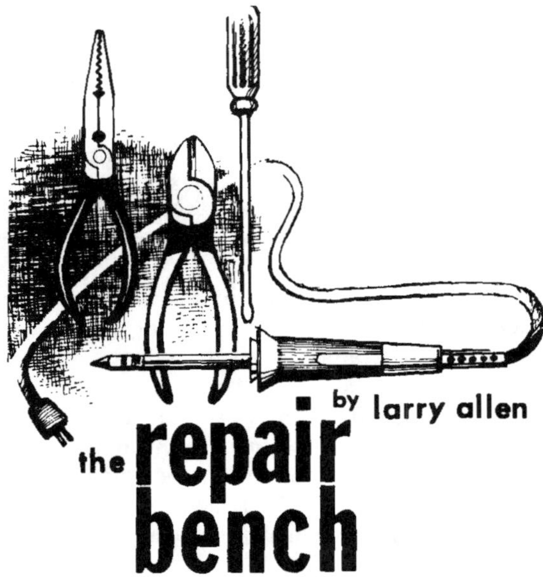

the repair bench

by larry allen

thinking your way through repairs

The editor sends me a lot of letters from fans of the *repair bench.* Many of them ask for help with this or that problem. Naturally, I can't answer them all. Yet, there should be some way to help you, and perhaps what I'm writing about this month will do it.

Very few hams are professional repair technicians. If they were, they'd probably get enough of electronics all day long and wouldn't fool with it as a hobby. But when something goes wrong with your gear, even knowing exactly how it's built and how it works doesn't always lead you to the defect very quickly. At those times you probably wish you were a technician trained in hunting down electronic faults.

I can't train you in a few magazine pages to be a technician. But I can show you a technique used by professionals in home-entertainment electronics.

There's one method that should be easy to learn. It's called *1-2-3-4 Servicing.* It was originated by Forest H. Belt in some books published by Howard W. Sams & Co. Although the books are not about ham gear, I could see immediately how easily the method applies to any kind of electronics. So here's how *1-2-3-4 Servicing* can be applied to equipment on your ham repair bench.

thinking in sections

The important thing about 1-2-3-4 Servicing is the way you think. You have to think of electronic equipment in an orderly fashion. For that purpose, all electronic devices are split up into four logical divisions. They are sections, stages, circuits, and parts.

First, you think *sections.* Take a ham receiver, for example. It has an rf section, i-f section, audio section, and power supply section. These are sketched for you in **fig. 1.** You might be tempted to break it down further, but these are the breakdowns you use for 1-2-3-4 Servicing.

A transmitter divides up into the speech or audio section, rf generating section, modulated rf section, rf power section, and power supply section. If the transmitter is ssb, there is also a frequency-change section. Transmitter sections are illustrated in **fig. 1,** too, and the comparison with the sections of a receiver may help you see what sections really are.

The distinction that identifies a section is simple: *A section handles only one kind of signal.*

In a receiver, for example, the rf section handles incoming stations. The i-f section amplifies the station signal only after it has been heterodyned down to the intermediate frequency of the receiver. The audio section handles voice signals alone, after they have been demodulated from the i-f signal.

In a transmitter, as long as audio (voice or speech) is by itself the audio section handles it. The rf signal, as long as it's alone, is in the rf-generating section. Once modulated, the signal is handled by a different section. If the transmitter is ssb, a frequency-changing section takes the modulated signal and changes its frequency to the transmitting frequency. Once the modulated signal is at the output frequency, an rf power section boosts it.

That's how you can identify sections in any kind of electronic gear. When the signal changes character, consider it as going into a different section.

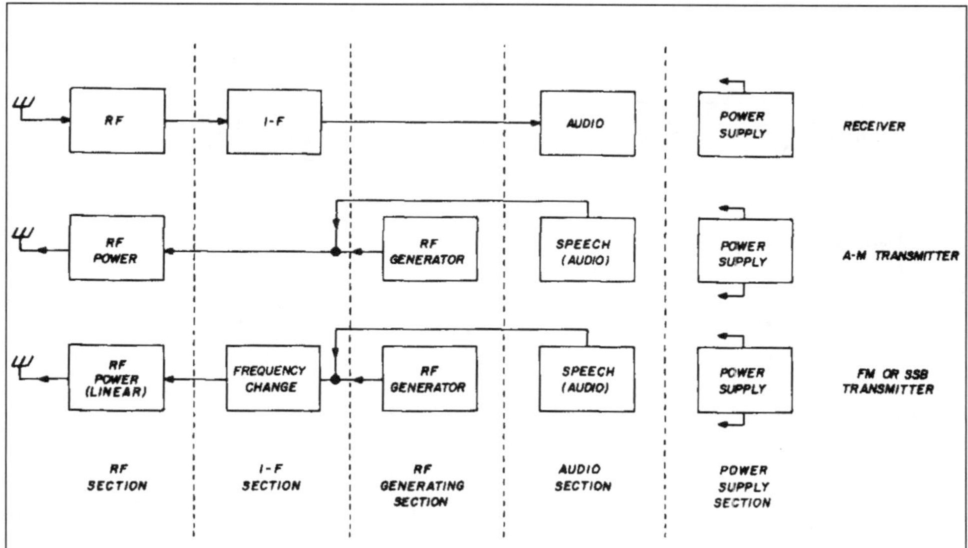

fig. 1. Sectionalizing ham equipment helps track down faults. Sections of different kinds of equipment are amazingly alike — if you learn the sections of one it's easy to divide up other electronic gear in this way.

stage by stage

The second division you think of for 1-2-3-4 Servicing is *stages.* Sections are divided into stages.

For example, the rf section of a ham receiver has at least three stages. They're shown in **fig. 2A.** There may be more than one rf amplifier stage. A stage usually comprises one tube or one transistor and the parts that go with it. A stage either generates or amplifies or alters a signal.

Inside this rf section, the oscillator stage produces a plain rf signal. The rf amp gives a boost to the signal picked up by the antenna (which can also be considered a stage). Both signals go to the mixer. That stage changes the character of the rf signals, mixing them together and coming up with a signal at some intermediate frequency (their difference).

The i-f signal still has the station modulation that was on the rf signal. But the character of the signal has changed. It takes a different kind of stage to handle it. So the mixer stage has introduced the signal from the rf section into the i-f section.

The mixer stage, though officially part of the rf section, is an *interface stage.* It interfaces the rf section to the i-f section. The stage that creates the character change in the signal is an interface stage.

The detector or demodulator following the i-f amplifier stages is an interface stage. It extracts the voice signal that is part of the i-f signal. Once demodulation takes place, there is no more i-f signal — only the voice or audio signal. So the detector stage interfaces the i-f section to the audio section. Amplifier stages in the audio section handle only audio signals.

If you want to carry the idea further, the speaker is a stage of the audio section. Actually, it's an interface stage between the audio section and the air. It converts the electronic audio signal into audible sound waves. That's what any interface stage does: change the nature of the signal between two sections.

An interface stage is always considered part of the section preceding. Sometimes it takes signals from more than one stage, as the mixer does in the rf section of a receiver.

Or, a stage may interface more than two sections. **Fig. 2B** has an example of this. This is the stage-by-stage division of an ssb transmitter. Stages are grouped

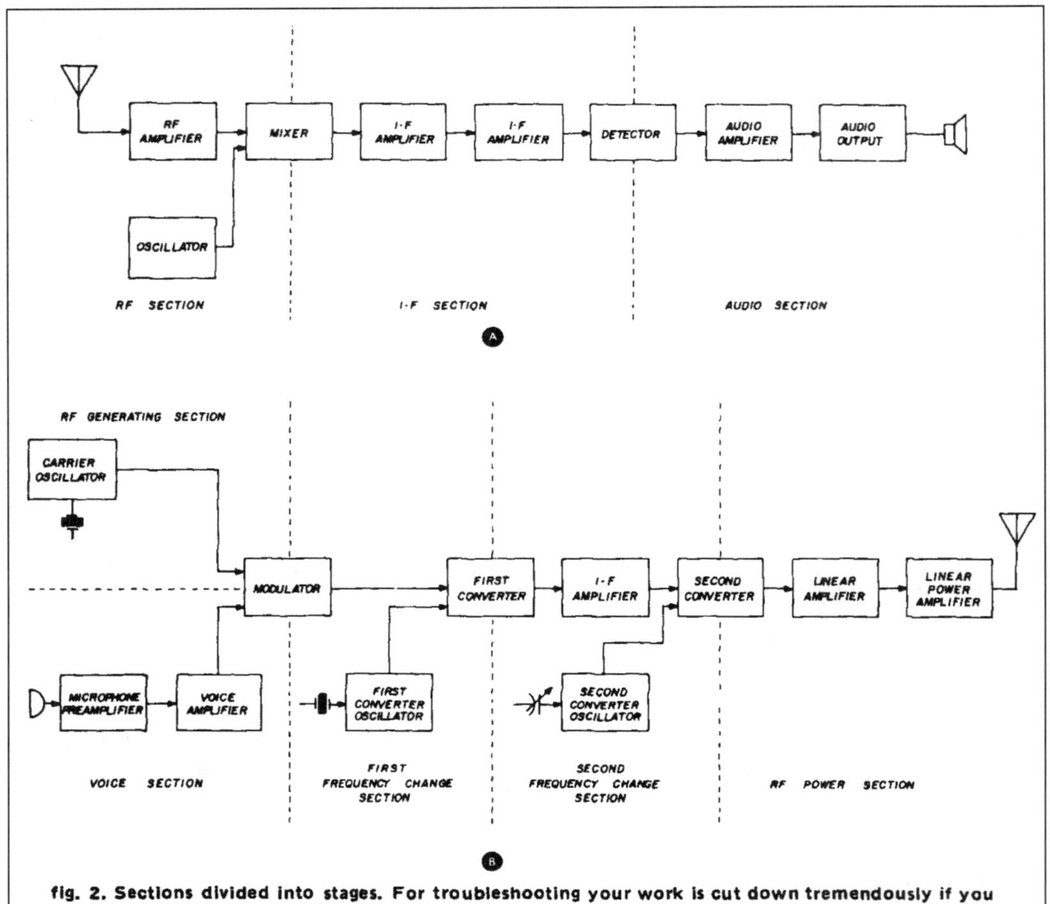

fig. 2. Sections divided into stages. For troubleshooting your work is cut down tremendously if you only have to test the stages in one section.

into sections so you can see the relationships. The voice section and rf-generating section are interfaced with the first frequency-change section by the modulator stage.

The voice section has two stages. The rf generator section (which can be called the *carrier* section in an ssb transmitter) also has two: the carrier oscillator and the modulator. It's customary to consider the modulator as an rf stage rather than a voice or audio stage, even though it involves both kinds of signals. The modulator takes the two signals and transforms them into a modulated rf signal. That calls for a different section, so the modulator is an interface stage.

The first converter stage interfaces the first frequency-change section to the second. This ssb transmitter uses double conversion to arrive at the output frequency. The amplifier stage in the second frequency-change section is called an i-f amp. (In a few transceivers, it's the same i-f amp used during receiving.)

The modulated i-f signal goes to the second converter, the interface stage between the second frequency-change section and the rf power section. A second conversion oscillator stage furnishes the unmodulated signal for conversion. The output of the second converter is a modulated rf signal at the output frequency.

All that remains is to amplify the output signal to full output power of the transmitter and feed it to the antenna. The rf power amps and antenna are stages of the rf power section.

what a circuit really is

That should give you a good idea how

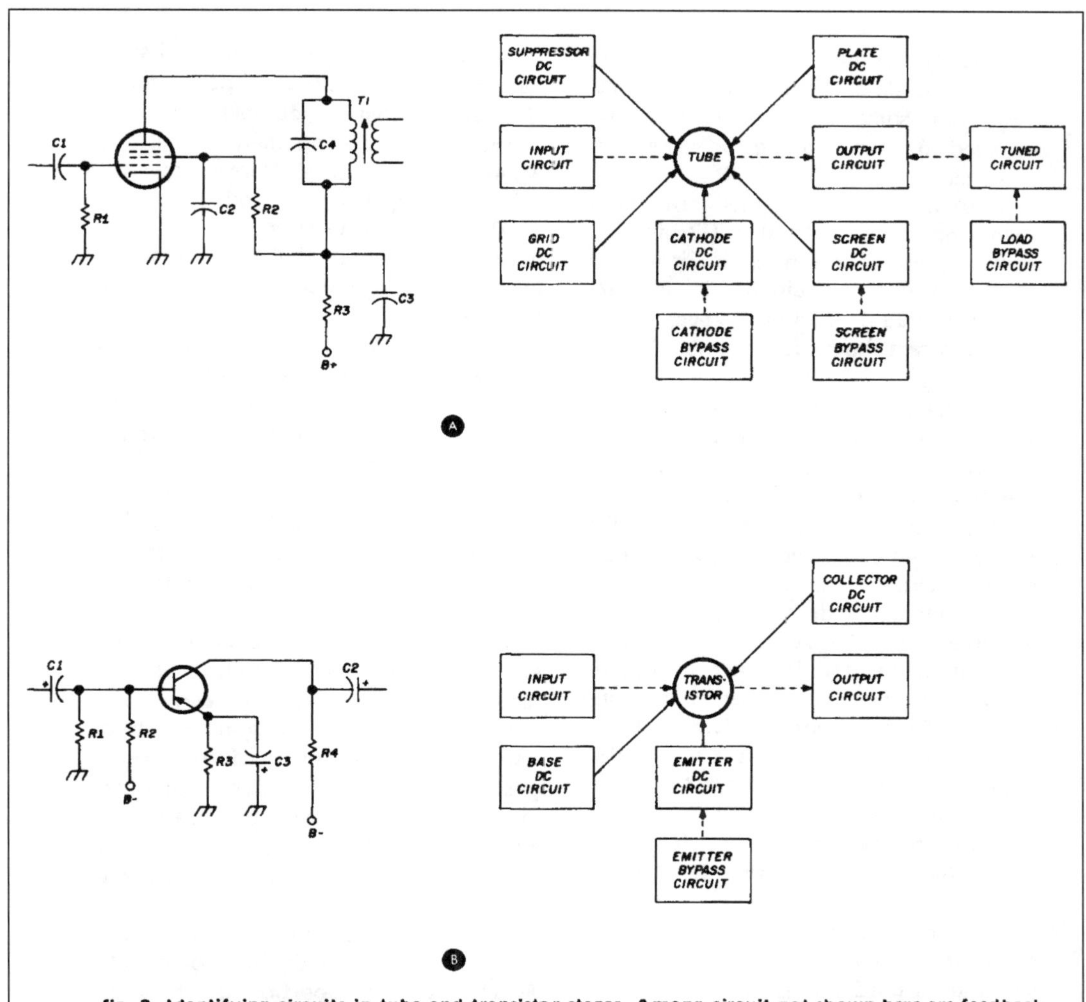

fig. 3. Identifying circuits in tube and transistor stages. Among circuit not shown here are feedback circuits (such as in an oscillator) and power-supply decoupling circuits (which aren't always drawn on schematics).

to think in sections and stages. Be sure you understand these divisions, because they're important in applying 1-2-3-4 Servicing.

The third logical division of electronic equipment is the *circuit*. Many hams — and a lot of technicians, too — think of circuits as being the things just described as stages. That's confusing. Actually, stages are made up of circuits.

Remember, a stage is a tube or transistor and the parts that go with it. Those parts are connected into *circuits*. There's an input circuit and an output circuit; a grid circuit, plate circuit, cathode circuit, and perhaps a screen circuit; a base circuit, emitter circuit, and collector circuit; there may be a feedback circuit, a tuned circuit, and so on. Any given tube or transistor stage has several circuits.

What you name the circuits depends on how you're considering the stage. For example, the tube and transistor stages in **fig. 3** have both dc and signal circuits identified for you. Solid arrows are for dc, and dashed arrows for signals.

The input circuit in both stages is C1 and R1. C1 is the input coupling capacitor, R1 the input load.

The output circuit for the tube is T1, C4, and C3. T1 and C4 are a tuned circuit that is the main load, and C3 is the load decoupling or bypass part of the circuit.

At the same time, the plate dc circuit

includes the primary winding of T1. The whole plate dc circuit is through R3 and T1. Thus some components must be involved in your thinking for more than one circuit.

Indeed, some components are in more than one circuit. Resistor R3, for instance, is in the screen dc circuit as well as in the plate dc circuit. In troubleshooting, you'd naturally expect a bad R3 to affect both dc circuits. And it would.

By the same token, R1 in the transistor stage is part of the input circuit for signals. Yet, at the same time it's part of the base dc circuit; it's part of divider R1-R2 which sets base bias. Resistor R4 is the output load, and is also in the dc collector circuit.

Two bypass or decoupling circuits are invisible in **fig. 3B.** Anytime a B-plus or B-minus connection is shown in a schematic, you must remember that it includes a filter capacitor, which is usually in the power supply. But it is part of the signal circuit of any stage fed from that leg or branch of the power supply.

When troubleshooting, don't ignore this bypass circuit. In **fig. 3B** a filter capacitor is part of the output load circuit, forming the ground return for signal at the bottom of R4. The bypass capacitor at the B-minus end of R2 makes R2 a part of the input load, in parallel with R1.

what makes up a circuit?

Of course the answer to that question is *parts.* You've already seen that, in my description of what a circuit is. And parts are the fourth division of electronic equipment.

One secret of 1-2-3-4 Servicing is this manner of thinking. You consider all electronic equipment in terms of these four divisions. You think of a receiver or transmitter as divided into sections. The sections, you think of as divided into stages. Stages are in turn divided into circuits, and those are divided into individual parts.

The four steps of thinking are, therefore: sections, stages, circuits, and parts.

Once you've learned to think of your ham gear as made up of sections, stages, circuits, and parts, you can use those divisions in an exceptionally logical way when trouble occurs. Here's how the 1-2-3-4 Servicing method works.

The first step is *diagnosis.* You diagnose which section of the faulty instrument has the trouble in it. There are three main helps to diagnosis.

One is to know the equipment. Be familiar with it. Study the schematic and read the instruction manual. Read books about it. Gain knowledge of it through experience. Once you're familiar with a receiver or transmitter, you know what sections it has. That aids diagnosis. For example, if a transmitter puts out rf, but not modulated, you know the trouble must be in the voice section.

Another approach to diagnosis is inspection. *Look* inside the unit. *Listen* to a receiver. Or listen to the relays of a transmitter. Or listen for some unnatural sound like a resistor or transformer frying or something arcing. *Sniff,* and sometimes *touch* – perhaps to sense overheating. In other words, use your senses to inspect the unit. They may help you diagnose which section has trouble.

And of course you diagnose by observing symptoms. Is the unit completely dead? Does some portion of it work? Some one function not work right? Do the operating controls work as they're supposed to? How about service adjustments? Such analysis of symptoms can point strongly to which section of a transmitter or receiver is bad.

the second step

Once you diagnose the faulty section, your next step is to *locate* the defective stage. Since you already know the faulty section, you have only a few stages to check. Instead of checking a whole setfull of stages, you check only two or three – those in the section you diagnosed.

You might locate the faulty stage by observing symptoms, or by noticing how certain controls react. But more likely you'll use instruments.

You already know there are two ways to look at stages. They are dc-operated, and they handle signals. You can test them for either kind of operation. You can inject a signal at the input of a stage and see how it travels through the remainder of the set. Or, you can use a tracer to see how far a signal gets, to see what stage stops it or distorts it or somehow fouls it up.

Or, you can measure dc voltages on stages in the section you've diagnosed as faulty. If you find a stage with dc voltages upset, you've located the faulty one. (One thing about this: if several transistor stages are dc coupled, the first one you measure may not be the truly bad one. You may have to "inject" dc voltages at strategic points to find out which stage is causing the trouble.

isolating the circuit

The third step of 1-2-3-4 Servicing is *isolating* the defective circuit. You may have done it already with some of your locating procedures.

For example, you may have traced a signal in a receiver as far as the collector of one stage but find it missing at the base of the next. You've located the faulty stage or stages; but, what's more, you've also isolated the faulty circuit. It's the circuit between the collector of one stage and the base of the next. That can only be the coupling circuit.

But even if you have no idea what circuit is bad, you have only a few to test. That's because you've eliminated those in other stages and sections by the first two steps. You now have to test only the circuits in one or at most two stages.

Signal tracing or injection both work for input, output, and coupling circuits. You can try adjusting tuned circuits; if they don't respond, they must be faulty. You can put a signal across decoupling and bypass circuits; the signal should be almost wiped out. Or, you can measure signal across them; it should be near zero.

You can use dc voltage tests. In a tube stage, you measure voltages in the plate, screen, grid, and cathode circuits. Keep in mind as you do that the grid affects all those others. If the voltages all are wrong, check for a problem in the grid circuit.

table 1. Basic steps to thinking 1-2-3-4 in servicing electronic equipment.

step	action	division
1	diagnose	sections
2	locate	stages
3	isolate	circuits
4	pinpoint	parts

In a transistor stage, the collector, base, and emitter dc circuits are the ones to check voltages in. Remember that the base controls collector voltage through its influence on current through the transistor.

If the transistor is a field-effect type, the voltages to measure are at the drain, the source, and the gate. The gate affects the drain voltage through its control on current through the channel.

pinning down trouble

The fourth and final step in 1-2-3-4 Servicing is *pinpointing* the faulty part. The job has by now been simplified almost to the point of no effort. Once you have the faulty circuit isolated, there are very few parts to think about. Even if you aren't sure which of two circuits is bad, the number of parts is small.

You can extend the tests you used in step three. You can make signal and dc tests that pinpoint the faulty part. You may have already done so during step three. For example, when you isolated the faulty coupling circuit, there's only one part so it's the bad one. That happens with other circuits, too.

But there are so few parts involved in this final step of 1-2-3-4 Servicing, you can freely succumb to individual parts tests. You've gained so much efficiency through the first three steps, this may be the quickest way to finish up. You can test most parts with your voltmeter or ohmmeter. (If you don't know how, I can tell you in a later repair bench.) Or, if you have a resistor-capacitor tester, a transistor tester, etc., use them.

Another fast way to test ordinary parts is by substitution. You can probably find what you need in your junk box.

table 2. Steps for 1-2-3-4 servicing.

Step	Method	Details
DIAGNOSE (section)	A. Know the equipment	1. The schematic 2. Instruction manuals 3. Experience 4. Study books
	B. Inspect — Inside and out — Off and operating	1. Look 2. Listen 3. Smell 4. Feel
	C. Observe symptoms	1. Dead? 2. Works in part? 3. Operating controls? 4. Service adjustments?
LOCATE (stage)	A. Observe symptoms	1. Dead or operating poorly 2. Controls and adjustments
	B. Signal tests	1. Injection — Signal generators — Finger (tubes only) — From similar set
	C. Voltage tests	2. Tracing — Oscilloscope — Vtvm and probe — Signal tracer
ISOLATE (circuit)	A. Signal tests	1. Input circuit 2. Output circuit 3. Bypass circuits 4. Tuned circuits 5. Feedback circuits
	B. Dc voltage tests	1. Tube — Plate, — Screen, — Cathode } Grid affects these — Grid 2. Bipolar transistor — Collector — Base — Emitter 3. Field-effect transistor — Drain — Gate — Source
PINPOINT (part)	A. Signal tests (parts in signal circuits)	1. Tracing 2. Injection
	B. Dc voltage tests	1. In-line voltage tests (Use Ohm's and Kirchhoff's laws) 2. Tests in nearby circuits (For parts connected between) 3. Remember interaction — Grid affects plate, screen, cathode — Base affects collector and emitter — Gate affects drain and source
	C. Individual parts tests	1. With special testers 2. With volts and ohms tests
	D. Substitution	(Limit to common, inexpensive parts)

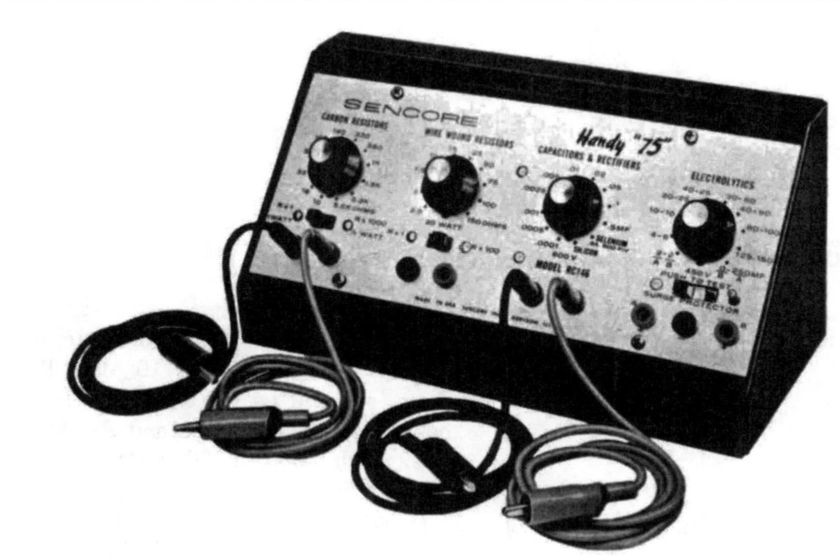

fig. 4. Commercially built parts substitutor saves carrying an inventory of small parts for test purposes.

Or, some manufacturers make substitution testers (**fig. 4**) that include capacitors and resistors of many values, some diodes, some electrolytics, and so on. Except for expensive parts, substitution is a good bet, even if you have to keep a small stock of typical values.

summing up

Now you can understand the chart in **table 1.** If you looked at it before reading how to apply 1-2-3-4 Servicing, it may not have made much sense. But it summarizes the steps of 1-2-3-4 Servicing as you apply them to troubles in your own equipment.

There's nothing mysterious about this method. If you're really efficient at finding trouble that occurs in your gear, you probably already use some version of this technique. If not, however, this logical approach can make troubleshooting easier than you ever thought possible.

To jog your memory for the method and how to apply it, you may want to cut out or copy the more complete chart in **table 2.** It gives you a thorough outline of the whole process and the means by which you can accomplish it. With it, you can apply 1-2-3-4 Servicing to any piece of ham gear you own.

next time . . .

Speaking of letters from you readers, I've really got a lot of them about using a sweep generator down in the low frequencies. In the very next *repair bench* I'll tell you how.

To get ready for it, I suggest you go back and review the earlier one about using a sweep generator and scope. I won't repeat much of the general information that tells how to set up the generator and scope for sweep alignment. That way, I spend most of my next *repair bench* explaining the techniques of sweeping at low frequencies and how to make use of it.

(Which all reminds me to remind you: If you have any other special ham repair problems you want covered in this department, it takes a letter to let me know.)

a stable, variable-output weak-signal source

Receiver alignment generator, frequency standard, neutralization signal source are just a few uses for this instrument

Bruce Clark, K6JYO, 1019 El Dorado Drive, Fullerton, California 92632

A necessary adjunct to any test-equipment bench is a stable weak-signal source. This device is basically a crystal-controlled signal generator with an adjustable output. Its applications are analogous to the tunable signal generator, but it can be a much more useful tool than either a signal generator or a crystal calibrator by combining the properties of both. Several vhf signal sources have been described in *ham radio*[1,2] and *VHFer.*[3] All either lack frequency stability as the output is varied, or else they have too low output at the higher harmonic frequencies. The unit shown in **fig. 1** uses a pair of inexpensive RCA 40673 dual-gate mosfets to eliminate these shortcomings.

circuit description

The addition of a crystal-selector switch, S1, allows the choice of frequencies from 1 through 10 MHz. Presently, my unit uses 1 and 3.5 MHz for the high-frequency range and 8.0 MHz, 8.222 MHz, and 8.333 MHz for frequencies from 50 through 1296 MHz. High harmonic output is assured by the use of an H-P 2800 hot-carrier diode multiplier. Usable harmonic energy up to 1296 MHz been observed on a spectrum analyzer. With the circuit shown, the output was -82 dBm at 1296 MHz, a readily detectable level. A plot of output vs. frequency is shown in **table 1.**

Basically, the circuit is an untuned Colpitts crystal oscillator driving a wideband buffer, which in turn drives the hot-carrier diode multiplier. Two pots are used to adjust the output level. R2 sets the voltage level of G2 in the oscillator. This control is set to either of two calibrated points depending on the application. For example, if weak signal testing is desired, R2 is adjusted to the point at which oscillation is barely sustained. If a strong signal is desired, R2 is set to maximum output. The output-adjust pot, R7, sets the voltage of G2 on the 40673 buffer. The pot shown gives a -40 dB change in output level. This adjustment range can be calibrated in conjunction with a known attenuator and your receiver S-meter. Another approach is to mark off a 10-dB range in 1 dB steps on R7's scale and use an external attenuator for coarse adjustments.

construction

I used double-sided copper circuit board soldered together to form a 3 x 6 x 1 1/2 inch box. An LMB or Bud aluminum box of equivalent size could be used and the circuit built on a copper

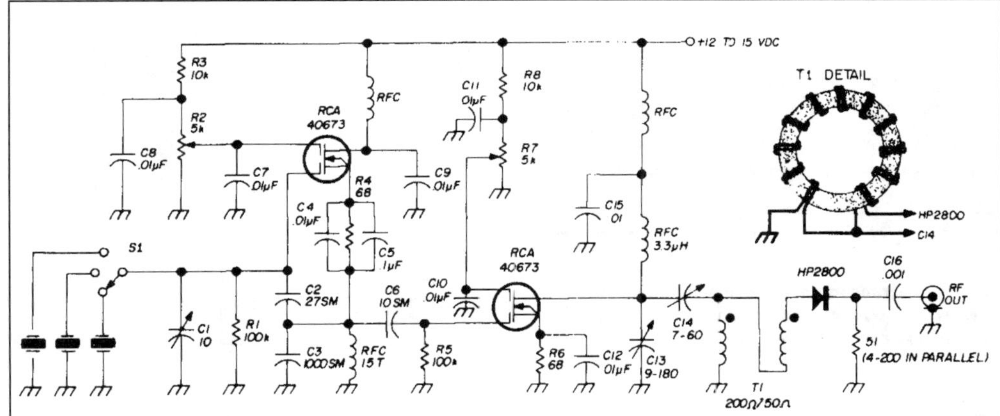

fig. 1. Schematic of the weak-signal source. T1: 7 turns no. 28 or 30 wire, bifilar-wound on no. 3E2A Ferroxcube toroid form. Rf chokes: 10 turns no. 24 or 26 on same form as T1 unless specified otherwise. C13: Arco 423. C14: Arco 404.

sub-chassis inside as shown in **fig. 2.** However, rf leakage may be a problem at low levels using this construction. Extra screws must be used to keep the box rf tight. This will minimize leakage for most applications.

A simple ac supply capable of supplying 12-15 V at 25 mA may be built in the box; or if desired, a 12-15 V battery supply may be substituted. If an ac supply is used, the line cord should be well bypassed with feedthrough capacitors to eliminate rf leakage from this source.

applications

The uses of the unit are many. With a 1- or 3.5-MHz crystal, it makes an excellent frequency standard, band-edge marker, or receiver alignment generator.

table 1. Output vs frequency.

frequency (MHz)	harmonic	output level (dBm)
8	-	-2*
16	2	-3.4
32	4	-17.4
48	6	-24.4
64	8	-29.0
144	18	-44.0
216	27	-51.2
432	54	-63.2
1296	162	-82.3

*0.625 mW across 50 ohms

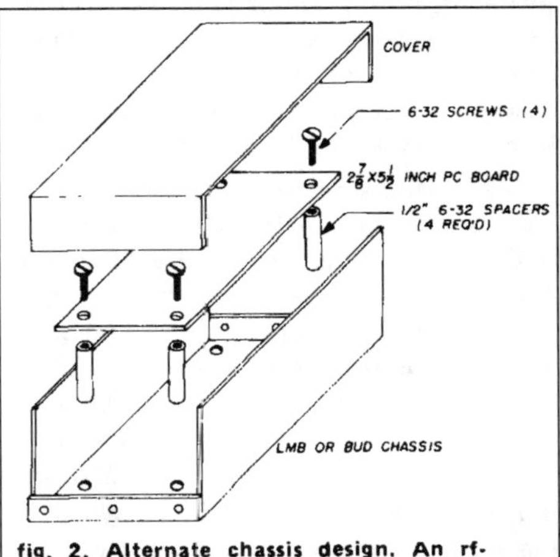

fig. 2. Alternate chassis design. An rf-tight enclosure is a must (see text).

With an 8-MHz crystal, it can serve as a weak-signal source for vhf front-end alignment. In its maximum power output mode, the unit becomes a signal source for neutralizing converters and high-power rf amplifiers.

Other applications include calibration of receiver S-meters and determination of receiver dynamic (agc) range. Also, you could probably replace the harmonic generator with a tuned circuit, key the buffer, and go super QRP!

vhf converter alignment

A word of caution about the use of the signal source for vhf front end align-

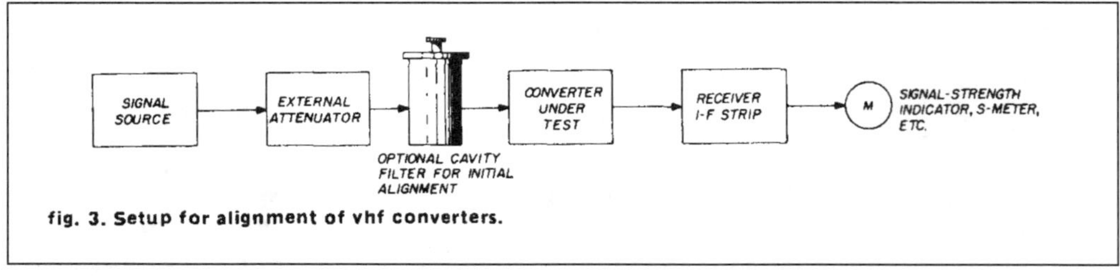

fig. 3. Setup for alignment of vhf converters.

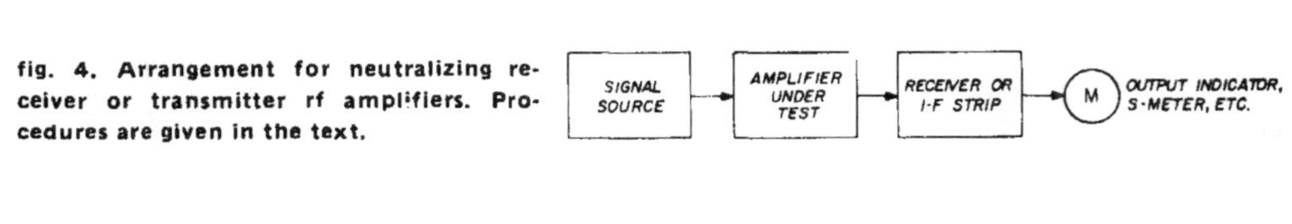

fig. 4. Arrangement for neutralizing receiver or transmitter rf amplifiers. Procedures are given in the text.

ment: if the device under test (**fig. 3**) has little front-end selectivity or is badly mistuned, it may be possible to tune it up on the wrong harmonic of the signal source; e. g., a 432-MHz converter with a 28-MHz i-f should be tuned up on its 54th harmonic (assuming an 8.000-MHz crystal in the source). However, the 47th harmonic of 8.00 MHz is 376 MHz, which is the image frequency of the converter. Since this is a lower-order harmonic, it contains more energy than the desired one and could be the source of a lot of frustration. A similar situation exists with a 144-MHz converter using a 28-MHz i-f where the 18th harmonic is 144 MHz and the 11th harmonic is 88 MHz, which is the image. The solution in these cases is to include some selectivity between the signal source and the unit under test, at least until you're sure the converter is properly adjusted to the desired frequency. A second method is to offset the crystal frequency a few hundred Hz high using C1. The lower-order harmonic on the image frequency will then fall below the desired harmonic and can be easily identified.

neutralization of rf amplifiers

Fig. 4 shows the arrangement for neutralizing a receiver or transmitter rf amplifier. The procedure for receiver amplifier neutralization is as follows:

1. With B+ off disable the amplifier by opening the source bias resistor or opening a filament lead.

2. With signal source at maximum output, tune in signal on receiver.

3. Peak input and output circuits, then adjust neutralizing control for a null. Repeat several times.

4. Reconnect source bias resistor of filament lead.

For transmitter power amplifiers:

1. For tube rigs, *turn off B+ and screen voltage.* For transistor transmitters, leave B+ on and open emitter lead.

2. Connect weak-signal source to input. Adjust to maximum signal level.

3. Connect receiver to rf output of amplifier.

4. Tune in signal source and peak amplifier input and output controls for maximum S-meter indication on receiver.

5. Adjust neutralizing capacitor for minimum signal feedthrough. Repeat several times for maximum null.

references

1. Del Crowell, K6RIL, "A 432 and 1296 MHz Signal Source," *ham radio,* September, 1969, p. 20.

2. J. W. Brannin, K6JC, "A Stable Small-Signal Source for 432 MHz," *ham radio,* March, 1970, p. 58.

3. S. Freeley, K6HMS, "A 144 MHz Weak-Signal Source," *VHF'er,* July, 1965.

low-cost

RX impedance bridge

Complete calibration and application information on the simple W2CTK impedance bridge

Bill Wildenhein, W8YFB, 41230 Butternut Ridge, Elyria, Ohio 44035

In the September, 1970, issue of *ham radio*, W2CTK described a simple impedance bridge capable of independent readout of R and X values.[1] Readers familiar with laboratory bridges will appreciate the potential of this instrument, but may doubt the accuracy of such a simple device. Less experienced readers may have shied away from this bridge since the original article showed only the use of Smith charts to interpret the bridge readings.

This article is written to compare the accuracy of W2CTK's bridge with a Boonton 250A. Some construction points are clarified, techniques are presented to substantially improve measurement accuracy, a wider range of applications is presented, and a step-by-step procedure is included for those who are unfamiliar with bridge calculations.

construction

First, let's tackle construction: **Fig. 1** shows the complete schematic of the impedance bridge, but only a part of the schematic will affect accuracy. **Fig. 3** shows the basic bridge. This portion is the heart of the unit and should be wired with short, heavy leads. A suggested layout is shown in **fig. 4.** The photograph may help with parts placement. Notice that the rear of the panel around the basic bridge has been stripped of paint to insure good grounding of bridge components. Be sure to use a *composition*-type potentiometer; a wirewound pot is completely useless.

A pot with a linear taper will result in the most useful dial calibration. The 56-ohm resistor can be any value from 47 to 68 ohms with no affect on final accuracy, once the R dial is calibrated with that particular resistor. Of course, this resistor must also be a composition type. Some wirewound resistors look like composition types, but are identified with one extra wide color band. Some, made during the Great War and sold surplus for years afterward are, in fact, wirewound, but do *not* have the identifying color band.

The variable capacitor can be any value from 150 to 365 pF. The lower values will limit the X range of the bridge, but have slightly better dial readability. The upper ranges have the opposite characteristics. Any of the values will result in an accurate bridge, but a 250-pF capacitor is about optimum. The diode can be any germanium type, similar to a 1N34. The .005-μF disc capacitor and the diode should have very short leads. Heat sink the diode while soldering it in place.

calibration

For convenience I will go over the calibration procedure. First, connect the basic bridge as shown in **fig. 5.** This hookup allows calibration of the R pot with dc. W2CTK is correct in his statement that rf measurement accuracy will not be impaired with this procedure. The rather shallow null obtained with this bridge is the major source of error, and will limit you to about 5% of calibration accuracy on both R and X measurements. Total error consists of this 5% *plus* the percentage error of resistors and capacitors used to calibrate the bridge.

For convenience, the battery can be fed into the RF IN jack. The 10k resistor is included so you don't bang the meter hard if the bridge is badly unbalanced. Connect a known resistor across the LOAD jack. Lead length is of no consequence here, since this is a dc calibration. For the same reason, wirewound precision resistors are also permissible. Touch the end of the 10k resistor to the center pin of the RF IN jack. Adjust the 100-ohm pot to bring the meter down to zero. Near zero you can touch the battery end of the 10k resistor to the center pin to improve the sensitivity of the adjustment.

If you continue turning the pot in the same direction, it will go below zero. The correct point is where the meter needle just reaches its zero resting position. Mark the R dial with the value of resistance you used for reference. Repeat the procedure for various resistors, or combinations of resistors, to obtain as complete a calibration as you wish. If you use new 5% resistors, your probable accuracy of R readout will be 10%.

Best R range for this bridge is approximately 20-ohms to about 3000 ohms. Below 20 ohms you may be limited by available X dial range – even with an essentially resistive load. Above 3000 ohms the dial graduations become quite close together, limiting the accuracy of readout. My bridge is calibrated to 5000 ohms, but it is difficult to get precise readings at that point. If you wonder about the variable capacitor in the circuit during this part of the calibration, forget it. It merely forms a convenient connection to the load jack.

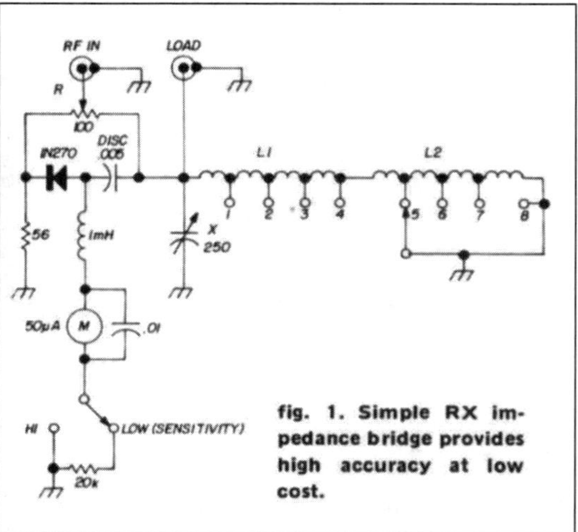

fig. 1. Simple RX impedance bridge provides high accuracy at low cost.

Homemade RX impedance bridge, right, and rf source, left.

After R dial calibration is complete, disconnect the lead from the variable capacitor to the 100-ohm pot. The variable capacitor is now the only component connected to the LOAD jack. The easiest way to calibrate the capacitor is to connect a capacitor bridge to the load jack and calibrate the capacitor in about 10-pF increments from 250 to 100 pF, and in 5-pF increments below 100 pF. There is no need to be concerned about stray capacitance here, so long as it is minimized. You are interested primarily in the difference between two readings, not so much the absolute value.

If you have no capacitor bridge available, you can get an excellent calibration using the external hookup shown in **fig. 6.** A coil is connected to the load jack with provision for placing reference capacitors across the LOAD jack. A source of rf power is link coupled to this coil, and a vtvm ac probe is connected across the coil

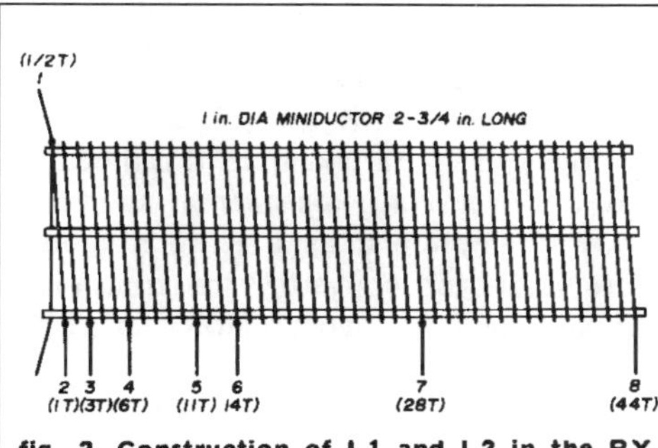

fig. 2. Construction of L1 and L2 in the RX impedance bridge.

to read the peak voltage when the coil and X dial capacitor is resonant at the rf input frequency.

The important thing here is to make a rigid setup, mechanically. The coil and coupling link must be quite rigidly held with respect to one another. The rf source must be stable. The entire setup must be held in a fixed position relative to the metalwork of the impedance meter to minimize variations in distributed capacitance.

The coil can be a 3-inch length of 1-inch diameter 16-pitch Miniductor. Later, this same material can be used for the bandswitched coil, so there is no loss. Cut the coil wire about 3 turns in from one end. Heat the ends and push them out of the support bars to provide connecting leads an inch or so long. Do not cut the plastic support bars between the two coil segments. They will help maintain the coupling. Mount the coil as shown in **fig. 6**.

With the setup of **fig. 6** assembled on a scrap of wood and connected to the LOAD jack of the impedance meter, first set the 250-pF X dial capacitor to maximum capacitance. Mark the dial accurately to indicate maximum. (It is advised to do the same for the R pot. This gives you a reference point in case the dial setscrews loosen and it is necessary to reposition them without re-calibrating.)

Feed an 80-meter rf source to the link and adjust the coil tap for a peak reading with the X dial capacitor at maximum. When this point is found, it must be maintained for the balance of the capacitor calibration. Solder it carefully in place. Connect, say, a known 100-pF capacitor to the binding posts. Re-peak the vtvm with the X dial capacitor. That point is 250 minus 100 pF = 150 pF. With only two 100-pF, one 50-pF and one 25-pF mica capacitor you can calibrate a 250-pF bridge capacitor in increments of 25 pF throughout its range. With care, you can estimate the intervening 5- or 10-pF increments with acceptable accuracy.

It would be worthwhile to buy this group of capacitors new in 5% tolerance or better. (These are available in silver mica to 1% quite reasonably.) If you use a grid dipper for this calibration you will find the dipper frequency pulls as the X dial approaches resonance, making it difficult to maintain exact frequency. For that reason it is better to use an 80-meter crystal-controlled oscillator.

This completes the calibration. You can now complete the wiring of the bridge. **Fig. 2** shows a suggested coil made from the Miniductor coil stock. If you used other than the suggested 250-pF variable capacitor the tap points will have to be changed. The only criterion for selecting tap points is this: On each amateur band from 3.5 to 30 MHz set the taps so that the band can be tuned at near maximum (about 200 pF) and near minimum (about 50 pF). This is necessary to get best X dial range in measurements. The easiest way is to unsolder one end of the wire from the 100-ohm pot to the variable capacitor, install the coil, then set each tap using a grid dipper. (The 100-ohm pot would swamp the tuned

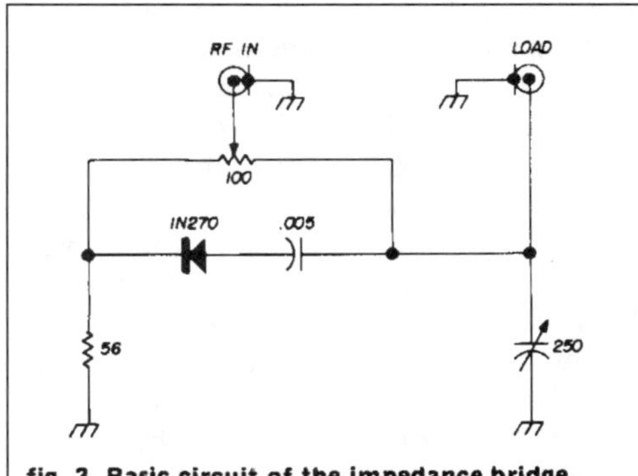

fig. 3. Basic circuit of the impedance bridge.

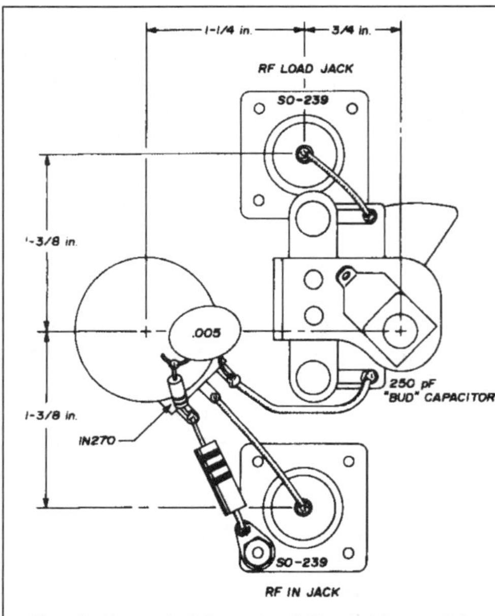

fig. 4. Suggested layout of the bridge resistor and capacitor.

circuit if left in, and would make it difficult to find a dip.) Unsoldering and re-soldering this lead will not affect bridge calibration if the wire is carefully repositioned.

rf source

The next concern is a source of rf to drive the bridge. Most solid-state dippers will lack the power necessary to do a good job. Some vacuum-tube grid dippers will also be marginal. In any case, a dipper has poor stability, so I would advise building an all-band crystal oscillator. The one shown in **fig. 7** covers 3.3 to 33 MHz with fundamental FT-243 surplus crystals. It will operate on the fundamental, second, third or fourth harmonic and provide ample drive for the impedance bridge. In my case I use a pair of 80-meter crystals and a pair of 40-meter crystals to do 90% of my work.

Layout is not critical. Just keep your rf leads short and direct, and your bypass capacitors installed with short leads. It is worthwhile to use up some of the old vacuum-tube gear in the junkbox on projects like this. Take an evening to do the metalwork nicely! It can be housed in a wrap-around made from a discarded piece of galvanized iron flashing. A spray can of gray lacquer will give it a professional touch. This oscillator will find many other bench uses other than as a bridge driver. For example, it can be used as a driver for experimental solid-state power amplifiers, as a temporary local oscillator for converters or transmitter mixers, as a frequency marker, as a QRP

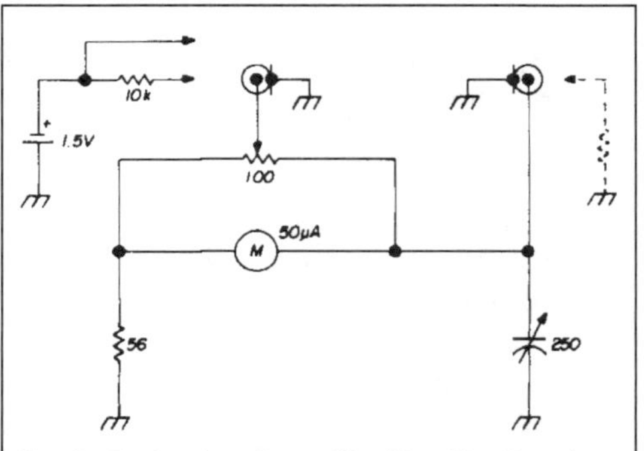

fig. 5. Test setup for calibrating the R potentiometer.

CW transmitter, etc. I find it useful as an rf source for calibrating wavemeters for novices and as a reference antenna driver in antenna tinkering. If you have an old set of plug-in coils, they can be used with somewhat improved efficiency.

bridge operation

Now, let's put the bridge to work. The following abbreviated procedure will be followed with some notes on methods of improving accuracy and interpretation of the results.

initial balance

1. Set the R dial to maximum resistance.

2. Feed rf at the desired frequency into the RF IN jack.

3. With nothing connected to the LOAD jack, alternately vary the R and X dials for best null.

4. Note the exact reading in pF on the X dial.

measurement

5. Connect the unknown load to the LOAD jack.

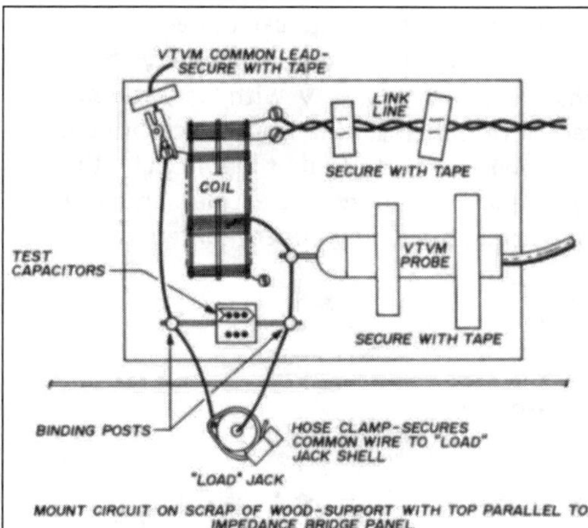

fig. 6. Test setup for calibrating the X capacitor.

6. Again adjust the R and X dials for best null.

7. Note the exact readings of the R and X dials.

8. The R dial reading obtained in **step 7** is the actual resistive component of the load. The *difference* between the X dial reading of **step 4** and the X dial reading of **step 7** is the reactive component of the load (C_p).

9. If both X dial readings are the same the load is purely resistive.

10. If the X dial reading of **step 7** is a higher capacitance than in **step 4,** the load is *inductive*. If the **step 7** reading is lower, the load is *capacitive*.

A little trick I use to get the best null is to view the meter from an angle so a little slit of white background shows through between the meter needle and the zero mark on the meter. In this way very small meter variations can be seen. You will find that by using a crystal-controlled oscillator for the bridge driver the X dial reading in **step 4** will be very accurately repeated, so once you know that number of pF for a given amateur band on a given bandswitch position, you can use the same crystal and skip the initial balance procedure. This is a big help when many measurements are necessary.

For example, I had an experimental tilted longwire antenna I wished to prune for best compromise on a number of bands. Since each pruning required a walk down the valley to the willow tree and back, it could be a full afternoon of work. However, knowing accurately what the X dial would be for initial balance setting it was possible to make a fast measurement on all bands from 80 through 10 meters and really see where the antenna was resonant. Inside of an hour I had it accurately set.

If, in **step 6,** you find it impossible to obtain a null, it means you ran out of X dial range. For example, if the X reading obtained in **step 4** were 25-pF and your load was *capacitive* by 50 pF obviously you would run out of dial range. So, switch to another bandswitch position, repeat the *initial balance,* and perhaps now the reading in **step 4** will be 200 pF. Again bridge the unkown load, and this time (assuming the load was 50 pF capacitive) the X dial will read 50 pF lower than the initial balance setting and you will have no trouble finding a null.

Of course, if the load is horribly reactive, it may be impossible to find any null. You could then grid dip the load to try to find where it was resonant, and this will tell you which way to go to get the load into the bridge range. If the load cannot be easily grid-dipped, you can

Construction of the RX impedance bridge.

make bridge measurements at other frequencies to find a frequency where it does fall within the bridge range. This same problem is common in laboratory impedance bridges. The Boonton 250A doesn't have nearly as much X dial range as W2CTK's simple bridge.

homebrew bridge and calculated it will look like 21.35 ohms. Try to approach that kind of accuracy with an *antennascope* or other simple impedance bridge.

However, I felt that, although the difference in impedance measurement was negligible, many experimenters might

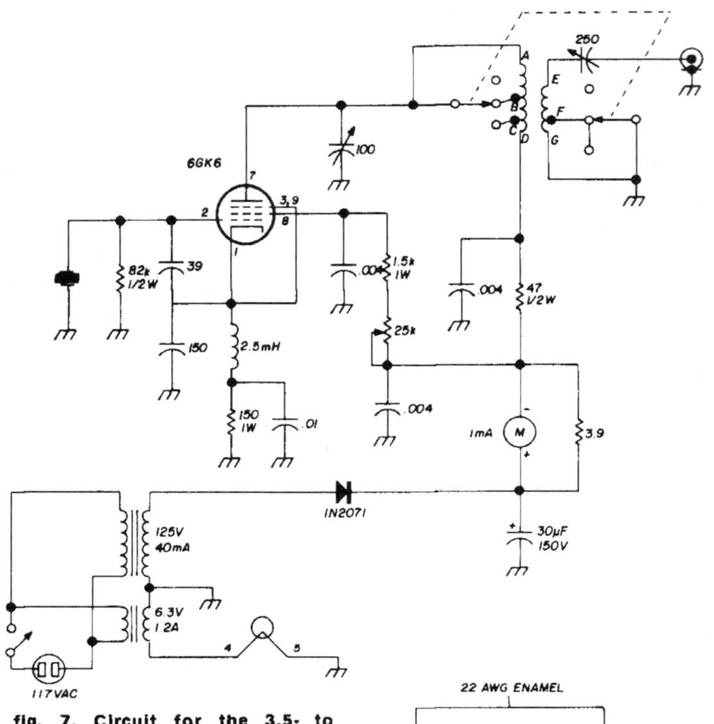

fig. 7. Circuit for the 3.5- to 33-MHz oscillator/multiplier used as an rf source with the RX impedance bridge.

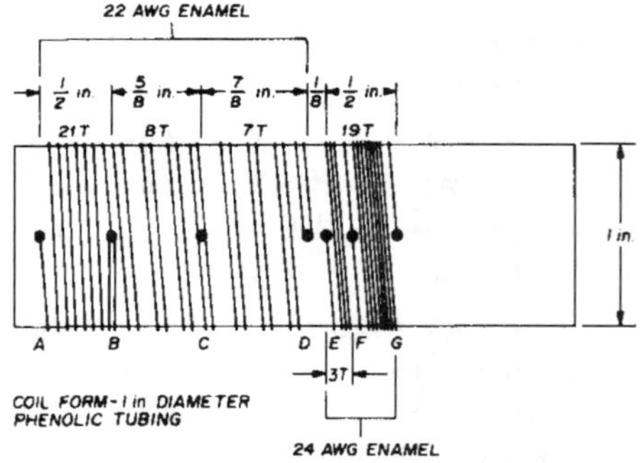

From a point approximately 150 or 200 ohms and higher on the R dial the foregoing procedure will yield excellent accuracy. Below about 150 ohms an error creeps in which will add approximately 5% additional error, at worst. The X dial does not follow a Boonton 250A. At first glance the error appears to be large, but it is not. For example, a 22-ohm resistor measured with the Boonton measured 60-pF inductive. With the homebrew bridge it measured 130-pF inductive. From the worked-out formulas in **table 1** you see that resistor (as measured by the Boonton) results in an actual impedance of 21.82 ohms. As measured by the

like to correct the bridge to read out very closely with a laboratory bridge. In this way a low-impedance load could be measured and the nearly correct reactance could be found that would make the network a pure resistance. In **fig. 8** resistors of different manufacturers, in 1/2-, 1- and 2-watt sizes and all values from 18 to 2200 ohms were checked on the Boonton and on the homebrew bridge. In all cases lead length was held to one-half inch. The line marked *A* is the results with the Boonton. Incidentally, the slope of the lines implies a constant series inductance of .0268 Microhenry; this is the resistor lead inductance.

You can construct a similar correction graph for your bridge if you wish. On a piece of 2x2 cycle log graph paper lay out the curve *A* line from a point 16.5 ohms on the top line to a point at 165 ohms on the bottom line. Now, select a number of composition resistors from your junkbox – values from 22 to 100 ohms. Carefully measure these on your bridge at 14 MHz using half-inch leads protruding from points of contact with the load jack. The C_p values are the values of capacitance which are the *difference* between the *initial balance* value and that obtained in actually measuring the resistor. The R_p values are those read off the R dial when bridging the resistor. Plot these values for a number of resistors. Then draw a line parallel to the *A* line and through the approximate average point of your values.

To use the chart refer to **fig. 8:** That chart says that at 55 ohms the Boonton would read 9 pF, but mine would read 22.5 pF. If my R_p reading is 55 ohms, my C_p reading will probably be about 13.5 pF too high. Incidentally, as you tinker with this bridge, perhaps measuring the same resistor on each band, you may suspect something amiss when you consistently get the same value for C_p. You might suspect this number to be frequency dependent, but it is not. If you ran that same resistor through the range on a Boonton, you would find a slight downward shift in R_p as you go from 3.5 to 28 MHz (perhaps shifting from 28 to 27 ohms) but the C_p value would consistently hang in there.

A load which is not nearly pure R (as these are) will show the expected variations in C_p with frequency. If you feel that the half-inch lead length is gilding the lily, first bridge the resistor with full lead length, then repeat the measurement with half-inch leads. Don't be surprised if the C_p value of a 27-ohm resistor shifts as much as 50 pF. This effect is most noticeable at low resistances (50-ohm lines, etc.) and should serve as a warning of what can happen with sloppy clip lead lashups. I've seen otherwise competent engineers and technicians completely blow an rf measurement due to lack of feel for these little details.

If you are an engineering student, you will find this little bridge a valuable help in getting acquainted with rf measuring techniques. The Boonton RX meter goes all the way to 250 MHz, so you can appreciate the fact that it can get pretty hairy at times to get valid measurements.

using the bridge

Now, let's look at the business of making sense out of the numbers you get from your bridge. **Table 1** is a set of formulas with worked examples to give you another method beside Smith charts as explained by W2CTK. The formulas assume F = MHz, R, X and Z in ohms, C in pF. you may wonder what ω represents. It is $2\pi F$. A handy little trick in amateur work is to remember that, for the low end of the bands, on 3.5 MHz, $\omega = 22$, at 7 MHz, $\omega = 44$, at 14 MHz, $\omega = 88$, 21 MHz, $\omega = 132$, 28 MHz, $\omega = 176$.

The homebrew bridge reads a load in *parallel* values. The example in **table 1** is for an almost pure resistance. It will be seen that the bridge read a 22-ohm resistance with a 56-pF capacitor in parallel. Actually, the 56 pF was in an *inductive* direction. This is the same as saying, "What coil in parallel with 56 pF will be resonant on 14 MHz?" You could find this on an ARRL Type A Lightning Calculator in less than 5 seconds. In the same length of time the calculator also says it would be a half-inch diameter coil, 1-3/4-inches long, 28 turns. You *know* a coil of those dimensions wouldn't likely be inside that resistor! If you are completely unfamiliar with the calculations you might be inclined to think it is a wire-wound resistor.

However, look at **table 1** and find the

table 1. Mathematical formulas which may be used with the impedance bridge if you do not use a Smith chart. In all examples f = 14 MHz, Rp = 22 ohms, Cp = 56 pF and $\omega = 2\pi f$.

1. Parallel reactance $Xp = \frac{10^6}{\omega Cp} = \frac{10^6}{6.28 \times 14 \times 56} = \frac{1\,000\,000}{4928} = 203$ ohms

2. Parallel inductance $Lp = \frac{Xp}{\omega} = \frac{203}{88} = 2.30$ microhenries

   or $Lp = \frac{.159160\ Xp}{F} = \frac{.159160 \times 203}{14} = \frac{32.3}{14} = 2.30$ microhenries

3. $Q = \frac{Rp}{Xp} = \frac{22}{203} = 0.1083$

4. $Q^2 = 0.0117$

5. Absolute impedance $|Z| = \frac{Rp}{\sqrt{1 + Q^2}} = \frac{22}{\sqrt{1.0117}} = \frac{22}{1.008} = 21.82$ ohms

6. Equivalent series resistance $Rs = \frac{Rp}{1 + Q^2} = \frac{22}{1.0117} = 21.74$ ohms

7. Equivalent series reactance $Xs = \frac{Xp\ Q^2}{1 + Q^2} = \frac{203 \times 0.0117}{1.0117} = 2.347$ ohms

8. Equivalent series inductance $Ls = \frac{Cp\ Rp^2}{(1 + Q^2)10^6} = \frac{56 \times 22^2}{1.0117 \times 1\,000\,000} = 0.0268$ microhenry

9. Absolute impedance $|Z| = \sqrt{Rs^2 + Xs^2} = \sqrt{21.74^2 + 2.347^2} = \sqrt{478.1} = 21.85$ ohms

10. Equivalent series capacitance $Cs = Cp\ (1 + \frac{1}{Q}2) = 56\ (1 + \frac{1}{.0117})$

    $= 56\ (1 + 85.4) = 56 \times 86.4 = 4838.4$ pF

    or, $Cs = \frac{159160}{F\ Xs} = \frac{159160}{14 \times 2.347} = \frac{159160}{32.858} = 4840$ pF

parallel reactance; it is 203 ohms. Such a large value of reactance in parallel with such a small resistance will hardly affect the resistance. It is the same as hanging an rf choke across the end of a pi network for safety. The pi net sees only the 50 ohms. If you continue down to the first calculation for impedance you see the total impedance of the 2.3 microhenry inductance in parallel with a 22-ohm resistance is 21.82 ohms. The coil hardly made any change.

Any parallel network can be converted to an equivalent series network or vice versa. From **table 1** you see that an equivalent series resistance of 21.74 ohms in series with an equivalent series inductance of 0.0268 microhenries results in an impedance of 21.85 ohms. Had you carried out all your calculations to a number of significant figures both impedances would have come out identically. This last set of values sounds more reasonable! In fact, you actually do have a 21.74-ohm resistor in series with a pair of tiny coils (the resistor leads).

Now, suppose you had been using this bridge to measure a wide-band balun for your antenna. Suppose the balun was supposed to look like 22 ohms. Unless you have a feel for the effect of reactance you might have felt the balun was junk, while, in fact, it was almost a perfect match! For that reason it is worth taking the time to work out enough examples to get familiar with the arithmetic. Then, since the little bridge is quite accurate, you can be very sure of your results.

The impedance was figured in two ways to provide you with a means of back-checking your work. Whenever you convert from parallel to series values it will be worthwhile to use this as a proof that the series values were correctly figured.

How else can you use these numbers? Suppose you made a mobile antenna out of some odd scrap. You could end up with somewhat similar numbers to those

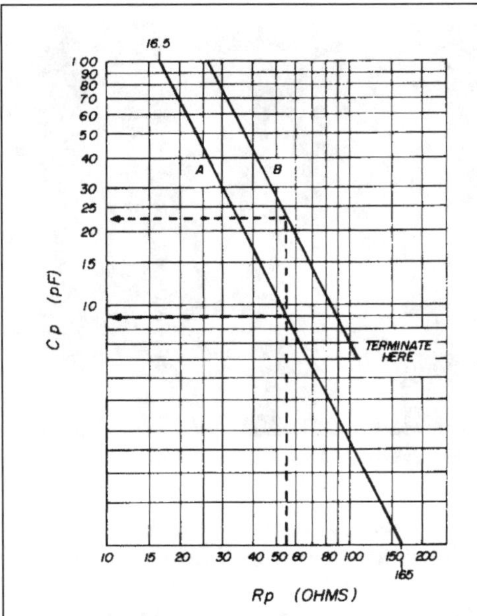

fig. 8. Calibration curves for the homemade impedance bridge (B), and a commercial Boonton 250 RX meter (A). These same curves can be plotted for your own bridge. See text.

in the example. If the antenna looked like 22 ohms in parallel with 56 pF *inductive,* you could make it 22-ohms resistive by merely hooking a 56-pF capacitor from whip to ground. If it looked 56-pF *capacitive* you could hook a 2.3-microhenry coil between whip and ground. Or, you could use the equivalent series values. If it showed 56-pF *inductive* on your bridge you could put about 5000-pF in series with the lead from the whip to the coax. If it looked *capacitive* you could put a 0.0268-microhenry choke in series from the whip to the coax. Of course, this example is pretty ludicrous, since it would be foolish to correct such a tiny inaccuracy. However, it gives you an idea of how very close you can work in antenna matching with this simple device.

applications

Another application would be to transform that 22-ohm whip to 50 ohms to match a transmission line. You might look up the arithmetic of L networks in the ARRL Handbook, but perhaps you are a bit unsure that the values you figured were correct. In such a case you can use the bridge backwards: Connect the L network to the whip, make the initial balance on the bridge, and connect it to the 50-ohm side of the L net. Set the R dial of the bridge at 50 ohms. *Don't touch the bridge!* Diddle the L network until the bridge nulls.

Perhaps you have an 80-40 dual dipole and want to include a balun to divorce your line from the antenna, but you can't afford the cost of the big toroid. Take a fat loopstick from a junked BC set and wind your 1:1 balun on it. Of course, you have no idea of the characteristics of the loopstick material. Connect one end of the balun to the bridge and solder a 51-ohm composition resistor to the other end. Vary the number and spacing of turns until it looks good. Remember that with a 51-ohm impedance it takes a fair amount of reactance to ruin the final impedance. Once the balun is done, put a dab of epoxy on the wire ends to hold them in place and you are in business at practically no cost.

Perhaps you built a little synchrodyne receiver and used an unmarked surplus toroid of unkown characteristics for an antenna coil. The set works fine except that when it's connected to a 50-ohm line it went dead. Don't spend a lot of time guessing the proper turns ratio! Perhaps the IC you were using called for 400 -ohms for best noise figure. Put a 390-ohm resistor across that winding, then bridge the antenna winding. When you get it around 50 ohms, non-reactive, put the receiver on the 50-ohm line and it will perform.

Perhaps you built a kilowatt linear from your junkbox. You carefully designed the pi network, but aren't sure the taps are correctly set on the coil to provide the proper plate load to the tube. Leave the linear turned off. Connect a composition resistor of the value you chose for the correct plate load from plate to ground. If the pi net is far from frequency you might not be able to bridge it, so set if for resonance with a grid dipper. Now look into the 50-ohm end of the pi net and adjust it to see if it will show 50 ohms with the capacitor values you calculated. If it does, you can be sure the plate sees the same load as the one you used from plate to ground. This will insure that the efficiency is up to the point you had calculated.

Of course, the impedance bridge is really useful in antenna work, but there are a few odd things you might run into. In the first place, if you bridge the coax at the transmitter end, it doesn't say that the antenna impedance is the same. Only if you take the time to bridge a good half-wave line (or multiple of half waves) and connect the end of that line to the antenna – will you get a valid reading.

W2CTK describes bridging a half-wave line. Essentially, make up one end of the coax, couple it to the bridge, terminate the far end with a 51-ohm resistor, check and snip, reterminate, check and snip until it reads 51 ohms, non-reactive point. Snip tiny pieces off so you don't whiz by!

If you bridge high-impedance antennas, you may run into certain quirks. My long wire showed the usual. The bridge was connected securely to the ground line coming in the basement window. This is a piece of number-4 wire about 3-feet long going to a ground rod outside the window. At that point it branches out with about 1000-feet of buried radials plus another hunk of number-4 to a second ground rod about 10-feet away. On 10 and 15 meters the bridge showed a lot of hand capacitance, and the null was quite dubious. The solution was to take the bridge outside the house, lay down a few strips of broiler foil on the ground, clamp them to the ground stake, clamp the bridge to the foil as close to the stake as possible, then lie down on the foil to put myself as near ground potential as possible. Now I had good nulls and accurate measurements. That 3-feet of number-4 lead was enough to put the bridge *above ground.*

Another odd factor showed up on 80 meters. The null would not go to zero and it wasn't hand capacitance. I shut off the bridge driver and the meter still read upscale. I soon realized that this antenna was quite nearly a quarter-wave long at the frequency of the local BC station. Measuring with my rf voltmeter showed a steady 200,000 microvolts of signal. *That* antenna makes a good one for checking cross modulation of receivers.

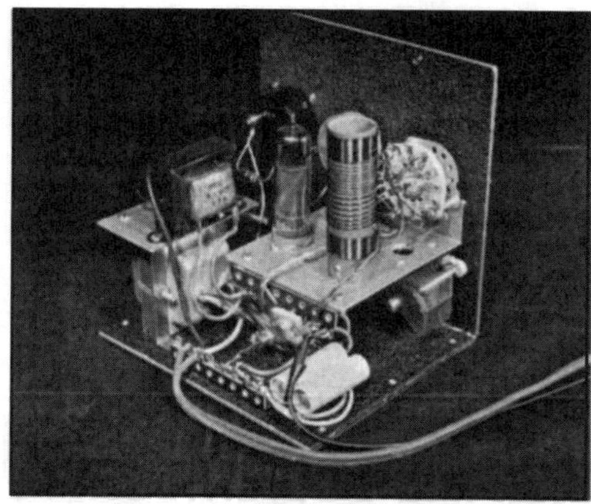

Construction of the rf power source.

conclusion

In conclusion, I feel that this instrument is an extremely useful device for the hamshack. With just a little care in construction and calibration it is remarkably accurate for its simplicity and low cost. I hope that the additional information presented here will encourage others to duplicate the device so that W2CTK can be properly credited with the importance of his little gem.

reference

1. Henry S. Keen, W2CTK, "A Simple Bridge for Antenna Measurements," *ham radio*, September, 1970, page 34.

low-power vhf dummy load

The tobacco load — a low-vswr dummy load that will handle 30 watts of rf

Dave Sprague, WB9DNI, 78 Paddock Street, Aurora, Illinois 60538

Here is a one-evening project that will provide a very useful accessory for your shack — a 50-ohm dummy load capable of dissipating more than 30 watts of continuous power with a vswr below 1.1:1 through 220 MHz and typically, below 1.5:1 at 450 MHz.

The rf load is neatly packaged in a 7-ounce pipe tobacco can, filled almost full with transformer oil. The load itself consists of ten 510-ohm (5%) 1-watt carbon resistors parallel-mounted between pieces of copper weatherstripping. An alternate design approach is to use 2-watt resistors for a 70-watt load. However, there is a tradeoff with this approach as the vswr at 220 MHz may be as high as 1.8:1.

construction

Two types of 7-ounce tobacco cans are readily available from your pipe-smoking friends. They differ in the type of lid. The older type uses a paint can type lid; the second, and newer, type uses a screw-on lid. Although either tobacco can will work, I used the screw type for this project.

fig. 1. Layout of the copper strips used in the construction of the 30-watt rf load.

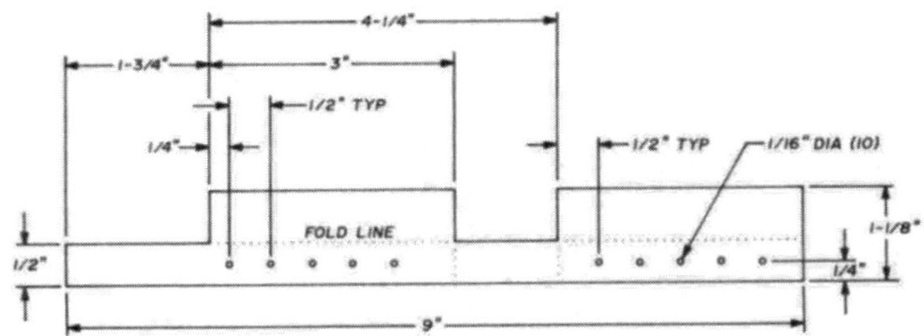

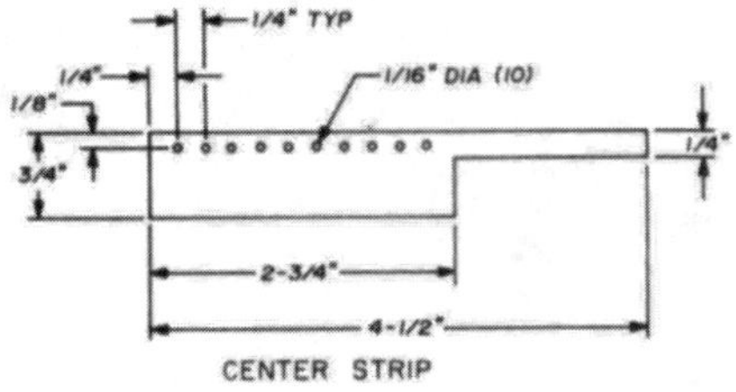

fig. 2. Copper strip used for mounting the 5% 1-watt resistors used in the rf load.

The metal which these cans are made from is soft enough that a scribe can be used to punch the necessary pilot holes when mounting the connector and vent tube in the lid. First, test the tobacco can for leaks by filling it with water. If any leaks are found, seal the seams of the can with a thin coat of RTV or similar oil-resistant sealer.

Completed rf load. Small tube to the left of the coaxial connector allows heat expansion of the cooling oil.

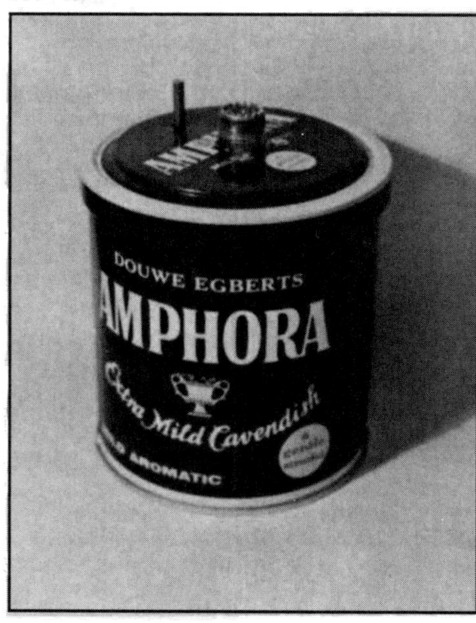

I tried several materials for the element mounting strips, but I found ordinary copper-clad weatherstripping to be electrically suitable, inexpensive and easy to form. Cut the strips to the specifications of **fig. 1.** Fold the larger strip as shown in **figs. 2** and **3.**

When the copper strips are completed, lay the smaller strip before you, as pictured in **fig. 1B.** Insert the first five resistors in alternate holes beginning with the hole closest to the right end. The end of the resistor body should be flat against the strip, but do not cut the soldered lead off at this time. A vise is a great help to keep the resistors properly aligned while soldering.

Next, turn the copper strip over and insert the remaining resistors in place with their bodies similarly aligned. Don't be stingy with the solder here – a good electrical connection is more important than looks. You should now have a unit which resembles **fig. 4.**

Cut the soldered lead of each resistor flush with the solder joint. Trim the free lead of each resistor to approximately ½ inch. Insert the center strip and resistors into the outer strip by carefully spreading the edges of the outer strip (see **fig. 5**) and guiding each resistor lead through the

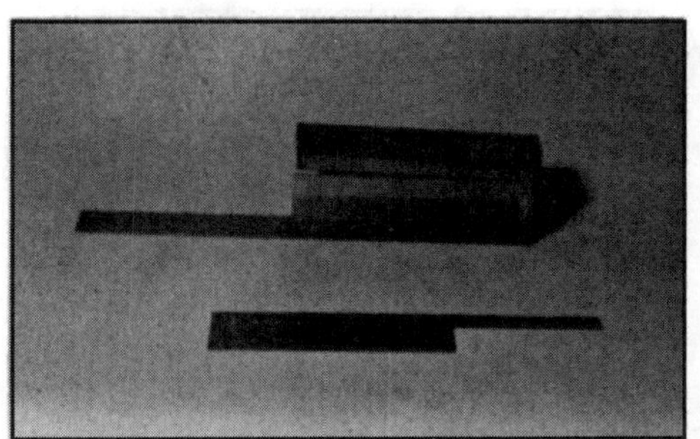

fig. 3. Copper strip is formed into a closed U-shaped box as shown here before installing the center strip and 1-watt resistors.

appropriate hole. Solder the free lead of each resistor to the outer copper strip, keeping the strip edge as close to the resistor body as possible. Trim the excess lead flush with the solder joint.

Solder the back and bottom edge of the outer copper strip as shown in **fig. 6.** When soldering, keep the center and outer strips separated. After verifying a resistance of 51 ohms between the strip tabs the load element is ready to mount in the tobacco can.

Mount a female SO-239 uhf coaxial connector through the lid of the can. I used pop-rivets to mount the connector but screws and nuts will work just as well if screw length is kept to a minimum.

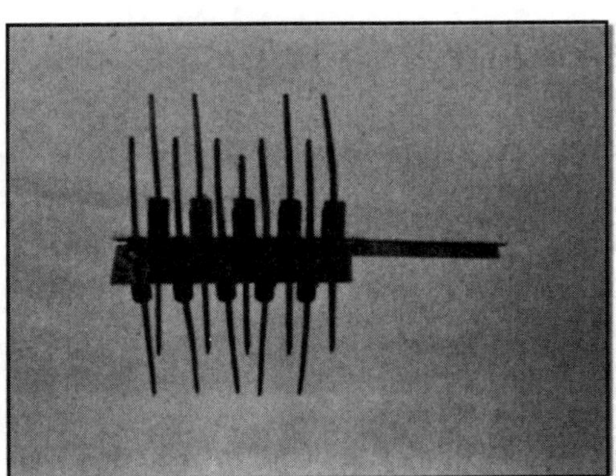

fig. 4. First, solder the resistors to the center strip. This center strip is then mounted in the U-shaped box as shown in fig. 5.

Solder a 3/4-inch piece of 1/8-inch OD copper tubing through the can lid so that the bottom of the tube extends to approximately 1/16-inch inside the lid. The load element should be mounted on the lid so that the bottom just touches the tobacco can bottom when the lid is secured (see **fig. 7**). There is sufficient tab length to fit almost any can type. Cut the excess from each tab, and solder the center tab to the SO-239 connector; solder the outer tab directly to the lid.

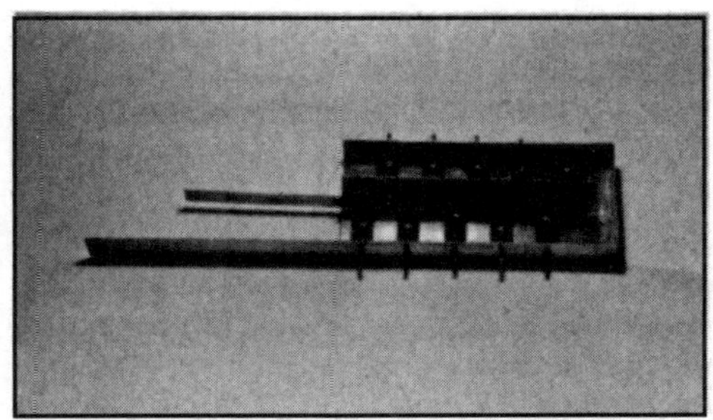

fig. 5. Install the center copper strip with resistors into the U-shaped box.

operation

Fill the tobacco can about two-thirds full of a transformer oil.* Insert the load element and secure the lid. Remove the lid and note the highest point the oil reaches on the resistance element. Add enough oil to just cover the entire element. This leaves sufficient space for expansion as the oil heats. Secure the lid on the can tightly, and tape the seam.

Use a through-line type rf wattmeter when testing the tobacco load. Or, use a vhf reflectometer between the transmitter and the load. Keep the power low for the initial trial. There is nothing to tune or adjust. If the vswr is high, recheck the

*A small quantity of transformer oil can usually be obtained from your local, friendly, public utility company. Use caution if a substitute oil is used — the dielectric constant of vegetable oil, for example, is not consistent.

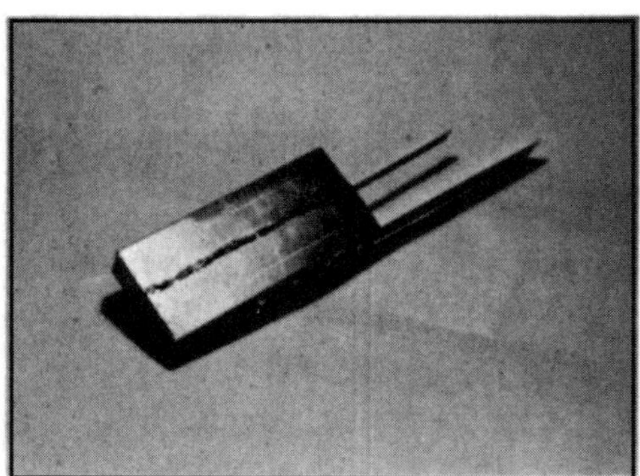

fig. 6. Solder the resistor leads and edges of the center copper strip to the U-shaped copper box.

dimensions of the load element and all solder joints.

I found that the tobacco load was able to withstand several rf overloads for short periods of time without damage to the load element. The thirty-watt rating is on the conservative side. With proper care this small load will pay for itself many times over.

fig. 7. An SO-239 coaxial connector is soldered to the lid of the tobacco can and the load element is installed.

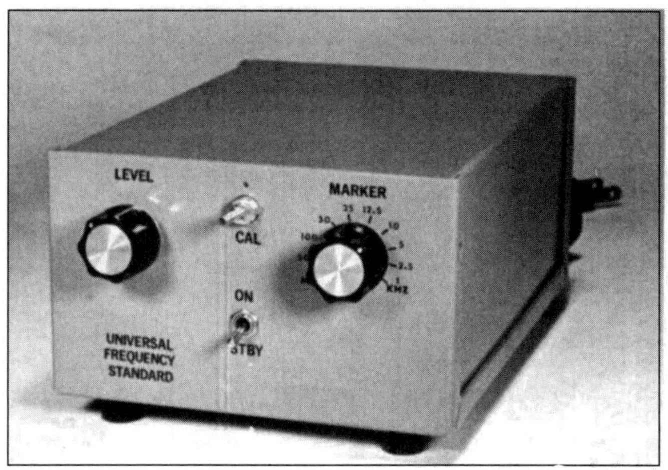

universal frequency standard

A precision frequency standard featuring a high-accuracy crystal, stable transistor oscillator and TTL logic

Bert Kelley, K4EEU, 2307 South Clark Avenue, Tampa, Florida 33609

This frequency standard has been designed to supply precision frequencies for several purposes. Its main use is with my station receiver as a precise and reliable frequency calibrator. It may also be used as a digital counter time base, a scope calibrator or to drive a digital clock. With all integrated circuit packages installed on the board there are no less than eighteen frequencies available extending from 2 MHz down to 1 Hz. Ten of these, of your choice, are connected to a rotary switch for calibrator use while the lower frequencies below 1 kHz that would not normally be used with the station receiver, are picked off terminals on the circuit board by a small clip lead when needed.

The design is flexible. If you do not need the versatility of the complete unit, the photo shows a frequency standard that will provide markers at 1 and 2 MHz, and at 500, 250, 100, 50, 25 and 12.5 kHz with the installation of only three digital IC packages. Provision is made on the circuit board for as many or as few output frequencies as are likely to be needed.

circuit features

Features include excellent frequency stability, front panel calibration and a precise, self-contained, regulated power supply. An adjustable level control is included so the calibrator output can be matched to incoming signals such as WWV for really accurate zeroing or advanced full on for strong, clear markers.

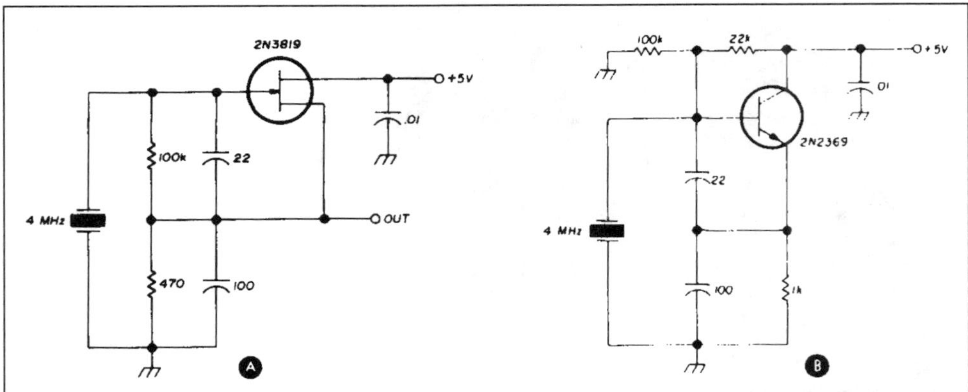

fig. 1. Solid-state crystal oscillator circuits. Fet circuit in (A) is considerably more stable than the bipolar circuit shown in (B). For a more stable bipolar oscillator circuit, see fig. 2.

With proper temperature compensation, the frequency standard will stay within 1 Hz of WWV at 10 MHz over an extended period of time with no adjustment.

This precision is not needed if the unit is used only to find band edges or set the receiver graticule. However, for frequency measurement or use as a time base for a counter or digital clock, you need all the accuracy you can get. The best reason for going first class is that it costs little more, and probably less, in this do-it-yourself project. The required stability can be obtained with inexpensive construction.

The oscillator circuit in this calibrator was first used as a stable time base for a digital clock and later in a receiver.[1] Several circuits evaluated in drift tests showed the Clapp-Colpitts to be most stable – not a new circuit, but seldom used in recent designs. While both the circuit used and the TTL logic is familiar, it is the combination of circuit, crystal and construction that makes this calibrator better. Stability comes when a few hertz drift is removed or greatly reduced from each of several sources. A 10-Hz drift caused by a trimmer might be tolerated, but when the drift from other sources is combined, the total becomes excessive.

Most oscillator circuits would cause even a perfect crystal to drift. Voltage and temperature changes cause changes in the semiconductor's internal impedance. Trimmers don't stay where set, wires move, capacitors change value. Solid-state circuitry has substantial advantages over vacuum-tube circuits; much heat is eliminated, components run cooler and the crystal is driven at the low levels recommended by the manufacturer. However, a substantial amount of drift can come from the semiconductor alone.

How much? Two circuits are shown. **Fig. 1A** shows a fet oscillator circuit using crystals in the 1- to 9-MHz range. The crystal ground for 32-pF load is sometimes brought on frequency with a small trimmer. If this circuit is built so the semiconductor is isolated, it can be heat cycled without affecting other components and the drift can be measured on a

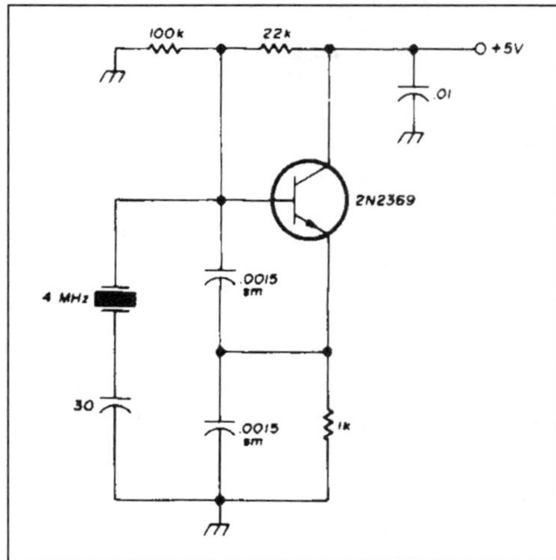

fig. 2. This transistor crystal oscillator circuit is stable because impedance changes are swamped out by the 1500-pF silver-mica capacitors.

Frequency standard and power supply are installed in a 3x5x7-inch (7.6x12.7x17.8-cm) Minibox.

digital counter. The best of several fets caused a change of 12 Hz at 4 MHz with a 5-degree ambient temperature variation.

Fig. 1B shows a transistor in a similar oscillator circuit, not recommended, but sometimes used. Although not as sensitive as the fet to temperature, a change of only one volt caused a 70-Hz frequency shift. If this circuit is modified as shown in **fig. 2,** there is a substantial improvement. Changes of ten degrees and one volt did not cause a frequency change readable on the counter. The transistor case could be heated with a soldering iron to the point where it burned the fingers with a 2 Hz change registered at 4 MHz.

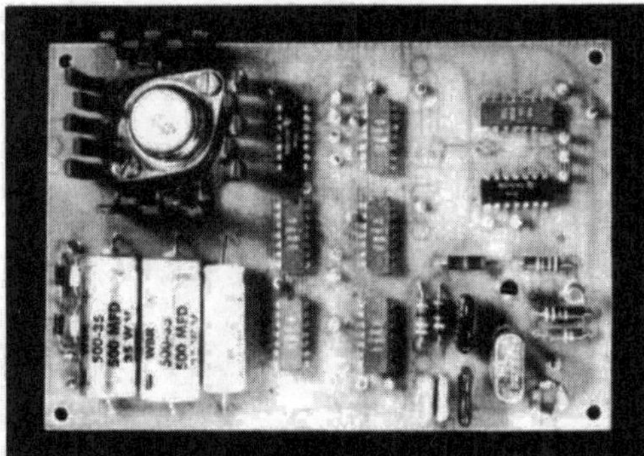

Printed-circuit board with all logic packages installed, as would be required for driving a digital clock or a frequency counter. One jumper is used.

These experiments show both the extent of drift that can be contributed by the semiconductor and indicates the solution. The more stable the capacitance used across the transistor, the better the stability. The 1500-pF capacitors have a very low reactance at 4 MHz, swamping out any other impedance changes in the circuit. This circuit will not oscillate with some transistors because there must be enough gain to sustain oscillation. This requirement is met by the Motorola HEP715, a pnp device with a typical beta of 120. Other transistors with similar current gain can be substituted.

the crystal

Some time ago while working with digital counter time bases[2] I noticed that the ordinary 100-kHz crystal was not as stable as it might be. A time base derived from the power line was nearly as accurate. Although used in amateur calibrators for many years, the 100-kHz rock is

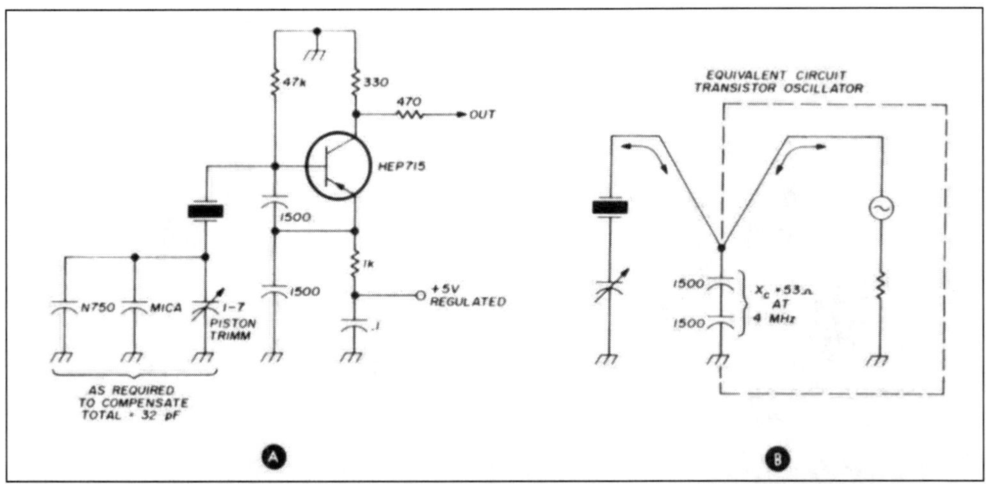

fig. 3. Stable transistor crystal oscillator circuit. Equivalent circuit in (B) shows how 1500-pF capacitors swamp out any internal impedance changes which would cause the output frequency to drift.

not a good choice as it must be stabilized by use of an oven; this adds bulk, expense and a heater supply.

A better crystal for a frequency standard is the high-accuracy 4-MHz crystal recommended by W6FFC.[3] High accuracy as used here means the crystal drifts less, and at a predictable rate. Although cheaper crystals in the 1- to 5-MHz range are better than those at 100 kHz, they are subject to random drift which is difficult to compensate. Therefore, it is important to obtain the better quality crystal. They are manufactured by both Sentry and International and cost less than $10.*

Note the drift characteristic curves for a typical AT-cut crystal in **fig. 4.** As the temperature increases, the crystal frequency decreases. Drift can be almost entirely eliminated at room-temperature operation which most amateurs are interested in by selecting the proper value of negative coefficient compensation capacitor. Curve A requires the most compensation while curve B requires little if any. A crystal cut for a 32-pF load allows about 30 pF for compensation with the piston trimmer adding the 2 or 3 pF needed to pull the crystal to the exact frequency.

*Write for their catalogs. Sentry Manufacturing Company, Crystal Park, Chickasha, Oklahoma 73018; International Crystal Manufacturing Company, 10 North Lee, Oklahoma City, Oklahoma 73102.

A new crystal might require anything from a maximum of 30-pF (N1500) to a 30-pF silver mica (NPO), depending on its temperature vs frequency characteristic. Since there is no way to know what will be needed, it is advisable to have a supply of different values of N750 and N1500 coefficient capacitors on hand, with small silver micas to pad the total to 30 pF, before starting any temperature compensation work. A piston trimmer is recommended because of the smoother adjustment and lack of drift. It is also easier to determine if capacitance is being added or removed, useful information when temperature compensating the calibrator.

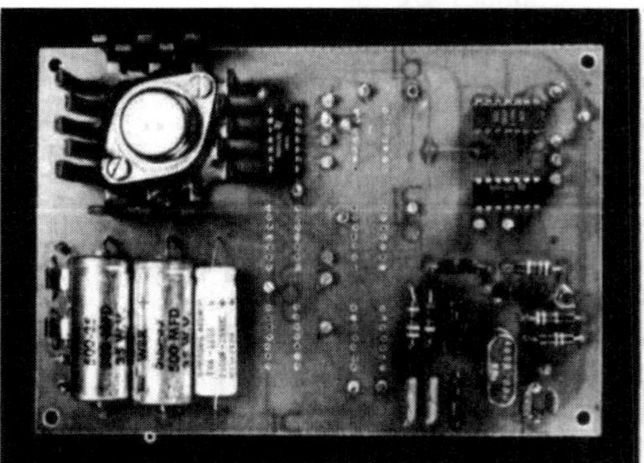

If you don't need the versatility of the complete unit, this photograph shows a frequency standard that will provide markers at 1 and 2 MHz, and at 500, 250, 100, 50, 25 and 12.5 kHz with the installation of only three logic packages.

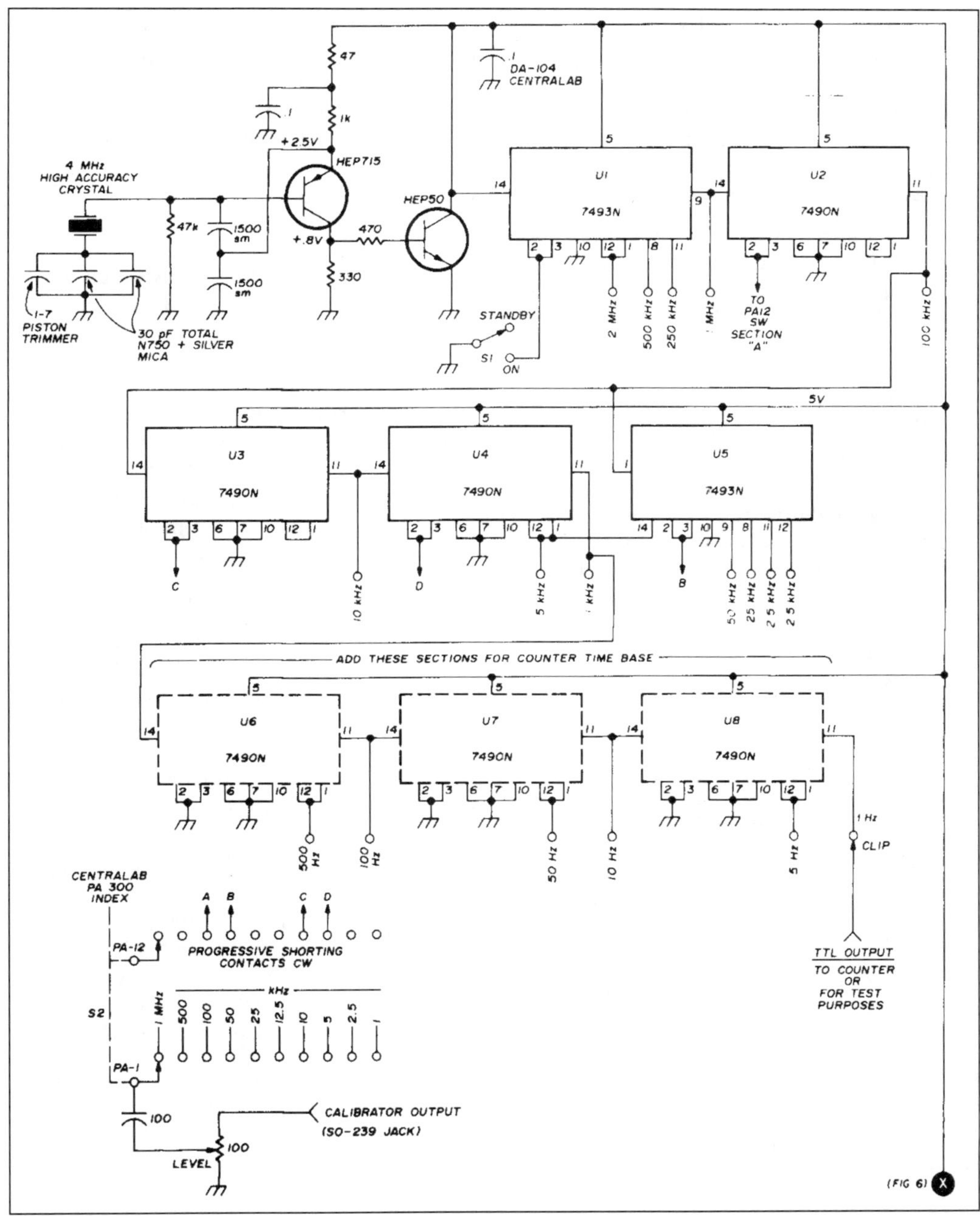

fig. 5. Complete schematic diagram for the universal secondary frequency standard. Unit uses high performance TTL logic ICs. A printed-circuit layout is shown in fig. 6.

the circuit

In fig. 5 the pnp oscillator transistor, Q1, is coupled to the TTL logic by transistor Q2. The 7493 binary dividers U1 and U5 are used to divide by factors of 2, with 7490 decade packages making up the remainder of the logic. IC U5 has two inputs for 5 kHz and 100 kHz. Reset pins 2 and 3 control operation of the logic, either by switch S1, the standby switch, or by progressively shorting contacts on the rotary switch.

This way, the oscillator runs continuously for best stability, and unused packages are disabled. It prevents some markers from leaking across the selector switch and being heard in the receiver. If this feature is not wanted, pins 2 and 3 should be jumpered to ground.

Board outputs and compensating capacitor terminals appear at convenient terminals at the top of the board made by forcing short lengths of bare number-12 wire into 5/64-inch holes. This facilitates exchange of compensating capacitors, or selection of a different logic output at some future time. After completion, it is difficult to work on the underside of the board without removing several wires.

The fast switching TTL logic has active transistor pull-up circuitry which is well suited to driving external loads. The 2900th harmonic of the 10-kHz marker is over S9 in the ten-meter band. In the unlikely event that a logic package fails, repair would be facilitated if Molex sockets are used.

The full current drain with all IC packages installed is 5 volts at 260 mA, easily supplied by a LM309K voltage regulator IC mounted on a heatsink (**fig. 6**). All power supply components except the power transformer are mounted on the circuit board. High temperature shutdown and overcurrent protection are provided by the regulator.

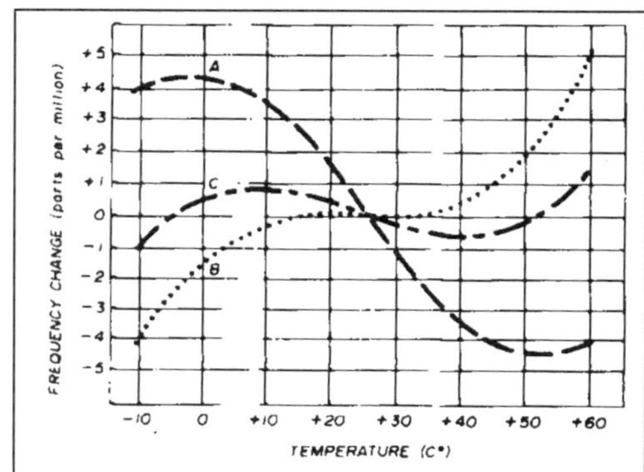

fig. 4. Frequency vs temperature chart for typical high-accuracy AT-cut crystals.

construction

This frequency calibrator is simple to build. The circuit board may be hand duplicated following the layout given in **fig. 7**, or an etched, plated epoxy board is available which speeds construction and minimizes errors.* Parts locations and identifications are screened on the board. It is only necessary to drill the IC holes with a number-60 drill, insert parts, and solder. The assembly is mounted in a compact 3x5x7-inch Minibox.

temperature compensation

Crystals should be ordered for 0.0005% tolerance, F-700 or SP7-P holder (depending on manufacturer), 32-pF load, 4 MHz at room temperature. New crystals should be operated for a time before starting any compensation work. You will need the previously mentioned supply of N750 and N1500 capacitors, and a receiver with an S-meter that will tune WWV.

Start with 15 pF in parallel with a 15-pF N750, allow an hour for the unit to stabilize, and adjust to frequency using the S-meter on the receiver as an aid to exact zeroing. Select a time when WWV is

*An epoxy, plated 4x6-inch printed-circuit board for this frequency standard is available from the author. $8.00, postpaid, in the United States.

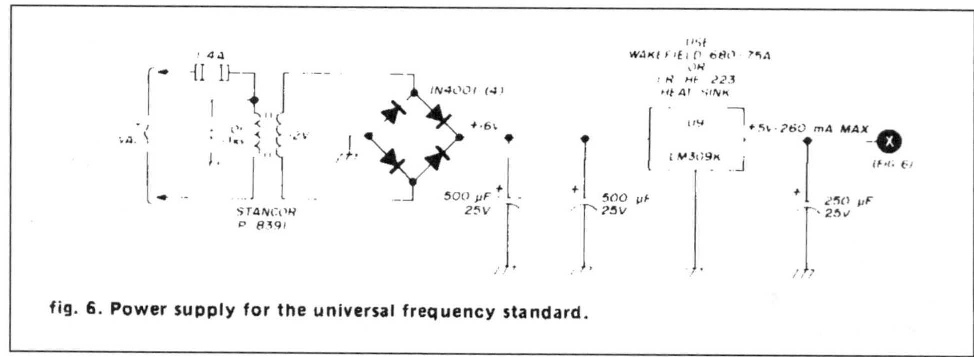

fig. 6. Power supply for the universal frequency standard.

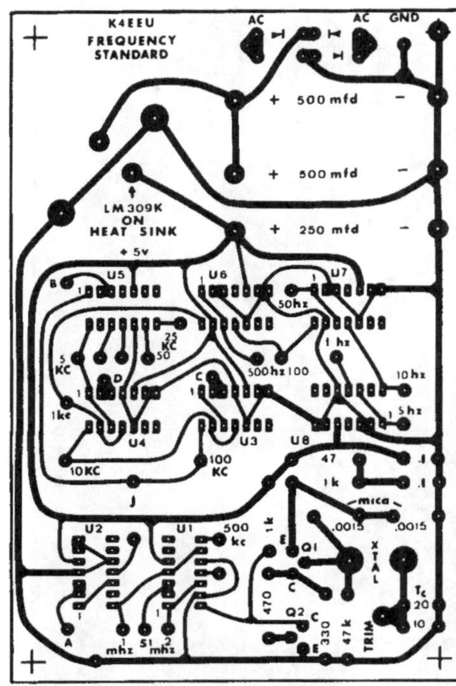

fig. 7. Printed circuit layout for the secondary frequency standard. Boards are available from the author for $8.00, postpaid.

moderately strong, with little fading, so the meter remains reasonably steady.

The standby switch, S1, is turned on and the calibrator *level* control advanced about halfway so both signals are heard. At first, the calibrator will probably be so far off frequency that an audible beat note will be noted, mixed with the WWV tone. As it is zeroed, the warble in the WWV tone will decrease in pitch until it is no longer heard and the S-meter will swing, rapidly at first, then slower, as tuning becomes more exact.

The amplitude of this swing will maximize when the calibrator level is equal to WWV's strength. It should be easy to set the calibrator exactly in zero beat with WWV at 10 or 15 MHz. When this is finished, note the temperature of the room on a thermometer, and record it for reference. Recheck the frequency with WWV periodically, and if there is any drift note the temperature and the direction the trimmer must be adjusted – to add or remove capacitance. If this trimmer capacitance must be reduced for a temperature increase, more N750 or even N1500 capacitance is needed, always padding the combination to a total of 30 pF. after a few tries the exact value of compensation will be found.

Take your time during this work and be sure a trend is established before changing capacitors, possibly making two or more observations before proceeding. The temperature compensation is easier to accomplish than it appears and makes the difference between an ordinary and a precision instrument.

some uses

The photos show a frequency standard supplying pulses for a digital clock. This clock controls nineteen slave clocks in a broadcast installation, so reliability and accuracy are important. The clock is made immune to momentary power-line failures by floating the dc supply across a nicad battery large enough to operate the logic until emergency power can be started or service is restored, whichever comes first.

Fig. 8 shows diodes used to drop the battery voltage to TTL requirements. The one-second pulses from this standard are so accurate that this clock stays on the tick with WWV for weeks with no correction.

In this circuit the 4-MHz crystal frequency is divided by a factor of 4 X 10^6 to obtain the one-second pulses. If this crystal drifted an extreme 1 Hz away from nominal, it would require four million seconds (or 46.3 days) for the clock to accumulate an error of one

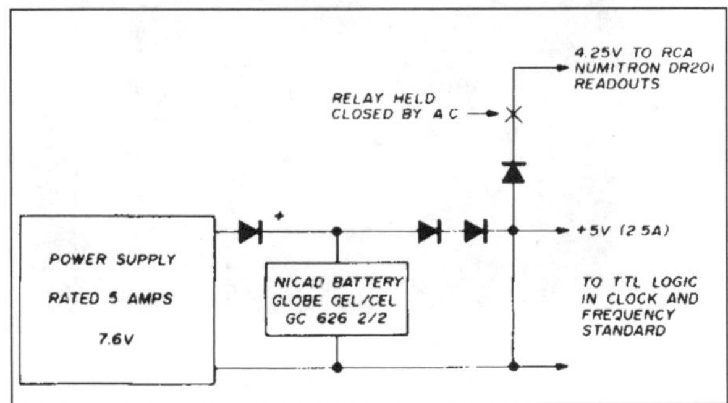

fig. 8. Failsafe power supply used with the digital clock described in the text.

Precision digital clock discussed in the text. The frequency standard board used to provide the 1-second driving pulses is located in the foreground on the upper deck. Nicad battery to the right is part of the failsafe power supply (see fig. 7).

second. But, since the crystal is compensated closer than this, and any minor drift is above and below the frequency, the average error is very small.

The secondary frequency standard would also make a good time base for a digital or Rec-Counter.[5] Readout kits with the tubes, storage latches and counter ICs are advertised in this magazine, so only the gating circuitry would have to be hand wired. Other applications for the standard include audio oscillator or signal generator calibration, or calibration of the sweep time base in oscilloscopes.

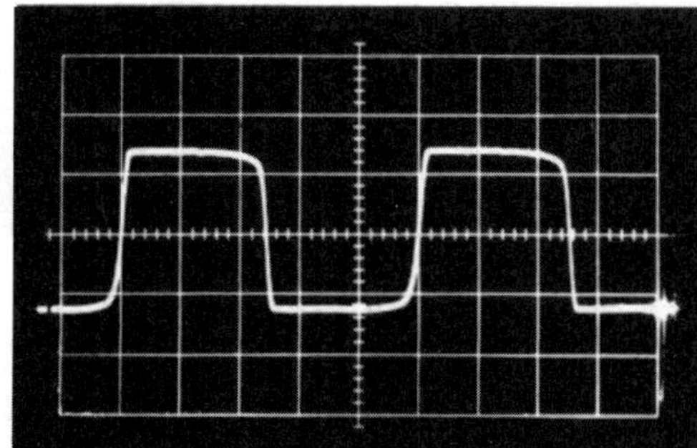

1-MHz output of the frequency standard as observed on a 10-MHz oscilloscope. Rounded waveform shows bandpass limitation of the scope. Horizontal scale is 0.2 microsecond per cm.

references

1. A.A. Kelley, K4EEU, "RTTY Monitoring Receiver," *ham radio,* December, 1972, page 27.
2. A.A. Kelley, K4EEU, "15-MHz Digital Frequency Counter," *ham radio,* December, 1968, page 8.
3. Irvin M. Hoff, W6FFC, "The Mainline FS-1 Secondary Frequency Standard," *QST,* November, 1968, page 34.
4. A.A. Kelley, K4EEU, "How to Make your Own Printed-Circuit Boards," *ham radio,* April, 1973, page 58.
5. Kenneth Macleish, W1EO, "A Frequency Counter for the Amateur Station," *QST,* October, 1970, page 15.

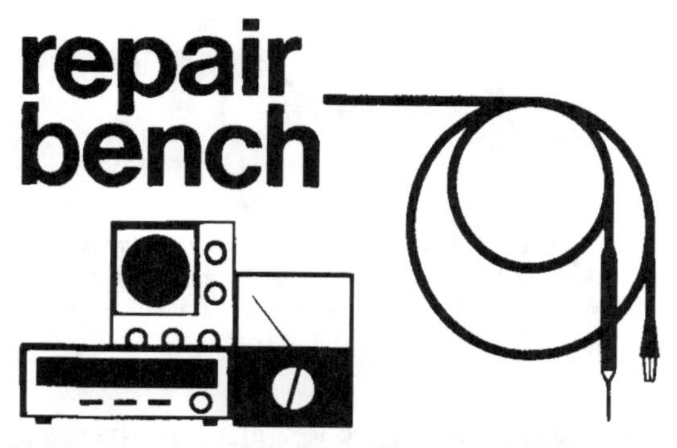

Michael James

basic troubleshooting

Troubleshooting and equipment repair are two of the toughest problems faced by radio amateurs today. Part of the difficulty is due to the fact that modern ssb equipment is much more complex than the old a-m and CW gear of twenty years ago, but perhaps more important, few amateurs build major pieces of their station equipment anymore so they are probably not as familiar with its circuitry as they should be. When your receiver or transmitter starts giving you trouble, more than likely it will be returned to the factory to be repaired. If the problem isn't too severe, you may avoid using that function which is affected or overlook it altogether. In some cases you may not even be aware of a problem unless another amateur brings it to your attention (distorted speech, poor sideband suppression or splatter, for example).

Although there *may* be some equipment repair problems that are best sorted out by the factory, in most cases you can save yourself a lot of time and money by fixing it yourself. Once you send your gear back to the factory, you may have to wait a month or more until you can get back on the air. In addition, you will probably have to pay the factory ten dollars an hour or more for their technician's time.

Troubleshooting electronic equipment is not difficult, nor does it require a bench full of test equipment. A large selection of test equipment may simplify the task, or allow you to solve a problem more quickly, but 90 per cent of all troubleshooting can be accomplished with a volt-ohmmeter and other simple test equipment you already have on your workbench. In those cases where you need a calibrated signal generator or an oscilloscope, you can often borrow one from your local radio club or from an amateur who lives nearby.

In the coming months this column will be devoted to troubleshooting techniques and how you can use them to fix your own equipment. Although much of the initial discussion will be in general terms that are applicable to practically any electronic equipment, future columns will discuss specific pieces of equipment and unique or unusual circuitry that requires a somewhat different procedure. If you have solved a particularly difficult equipment problem, we would like to hear about it. There may be others who will be helped by your success.

basic troubleshooting

There are three basic troubleshooting techniques which can be used to locate and fix circuit malfunctions: signal tracing, resistance measurements and voltage measurements. In receivers and transmitters the problem area is usually located with signal tracing, then pinpointed with resistance and/or voltage measurements. Although some electronic circuits such as gain-control circuits don't lend themselves to signal tracing, the majority of receiver and transmitter circuits can be quickly checked with this method. Once you know how to use signal tracing, in fact, you will probably agree that it's one of the quickest ways to track down a circuit problem.

Basically, signal tracing consists of injecting a signal at the input to a piece of equipment and checking its path through the equipment. If the signal appears at the input to a stage, but not at the output, that stage is the culprit. It may not be the only culprit, but once it's been fixed, you can locate other problem areas along the signal path.

The signal tracer is essentially a very quiet, high-gain audio amplifier with headphone or speaker output. One commercial version which is available at modest cost is shown in the accompanying photograph. If you wish, you can build a simple high-gain audio amplifier around an op amp IC as shown in **fig. 1**, and in a pinch you could even use one channel of your stereo system. This is all you need if you're working with audio systems, but if you're troubleshooting rf and i-f stages, you will also need a simple demodulator probe such as that shown in **fig. 2**. The one I use is built into a discarded plastic ballpoint pen. You can also use one of the rf probes which are available for vacuum-tube voltmeters.

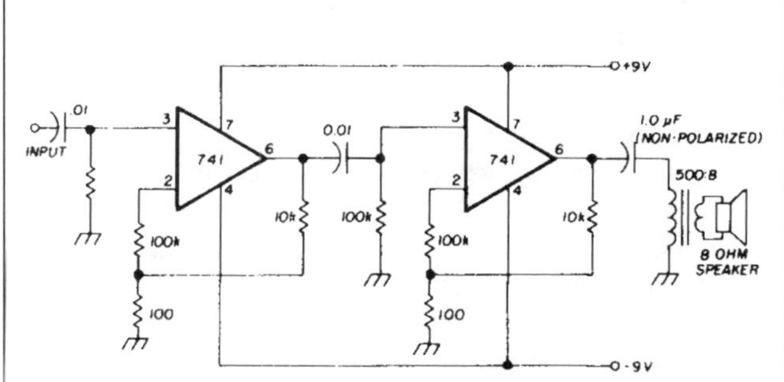

fig. 1. Signal tracer circuit which is based on the 741 op amp ICs. Gain is about 80 dB; audio power output is approximately 40 mW, sufficient for most signal tracing.

In addition to the signal tracer (audio amplifier and rf probe) you will also need a signal injector – a device which has a broadband signal output from audio through vhf. There are several pencil-sized signal injectors on the market for less than ten dollars. Most consist of a simple 1 kHz multivibrator which has high harmonic content well above 30 MHz. The circuit in **fig. 3**, which uses inexpensive high-speed switching transistors, can be used for signal tracing through at least 50 MHz. Built on perfboard, this unit is small enough to fit inside the aluminum cases in which expensive cigars are sold (you could also use a plastic pencil holder or toothbrush case).

signal tracing

Whatever kind of signal tracer you decide to use, you'll want to get the most out of it. Many people who already use signal tracers seem to think that signal tracing is limited to localizing trouble in one section of a receiver or transmitter. However, as will be shown later, the signal tracer can also be used to pin down defective components. All you have to do is know how to use it.

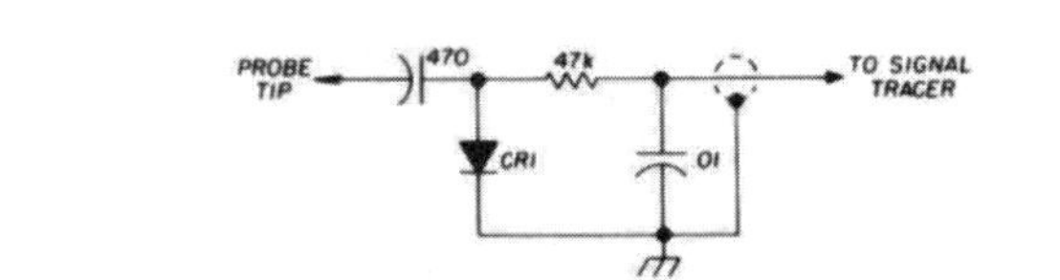

fig. 2. Demodulator rf and i-f probe for signal tracing. Diode CR1 can be practically any signal diode such as 1N34A, 1N60, 1N67A, etc.

Fast troubleshooting with a signal tracer demands logic, and you'll have to supply that. But even if you haven't done any troubleshooting before, you'll be amazed at how quickly you can track down a faulty circuit with a signal tracer. Fixing the bad circuit after you've located it may be another story, but if you use logic, and the resistance and voltage measurements we will discuss in future columns, you can probably repair any electronic circuit ever built. Things are simplified tremendously if you have a copy of the schematic or the manufacturer's maintenance instructions, but even without these you can, with persistence, be successful.*

As a starter I'll show you how to use a signal tracer to troubleshoot the sophisticated amateur communications receiver shown in **fig. 4**. This block diagram is fairly typical of modern superheterodyne receivers although some models may use only one frequency-conversion stage, or may not be equipped with a crystal calibrator or a separate a-m detector. I should also point out that it doesn't make any difference if your receiver uses vacuum tubes, transistors, or some combination of these -- the basic signal tracing technique is the same.

First of all, take a look at the schematic and mentally divide it up into blocks representing each stage or function in the set. **Fig. 4** has been divided into four basic sections: rf, high i-f, low i-f and audio. In some cases you might want to consider the detectors separately, but they are usually included as part of the last i-f.

First, the rf section. When signal tracing here you'll have to use the demodulator probe. If the receiver is connected to an antenna you will hear a mismash of incoming signals because most receivers don't have sufficient front-end selectivity to pick out any one signal – that's done further on, in the low i-f. If all the rf stages are normal, once you set the bandswitch all the signals within several-hundred kHz will be heard through the signal tracer. The collectors of the rf amplifier and mixer transistors (plate circuits in vacuum tube receivers) are the points to check with your probe. If you don't get any signal output from the mixer, something in the rf section is dead.

*Manuals for most amateur equipment manufactured between 1940 and 1965 are available from Hobby Industry, WØJJK, Box H864, Council Bluffs, Iowa 51501. Send self-addressed, stamped envelope for quote.

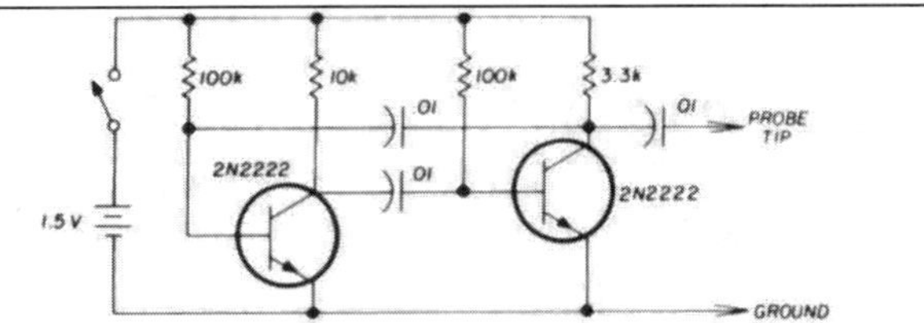

fig. 3. Signal injector is basically 1 kHz multivibrator which has high harmonic output. Circuit shown here has usable output up through 50 MHz. Practically any npn transistors will work in this circuit.

The high i-f processes the output of the first mixer and consists of a bandpass filter, the second mixer and the variable frequency oscillator. If any of the circuits in the high i-f isn't working properly, the signal picked up by your tracer at the output of the second mixer will reflect it. The low i-f includes the selective filter, i-f amplifier amplification and the detectors. You'll need your demodulator probe for the i-f stages, but the quickest test point for the whole section is after either of the detectors. Here you should hear a clear, undistorted audio signal without the probe. The audio section can also be checked without the probe. If the receiver is okay, you should hear a nice strong signal at the output of the last audio stage.

If the receiver isn't working properly, the quickest

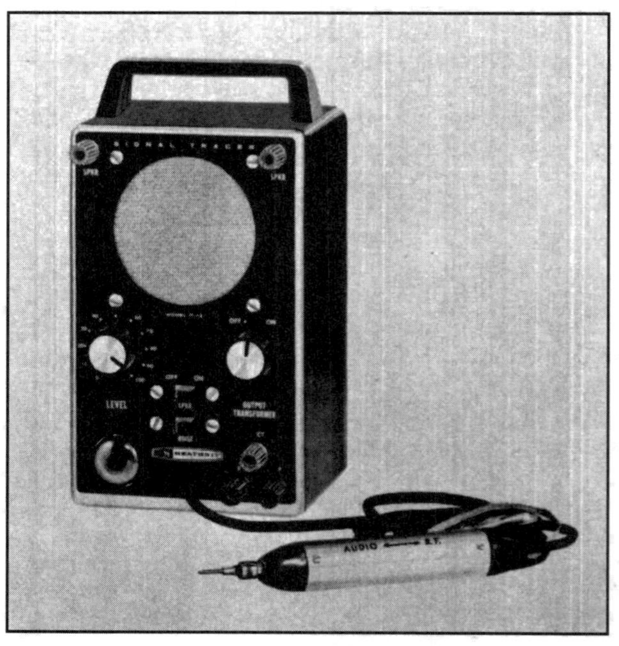

Heathkit IT-12 signal tracer has both visual (eye tube) and audio output. A switchable audio-rf probe is included with the kit, which sells for $32.95.

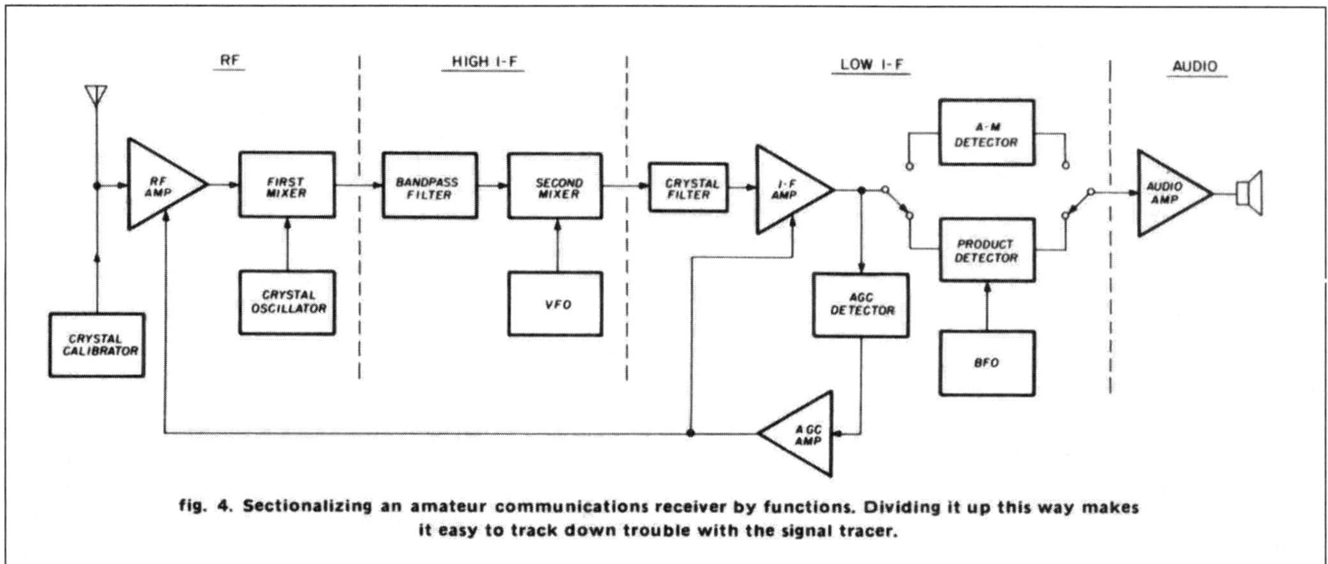

fig. 4. Sectionalizing an amateur communications receiver by functions. Dividing it up this way makes it easy to track down trouble with the signal tracer.

way to find the bad circuit is to check signal output about halfway through the set. A good point is the output of the second mixer. If the receiver is connected to an antenna the signal you hear should change as you tune the vfo (since the demodulator probe is an a-m detector, ssb signals will be unintelligible). If your receiver will tune to one of the WWV channels, this makes an excellent test signal, or you can use your signal injector. The pitch of the wideband injector signal, however, will not change as you tune the vfo.

If the signal is okay at the output of the second mixer, you have cleared the front-end circuits of any suspicion and can proceed to the last half of the set – the output of the low i-f amplifier is a good point. If you don't get an output from the second mixer, the low i-f and audio sections are probably okay.

Assume you get nothing at the output of the second mixer. Divide the front end roughly in half and use the tracer and demodulator again. The output of the first mixer is a good check point. If you have the proper signal there, there's something amiss in the bandpass filter, vfo or second mixer. If there's no signal output from the first mixer, the rf amplifier, crystal oscillator or first mixer stage must be at fault.

The last half of the receiver can be attacked with similar logic. If the signal was okay at the second mixer, the next logical dividing point is the output of the detector, which can be checked directly, without the probe. A signal in the tracer means that everything is okay up to there and the trouble is in the audio section. If you don't get a signal, check the output of the other detector. No signal means it has been blocked between the second mixer and the detector – the crystal filter or one of the i-f amplifiers is the problem.

Note that with only two signal tracer checks you have ioslated the problem to one small, functional section of the receiver. If the signal is okay at the input to a stage and not at the output, it's obvious the trouble is between those two points. It's a simple matter to check each of the individual stages within a section to pinpoint the offending one.

The divide-and-conquer technique of stage isolation works just as well for other symptoms as it does for a radio that is completely dead. You can hunt noise or hum, for example, tracking down the stage where the trouble first appears. It also works for distortion.

other checks

If the receiver is suffering from poor sensitivity, the problem can be signal traced by the "straight through" method. If reception is poor, the fastest way to determine which amplifier isn't doing its job is to check the gain of each stage by touching the signal-tracer probe to the input and output; if there is little or no increase in signal strength, the amplifier is weak. Although transistor mixer stages usually have some gain, vacuum tube mixers seldom exhibit gain and may often have a small signal loss, so keep this in mind. The filters introduce loss, too, but you can judge if it's too much after you have a little practice.

There are other little tricks of troubleshooting logic that make it easy to find troubles. If your receiver works alright on a-m but not on ssb or CW, for example, the difficulty is probably with the product detector or bfo –

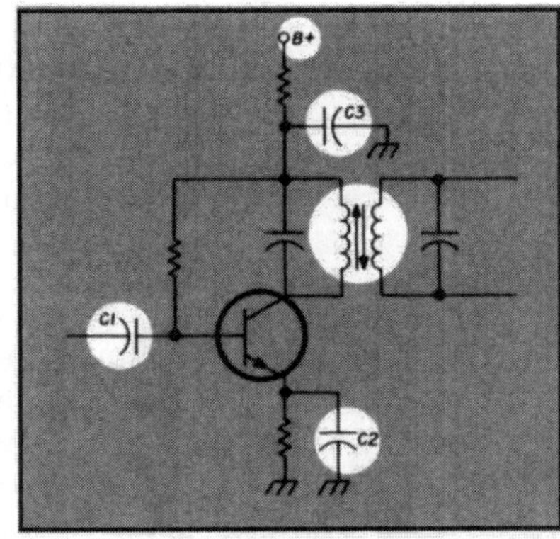

fig. 5. You can check these components with your signal tracer without even unsoldering them from the circuit.

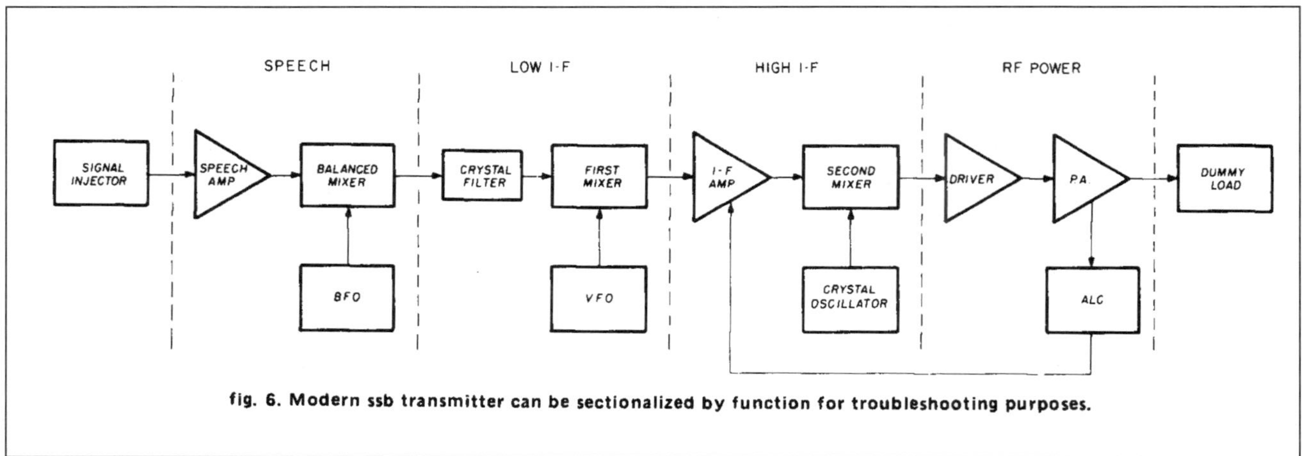

fig. 6. Modern ssb transmitter can be sectionalized by function for troubleshooting purposes.

they are the only stages which are not common to a-m. If weak signals sound okay, but strong ones distort, a good suspect is the agc stage which may not be controlling the rf and i-f gain as it should, letting strong signals overload the receiver. Likewise, frequency jumping or drift can usually be traced to the vfo; audio distortion eliminates all but the detector and audio stages; and poor selectivity is usually caused by a bad crystal or mechanical filter.

getting closer

After they've pinpointed the stage which is causing the problem, many technicians put away their signal tracer and reach for their voltmeter. However, the signal tracer can still tell you things about the circuit you can't find out with a voltmeter. In the amplifier circuit of **fig. 5**, for example, the highlighted coupling and bypass components can be tested right in the circuit without even unsoldering them.

The coupling capacitor, C1, and the interstage transformer, T1, should pass the signal along with very little attenuation. Whether they are large, as in audio stages, or small, between rf or i-f amplifiers, there should be about the same amount of signal on both sides. If there is any attenuation, it should be small. To check, touch the tracer probe to the input side of the component, then to the output side – if the output is much weaker than the input, the part is defective.

The bypass capacitors, C2 and C3, shunt the signal to ground and their values are chosen to short out practically all the signal at the emitter (C2) and at the power supply end of the interstage transformer (C3). The tracer should hear very little signal at either point. If there's any substantial signal the capacitor isn't doing its job. Even if the transistor is in good health, bad bypass capacitors at C2 or C3 will seriously degrade the gain of the stage.

Sometimes, when checking stage gain or components, you'll find that you don't have sufficient signal strength to determine if a component is doing the job it should. In this case it's helpful to place the signal injection directly at the input to the stage. This will bring the signal level up to the point where you can make meaningful measurements. You can also use the signal injector to quickly move through the receiver to determine which stage is causing the problem. Simply touch the probe of the signal injector to the input of each stage, starting at the audio output stage, and move back toward the front end, stage by stage. If everything is working properly you will hear the 1 kHz modulation through your receiver's speaker as you inject signal into each stage.

Finally, you can check the B+ line with your signal tracer for any traces of hum. Power supply filter capacitors are like any other bypass capacitors in that they should shunt all signal voltages to ground (power supply ripple in this case) and leave only pure dc. If one of the filter capacitors is weak, you'll hear a considerable amount of hum in the signal tracer. If the dc line isn't properly decoupled you may hear a whistling or hissing sound that is an rf or i-f signal if you could unscramble it. This can usually be traced to a bad bypass (decoupling) capacitor somewhere along the B+ line.

transmitter signal tracing

A modified form of signal tracing is also suitable for tracking down problems in ssb (and a-m) transmitters. In this case the signal injector is connected to the microphone jack and the transmitter is terminated in a dummy load as shown in **fig. 6**. Except that the position of the stages is reversed (audio front-end, rf output), the functions of the various stages in a modern ssb transmitter are not that much different than those in a superheterodyne receiver.

By using the signal tracer to track through the stages of the transmitter, you can quickly locate a stage which is blocking the signal (use the demodulator probe for the balanced modulator output and all following stages). The rf output from the final amplifier may be a little too much for the detector diode in the probe so don't connect it directly to the output – placing the probe tip next to the power amplifier compartment should provide enough signal for tracing purposes.

Although the signal tracer won't track down distortion, poor sideband suppression, or vhf parasitics in the transmitter, it's useful for quickly isolating a nonfunctioning stage or component. The signal tracer can also be used to eliminate hum and locate bad decoupling capacitors which are causing unwanted rf feedback. Other transmitter troubleshooting techniques will be discussed in a future column.

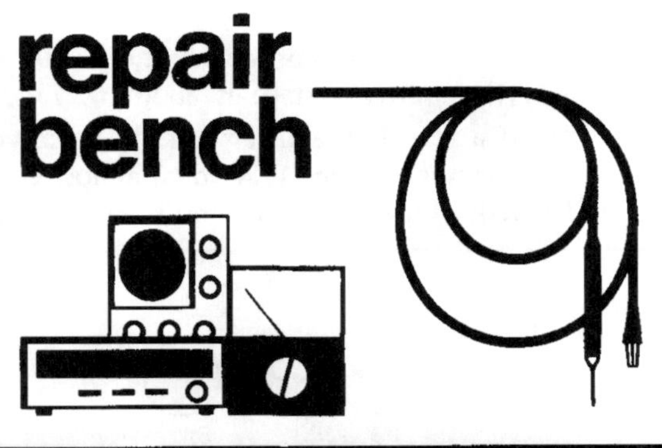

Joe Carr, K4IPV

troubleshooting transistor circuits

In this installment of repair bench I'll discuss troubleshooting of bipolar transistor circuits. The principal test equipment will be the vom or vtvm and, if you prefer, a transistor tester. First let's review some things about transistors that might affect servicing and trouble shooting.

preliminary considerations

In **fig. 1A** a stylized npn transistor is shown with its base-to-emitter junction connected across a power supply that causes it to be reverse biased. In this condition, charge carriers will be drawn away from the region at the junction, creating a relatively wide depletion zone. This allows only a very small current to flow across the junction. In **fig. 1B** the base-to-emitter bias voltage is changed so that the base-emitter junction is forward biased. Charge carriers are repelled by the battery polarity and are driven toward the junction barrier. Here they can combine with oppositely charged carriers from across the junction.

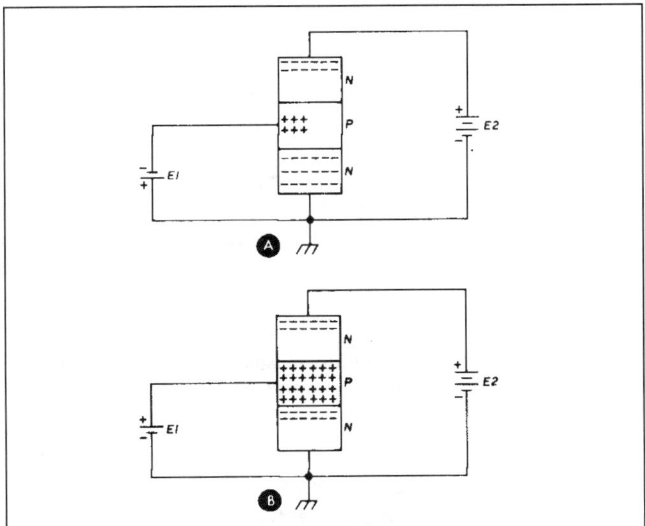

fig. 1. Relationship of charge-carrier propagation across the base-emitter junction in an npn transistor with reverse bias (A), and forward bias (B).

Figs. 2A and **2B** show proper forward-bias-voltage relationships for npn and pnp transistors respectively. In an npn transistor circuit the base is positive with respect to the emitter by approximately 0.7 volt for silicon and 0.2 volt for germanium types. Pnp transistors have about the same values of voltage drop across the base-emitter junction, but it is of opposite polarity. On these transistors the base is more negative (or less positive) than the emitter. (Keep in mind that these polarities are relative quantities).

table 1. Junction voltages to be expected in a normally operating transistor amplifier. Values for silicon devices are shown, followed by those for germanium (in parentheses). Readings 1 and 3 were taken with the minus meter probe on the emitter; reading 2 was taken with the minus probe on the transistor base.

junction	npn	pnp
1. base-to-emmiter	+0.7 (0.2)	-0.7 (0.2)
2. collector-to-base	++	- -
3. collector-to-emitter	++	- -

A pnp transistor, as in the example of **fig. 3**, is still correctly biased despite the fact that the voltages on the elements are positive with respect to ground. Since the base is at 9.3 volts and the emitter at 10.0 volts, the base will measure − 0.7 volt with respect to the emitter. **Table 1** gives voltage levels to be expected in a normally operating transistor amplifier stage. These voltages may not hold, however, in control circuits (such as squelch) or pulse circuits where the transistor may be reversed biased in one mode or another.

dc voltage checks

Any transistor stage has several dc voltages of interest to the troubleshooter. You want to measure the voltage drop across the emitter and collector load resistors and the base-emitter bias voltage. From these measurements you should be able to spot most faults. Of course, a circuit diagram that shows correct values would be of immense use, but it's not always necessary if you know what "ballpark" levels to expect.

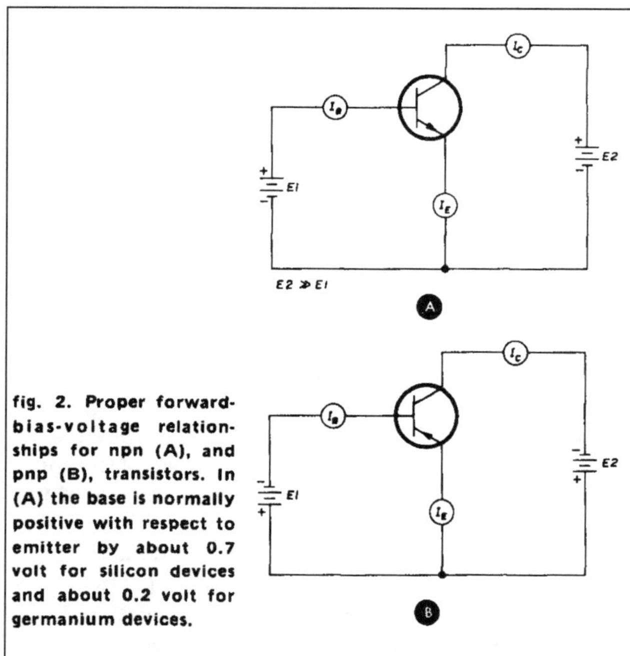

fig. 2. Proper forward-bias-voltage relationships for npn (A), and pnp (B), transistors. In (A) the base is normally positive with respect to emitter by about 0.7 volt for silicon devices and about 0.2 volt for germanium devices.

Perhaps the best method for learning the pattern of dc voltages that might be expected in the more common forms of transistor failure is the "case-history" approach. Assume that you have isolated the fault to a stage such as in **fig. 4** using signal tracing, signal injection or some other technique. Make the dc voltage measurements shown circled in **fig. 4.** Note that the emitter voltage is zero. Unless somebody has successfully repudiated Mr. Ohm, you're safe in assuming that the emitter current is also zero. You can also conclude that the collector current is either zero or at a very low value, because the collector voltage is close to the source voltage: a level about $V_{cc}/2$ would be normal. Measuring the base-to-emitter bias voltage, you find 10.7 volts instead of 0.7 volt. These symptoms usually point to an open base-to-emitter junction in the transistor. An ohmmeter or transistor checker will tell the tale if you're still in doubt.

Fig. 5 shows another common defect and its voltage relationships. The emitter voltage is about 0.27 volt, which results in a very low emitter current. Since you again have a collector voltage (17.4 volts) almost equal to V_{cc} (18 volts), you can say that no significant collector-emitter current is flowing. The base-to-emitter voltage, however, is almost normal, so this junction is probably all right. In this case the collector-to-base junction is open.

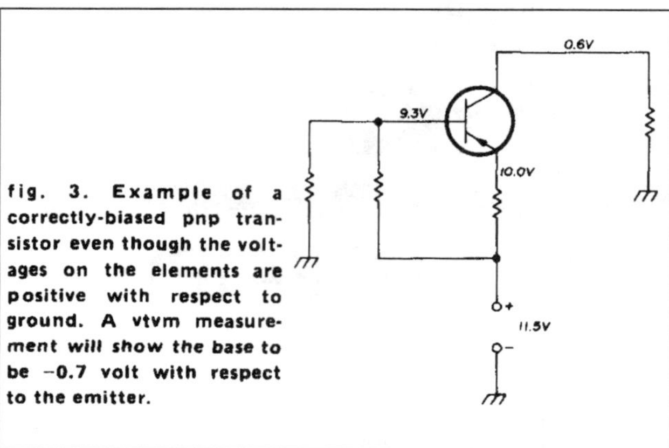

fig. 3. Example of a correctly-biased pnp transistor even though the voltages on the elements are positive with respect to ground. A vtvm measurement will show the base to be −0.7 volt with respect to the emitter.

A further common fault in transistor circuits is shown in **fig. 6.** Measurements show a low collector voltage equal or a voltage very close to the emitter voltage. This transistor probably has a collector-to-emitter short circuit. If enough power has been dissipated, this effect may cause collector and emitter resistors to burnout. This condition almost always occurs in class-A audio amplifiers.

transistor testers

Once you've decided that a particular transistor is suspect, you might wish to make further tests using one of the many transistor checkers on the market. Simple instruments that claim to measure transistor beta are available at low cost, both ready built and in kit form. Be aware that the really inexpensive checkers might be

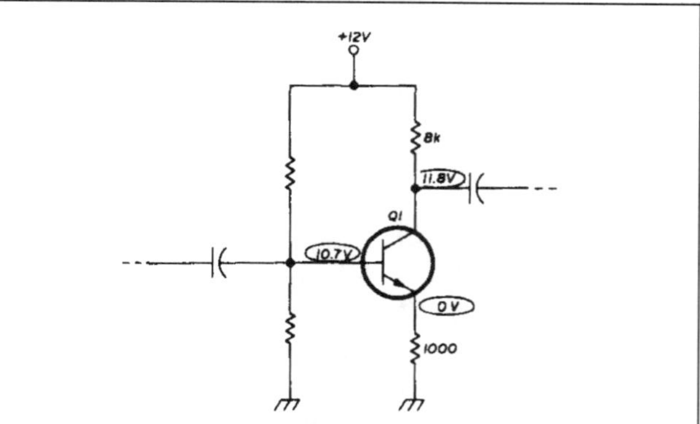

fig. 4. Example of the "case-history" approach to fault isolation in a transistor stage. Circled measurements indicate an open base-to-emitter junction.

too simple to give an accurate picture of the transistor's characteristics. The best transistor checker for ordinary service will provide some means for varying base current so you can determine whether the transistor base is able to control collector current.

An ohmmeter can be used as a transistor checker and can tell you quite a bit about the device under test. Certain precautions must be observed, however, as to the types of ohmmeters that are acceptable. Determine what type battery is used in the *ohms* section of your meter. If it's more than 1.5 volts, you may ruin as many transistors as you test! Some older instruments use battery voltages as high as 22.5 volts for the ohmmeter sections.

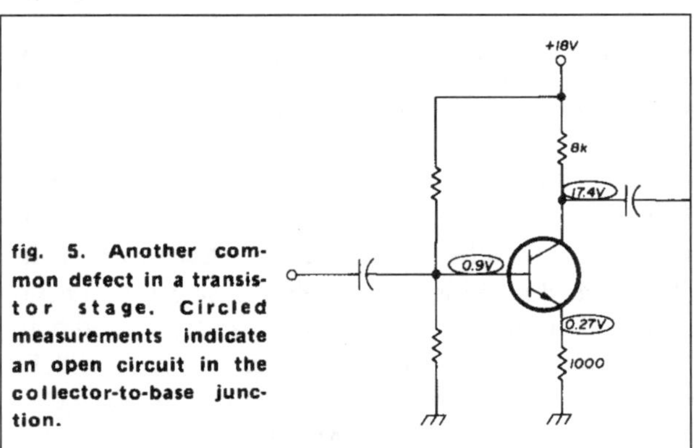

fig. 5. Another common defect in a transistor stage. Circled measurements indicate an open circuit in the collector-to-base junction.

Alternatively, some modern fet voltmeters have an ohmmeter source voltage that is too low for use as a "Transistor tester." This is a double-edged situation, because a desirable feature of such instruments is that they can be used to make in-circuit resistance measurements without removing the semiconductor devices. It is this very feature, however, that eliminates the instrument as a transistor tester. The dc voltage across the ohmmeter probes is too low to forward bias the pn junctions.

Test each junction separately by measuring its resistance twice (**fig. 7**). Measure the resistance the second time with the probes reversed from the direction used on the first try. Transistor junctions can be viewed as pn diodes and will not pass current in both directions under normal circumstances, so a very high resistance should occur in one direction and a much lower resistance in the opposite direction.

Normally, the reverse-forward ratio should be greater than 10:1. Check both base-to-emitter and collector-to-base junctions in this same manner. On power transistors use the RX1 scale, and on small transistors use the RX100 and RX1000 scales. If too low a resistance scale is used on those small transistors you may blow the junction.

Collector-to-emitter leakage may be checked in the same manner. Make two readings and take the higher one as the leakage resistance. The higher the better, and if it approaches your ohmmeter's idea of "infinity" so much the better. While making this test, you can also ascertain whether the base can control collector current. Connect the ohmmeter probes between collector and emitter. Next, short the base to the collector and note whether the resistance reading drops. If there is no response, reverse the probes and try again. If again there is no response, assume the transistor is dead.

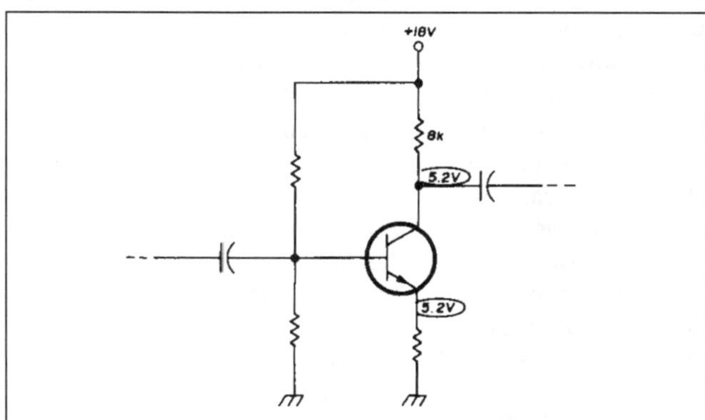

fig. 6. Circled measurements indicate a probable collector-to-emitter short circuit.

which stage is bad?

Signal tracing or signal injection techniques may provide the answer. However, a high-amplitude signal from a signal generator, or a transient generated when connection is made, may shock excite the defective transistor into normal operation. A far better technique would be to use dc analysis initially, then use one of the more traditional techniques only if the dc test fails.

Dc signal tracing requires only a vtvm or high-impedance vom. **Fig. 8** shows the transistor lineup in a

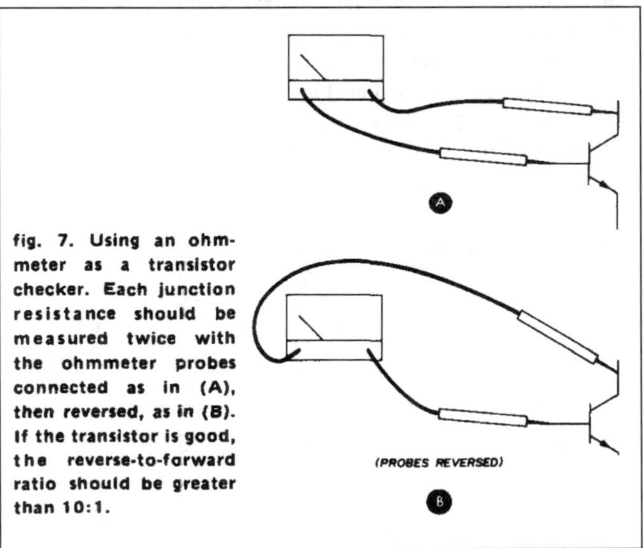

fig. 7. Using an ohmmeter as a transistor checker. Each junction resistance should be measured twice with the ohmmeter probes connected as in (A), then reversed, as in (B). If the transistor is good, the reverse-to-forward ratio should be greater than 10:1.

typical vhf fm receiver. This unit uses npn transistors, so ground the voltmeter minus probe, then use the positive probe to measure the voltage at each emitter in succession. Loss of emitter voltage indicates an open transistor or some defect that causes bias to be removed. The dc level tells you "this is the bad stage," and from there you can ascertain what's wrong. A higher-than-normal emitter voltage, on the other hand, indicates a leaky or shorted transistor.

Rf amplifiers and some gain-controlled i-f amplifier stages may yield false negative results when this test is used because of agc action. If you can't disable the agc, try rocking the tuning dial back and forth while noting meter readings. The agc-controlled stage will show voltage variations as the dial is tuned across incoming stations. Although the agc can foul you up when troubleshooting, it can nevertheless be used to advantage. If the emitter voltage in this stage (or the collector voltage in some pnp rigs) varies as the dial is tuned across an active band, you can be pretty sure the defect is *not* between the antenna input terminals and the point where the i-f signal is sampled for the agc drive.

An S-meter can be used instead of a voltmeter as an overall check. Note whether the S-meter deflects normally as the receiver is tuned across the band. If it does then look elsewhere; if no S-meter deflection occurs, a problem exists within the agc loop.

A similar technique may be used to troubleshoot receivers that use pnp transistors, with modifications in procedure to account for the difference in transistor polarity. (In both cases assume that negative grounding is used – a fair assumption in most mobile equipment, but one possibly tinged with errors in some home equipment). In a pnp stage, connect the voltmeter positive probe to the B+ line and use the minus probe to measure the emitter resistor voltage drop (**fig. 9**). Most receivers use a fairly hefty electrolytic capacitor to decouple the B+ line; this component may often be used as a point of identification if no schematic is available. As in the case of npn stages, the voltage drop at the emitter resistor can give clues as to device malfunction.

The oscillator stage can be checked using dc analysis techniques as an indicator of oscillation (but not of *oscillation frequency). Connect a voltmeter across the*

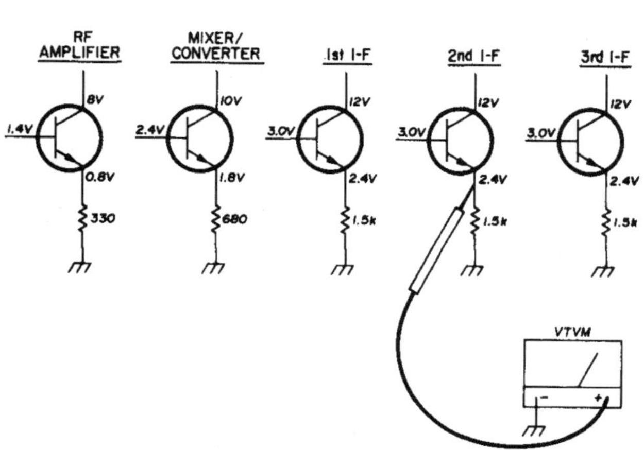

fig. 8. Dc signal tracing with a vtvm in a typical vhf-fm receiver using npn transistors. The probes should be connected as shown. Loss of emitter voltage indicates an open transistor or some defect that causes loss of bias.

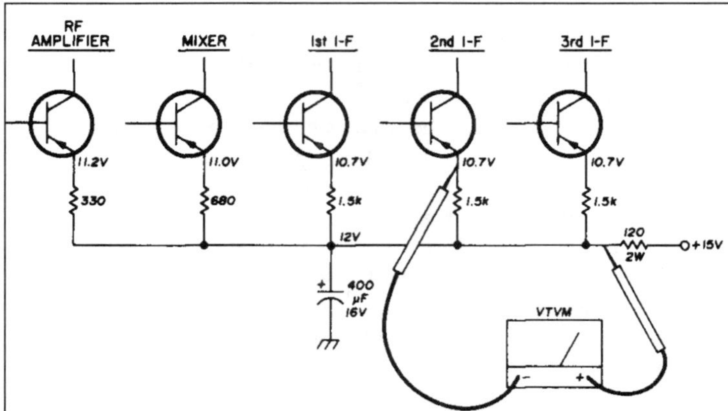

fig. 9. Using a vtvm for dc signal tracing in a vhf-fm receiver with pnp transistors. As in the circuit of fig. 8, the voltage measured at the emitter resistor will give clues as to device malfunction.

emitter resistor, using whichever procedure is applicable to the transistor polarity. Tune the dial from one end of the band to the other. The emitter-resistor voltage drop will vary as the dial is tuned. This change will be greater on general coverage-receivers than on amateur-band-only models; but, even in the latter, some change will be noted. In crystal oscillators, sometimes a change in emitter-resistor voltage drop will occur when the crystal is removed from the circuit. In either case, the change identifies oscillation.

caution note

Vacuum-tube test equipment is tolerant of some abuse, but transistorized equipment is not always so forgiving. Use of ungrounded ac-powered test equipment can easily generate both transient and steady-state voltage levels, which are quite capable of destroying transistors. Ground the cases of your test equipment. Two-wire power cords always identify ungrounded test equipment. For safety reasons it's wise to convert to three-wire power cords, in which the third wire is grounded to the chassis and cabinet. Of course, this only applies to equipment which is *not* ac/dc.

Grounded test equipment can create problems when troubleshooting, as in **fig. 9,** where the grounded case might be connected to the voltmeter minus probe. In that case, you must use either a battery-operated voltmeter such as a vom or fetvm, or a voltmeter which, although ac powered, isolates both input probes from chassis. This, incidentally, seems to be the way many modern digital voltmeters are being made.

conclusion

The material presented here is intended for the average amateur who likes to service his own equipment. I've tried to cover the most likely problems that may be encountered. The procedures given should prove useful and will allow you to get back on the air as soon as possible without spending a lot of money for a repair bill.

shirt pocket transistor tester

The transistor tester described here is useful when troubleshooting circuits. It's also great to take along while shopping the surplus stores for checking unknown devices. Large quantities of older transistors are available at low cost in surplus outlets, many of which are entirely suitable for many projects.

This tester will allow you to grade devices into npn/pnp and good/bad categories. It will also give you a pretty good idea as to whether the transistor is germanium or silicon. The best way to illustrate what the tester can do is to go through a typical test routine. The following procedure is based on the fact that the loaded terminal voltage of the battery in the tester is at least 7.5 volts.

using the tester

Suppose you wish to test a small-signal transistor that has no recognizable markings. Before plugging it into the tester, make sure the function switch is at the *short* position. This test must *always* be made first. Since nothing is known about the device, either npn or pnp can be chosen on the toggle switch. If the meter shows any reading at all, you'd normally reject the transistor without making any further tests. In this case however, if the toggle switch were set to npn and the device under test was a pnp (or vice versa), the meter would indicate a short circuit. (More about this later.)

Move the toggle switch to its other position and check for a meter reading. If the meter still deflects, we no longer care what the device is because it's definitely bad. Do not attempt to make any other tests on a bad device; it's bad for the meter! If the meter reads zero, then we've determined the device type and the fact that it's probably usable. The second step is to determine the quality of the device.

Turn the knob to test for I_{cbo} (collector-to-base current, emitter open). Most modern small-signal transistors exhibit nanoamperes of I_{cbo}. Many older types have microamperes of I_{cbo}. The former will give little or no reading on this meter, while the latter will produce small to moderate readings. We'd normally relegate a transistor with a high I_{cbo} reading to the trash can or give it to some curious youngster, so further testing at this point is really academic. In any case, a transistor that produces a moderate-to-high meter reading is probably a germanium device. Germanium devices typically have I_{cbo} values 10 to 100 times that of silicon devices, which is one reason why most transistors made today are silicon. Since I_{cbo} will increase with temperature and effectively increase the forward bias on the device in a circuit, a condition known as thermal runaway ensues. And if the I_{cbo} reading is moderate to high, be prepared for a high I_{ceo} (collector-to-emitter current, base open) reading on the next test. I_{ceo} always will be greater than I_{cbo} by a factor approximately equal to the current gain (beta) of the device. The modern silicon devices may not show an indication on the meter on this test either.

The final test is for beta. The switch selects a high beta range of 500 and a low range of 100. You won't find transistors with a beta of 500 very often, but they do exist. A transistor I like to use is a 2N3391, which typically gives beta readings between 300 and 400. The more usual case for the older types is a beta of 100 or less, and many of the types made for switching applications read quite low.

Having finished this test sequence, we've obtained some pretty useful information about the device under

Simple shirt-pocket transistor tester is packaged in small instrument box only 2½" (66mm) wide and 4" (85mm) high.

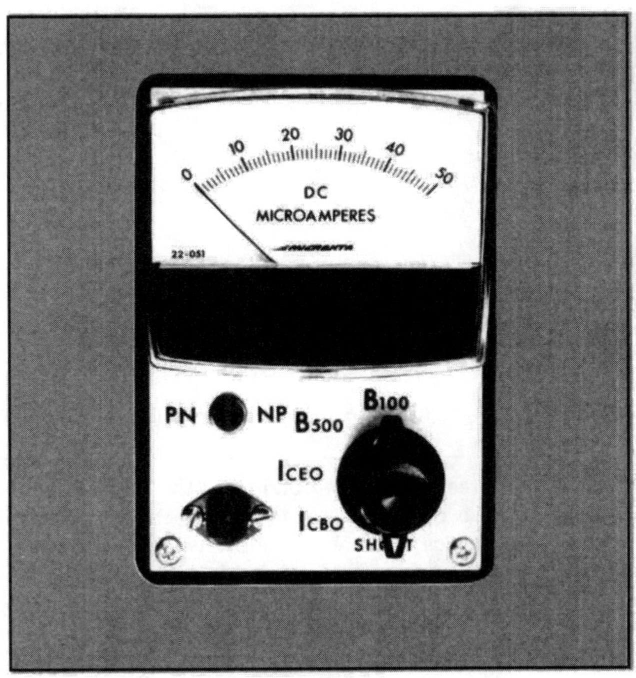

*A complete parts kit for this transistor tester is being made available in conjunction with this article. For ordering information and prices, write to G.R. Whitehouse & Co., 10 Newbury Drive, Amherst, New Hampshire 03031.

By Dave Cheney, WØMAY, 4808 N. Monroe, Loveland, Colorado 80537

test. Of course, the most important aspect of these tests is to determine if the device is suitable for our needs. Admittedly the tests don't tell us anything about frequency response, but this requires an entirely different tester. If you really must know whether a device is germanium or silicon, an additional test can be made. Set a volt-ohmmeter or vacuum-tube voltmeter to the one-volt range, and connect the meter test probes between base and emitter with the device plugged into the tester. Select npn or pnp, depending on the device type, and set the function switch to the *B500* position. A germanium device will read 0.2 to 0.3 volt while a silicon will read 0.6 to 0.7 volt.

The short test is really an I_{ces} test, as shown in **fig. 1A.** In this test the full-scale meter reading is 5 mA. Resistor R1 limits the current to about 6 mA, depending on battery condition. This resistor provides some meter protection and limits battery current drain during short tests. A meter reading will be obtained if the device has a collector-to-base or a collector-to-emitter short. To see why an npn device will show a short when the toggle switch is set to pnp (or vice versa), visualize the collector-base junction as a reversed-biased diode in normal operation. If the battery polarity is reversed, the collector-base junction becomes forward biased, which is the case when the toggle switch is in the wrong position.

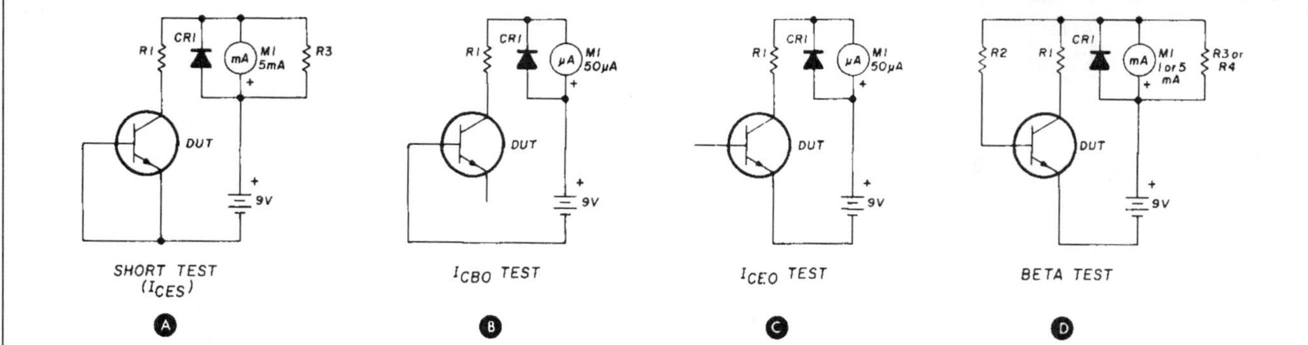

fig. 1. Circuits demonstrating the transistor tester operating principles. R1 is a current-limiting resistor and is in the circuit at all times. R2 sets the base current value for beta tests, and R3 is a meter shunt resistor, which is discussed in the text.

The I_{cbo} test circuit is shown in **fig. 1B.** The meter reads 50 μA full-scale, as no meter shunt is used. This value of current seems to be a good compromise for checking the collector-to-base leakage of most transistors. The same current range was used for the I_{ceo} test. Some transistors, especially germanium types with high I_{cbo} and beta, will pin the meter. Fortunately, this doesn't happen very often. If you desire a higher current range for this test and still wish to retain the meter calibration, a meter shunt resistor could be connected to set the full-scale value to 500 μA. The I_{ceo} test circuit is shown in **fig. 1C.**

The *B500* and *B100* test positions use the same circuit, except that different meter-shunt values are used. The *B500* test uses the same meter shunt used in the *short* test. On the *B100* test, a meter shunt resistor sets the full-scale current to 1 mA. The basic circuit is shown in **fig. 1D,** but only one shunt resistor is shown.

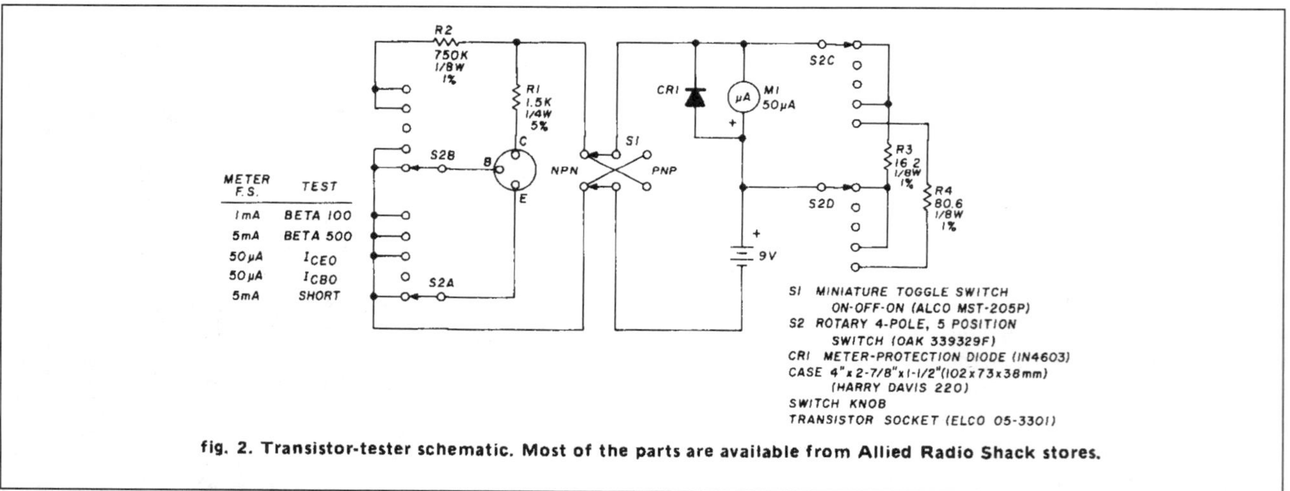

fig. 2. Transistor-tester schematic. Most of the parts are available from Allied Radio Shack stores.

The complete test circuit is shown in **fig. 2.** Note that collector resistor R1 is connected at all times. Its purpose has already been explained. Note, however, that this scheme does not adversely affect the test readings.

The diode across the meter is for over-current protection. Any small-signal germanium unit is suitable. The meter tracking accuracy doesn't compare with that of more costly units but is entirely satisfactory for this use.

The meter that maintains its zero position while in use has good damping. At the present price of about $7.95 it's a good choice for this kind of project. The meter isn't supplied with electrical data on either meter resistance or full-scale terminal voltage. Information such as this seems to be missing from many of Radio Shack's products. I suspect that the low price and lack of information means that the specifications aren't held to close tolerances.

The shunt resistors can't be calculated without having one or the other of the above specifications, so I measured mine. I applied current until the meter read full scale then measured the terminal voltage, which was 80 mV. (I didn't carry the test to its conclusion to see if the full-scale current was truly 50 μA.) Assuming the proper current value, an Ohms law calculation shows that my meter should have an armature resistance of 1600 ohms. If you're so inclined, you can measure your meter terminal voltage and recalculate the shunt resistor values for R3 and R4 using

$$Rs = \frac{I_m}{I_s} \times R_m$$

where: R_s = shunt resistance
R_m = meter resistance
I_m = meter current
I_s = shunt current

Note that I_s = circuit current minus meter current (I_m).

construction notes

The transistor tester is very compact and the battery is a tight fit. My original intention was to use an 8.4 volt mercury battery (E-146X), which is almost the same size as the common carbon-zinc 9-volt transistor radio battery. The problem is that the two batteries are not exactly the same size. The mercury battery is just slightly thicker and will not allow the front panel to fully close. I mention this because the mercury battery is the most suitable for this application. The mercury battery has a relatively flat discharge characteristic, so that the terminal voltage remains fairly constant until the end of its life. Such a battery will improve the long-term accuracy of the tester. I didn't realize the physical differences between the two batteries until it was too late, and the carbon-zinc cell is used at present. The battery is positioned between the bottom of the case and the back of the meter. The rotary switch in my tester was a junk box item and it's not as suitable as the one in the parts list. The recommended switch uses only two wafers, which provide more room inside.

Transistor tester parts layout. A slightly larger case than used here (see fig. 3) may be used to accommodate a type E-146X 8.4-volt mercury battery, which will improve the long-term accuracy of the tester.

Those who wish to use my layout may refer to **fig. 3** for the front panel dimensions. Note that this is the rear view of the panel, and that the meter-mounting screws are not symmetrically displaced from the center of the meter cutout.

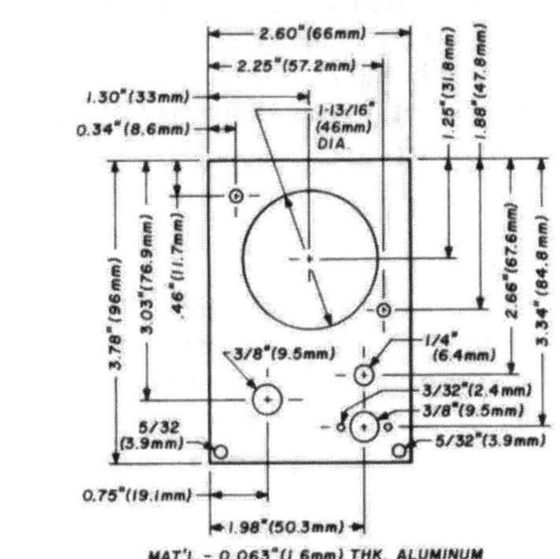

fig. 3. Front-panel layout viewed from the rear. The two 5/32-inch (3.9mm) holes at bottom are located 0.156 inch (3.9mm) in from each side and are for panel-to-case mounting screws.

alignment and test

No alignment is needed other than adjusting the meter mechanical zero. The battery should be replaced when its loaded terminal voltage reaches 7.5 V. If the voltage falls below that value, the *B500* test will read low. A simple battery check may be performed as follows:

1. Set the function switch to *short*.

2. Set the toggle switch to either position.

3. Insert a small bare wire in the test socket between the collector and emitter pins.

4. If the meter reads less than 50, replace the battery.

There you have it: a handy-dandy transistor tester you can take to the surplus store in your pocket. If you've read this far, you probably need one too.

integrated circuit base-step generator

A straightforward circuit which may be used in building a curve tracer

Several articles have appeared recently describing curve tracer adapters for oscilloscopes. These adapters provide a means of measuring and displaying the current-versus-voltage characteristics of two- and three-terminal semiconductors on an oscilloscope. All of these adapters have two basic sections: a collector sweep circuit, and a base-step generator.

The collector sweep circuit generates a sweep voltage which is applied to the collector of the device under test. This voltage is usually derived from stepped-down line voltage which has been halfwave (60 Hz) or full-wave (120 Hz) rectified. The magnitude of the waveform is controlled by either an adjustable resistive divider or by a variable autotransformer. The circuit also usually contains a resistor for sensing collector current, and/or current-limiting load resistors. The design of this portion of the curve tracer adapter is straightforward and has been adequately discussed in previous articles.[1,2]

base-step generator

To generate a family of curves for three-terminal devices, the adapter must include a base-step generator in addition to the collector sweep circuit. This circuit generates a series of voltage or current steps which are synchronized with the beginning of each collector voltage sweep. These steps are then applied to the base or gate of the three-terminal device under test. The accuracy of this portion of the adaptor is a major factor in the overall accuracy of the curve tracer adapter.

To achieve good accuracy and low parts count, the base-step generator circuit of **fig. 1** was designed using a digital-to-analog, integrated circuit approach.

The resulting circuit has the following features:

1. Calibration of the circuit involves only one adjustment.

2. Accuracy of the circuit is determined only by the accuracy of resistor ratios, not absolute resistor values.

3. A true current source supplies base current.

4. The number of current or voltage steps is selectable.

5. Parts count is low and therefore construction is easy.

6. It is easily adapted to the builder's individual requirements.

circuit

The heart of the circuit is Motorola's MC1406L, a six-bit, digital-to-analog (D/A) converter. Before introduction of this device, most D/A converters were too expensive to be used by amateurs in their projects. Now, a D/A converter is available that is within most amateurs' budgets.* The basic configuration of the MC1406L is shown in **fig. 2**.

A reference current, I_{ref}, is established and flows into pin 12. I_{ref} is given by

$$I_{ref} = \frac{V_{ref}}{R_{12}}$$

The output current, I_o, which flows into pin 4, is an accurate fraction of I_{ref}. This fraction is determined by the digital word present at the inputs of the MC1406L, pins 5 through 10. The output current is given by:

$$I_o = I_{ref}\left(\frac{\overline{A}_1}{2} + \frac{\overline{A}_2}{4} + \frac{\overline{A}_3}{8} + \frac{\overline{A}_4}{16} + \frac{\overline{A}_5}{32} + \frac{\overline{A}_6}{64}\right)$$

For example, if the digital inputs (A_1 - A_6) are of the

*Single quantities of MC1406L are available at the price of approximately $5.90.

By **Henry Wurzburg, WB4YDZ/7,** Motorola Semi-Conductor Products, Inc., Phoenix, Arizona 85008

form 101100, then $\overline{A_1}$ through A_6 would be 010011, and the output current, I_o, would be

$$I_o = I_{ref}(\frac{0}{2}+\frac{1}{4}+\frac{0}{8}+\frac{0}{16}+\frac{1}{32}+\frac{1}{64})$$
$$= I_{ref}(\frac{19}{64})$$

The output current can range from 0/64 to 63/64 of I_{ref} in increments as small as 1/64, depending on the state of the digital inputs. A set of guidelines is given in Appendix 1 on the proper operating conditions for the MC1406L IC.

If the output of a digital counter is connected to the inputs of the MC1406L, a series of current steps result on its output. This is the technique that is used in the base-step generator of **fig. 1**. Referring to the circuit, a train of clock pulses is derived from rectified line voltage. My curve tracer adapter used full-wave rectification of the line voltage for the collector sweep. Therefore, the ac voltage from which the clock pulses are derived is likewise full-wave rectified. This voltage is applied to Q1 at point A. The pulses from Q1 are inverted by U1A, and its output is applied to the clock inputs of U2 and U3. These J-K flip flops, along with U1B and U1C, form a synchronous divide-by-eight counter.

The outputs of this counter are then applied to the $A_1 - A_3$ inputs of the D/A converter, U5. I_{ref} is adjusted by potentiometer R1 to be 400 μA and the inputs A_4 through A_6 are tied to +5 volts. With this configuration, the output current into pin 4 is incremented 1/8 I_{ref} or 50 μA, each time the counter is incremented. Note that the minimum increment of current that can be obtained with this configuration is 1/8 I_{ref}, or 50 μA. If current increments other than 50 μA are desired, they can be obtained by appropriate adjustment of R1 and by using inputs A_2-A_4, A_3-A_5, etc. The guidelines for proper MC1406L operation, as set forth in Appendix 1, should always be followed when such a modification is done.

The output of U5 is then fed into a current amplifier composed of U6, Q2, and their associated resistors.

The basic configuration of the current amplifier is shown in **fig. 3**. The relationship between the input current I_1 and the output current I_2 is given by

$$I_2 = I_1 \ (1+\frac{R1}{R2})$$

An important feature of this configuration is that the output current is constant if the input current is constant, independent of the voltage between the output terminals. Note that the accuracy of the current gain is dependent only upon the ratio of R1 and R2, not on their absolute values. This affords the builder the opportunity of obtaining highly accurate current gains without the use of expensive precision resistors.

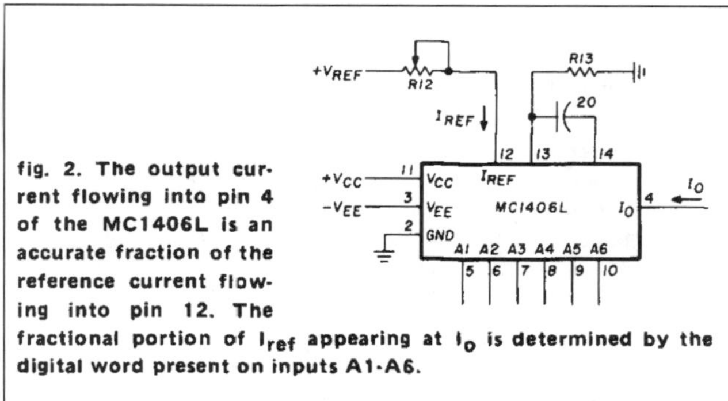

fig. 2. The output current flowing into pin 4 of the MC1406L is an accurate fraction of the reference current flowing into pin 12. The fractional portion of I_{ref} appearing at I_o is determined by the digital word present on inputs A1-A6.

This current amplifier configuration is used in the circuit of **fig. 1**. Transistor Q2 reduces the amount of current U6 must supply by a factor of Q2's beta. Resistor R9 converts the current steps into voltage steps for fet measurements. This resistor is the only precision part necessary in the circuit. U4 and S1 select the number of steps generated by resetting the counter section at the appropriate point in the count sequence. The circuit power supply is shown in **fig. 4**. At this point two words of caution are in order:

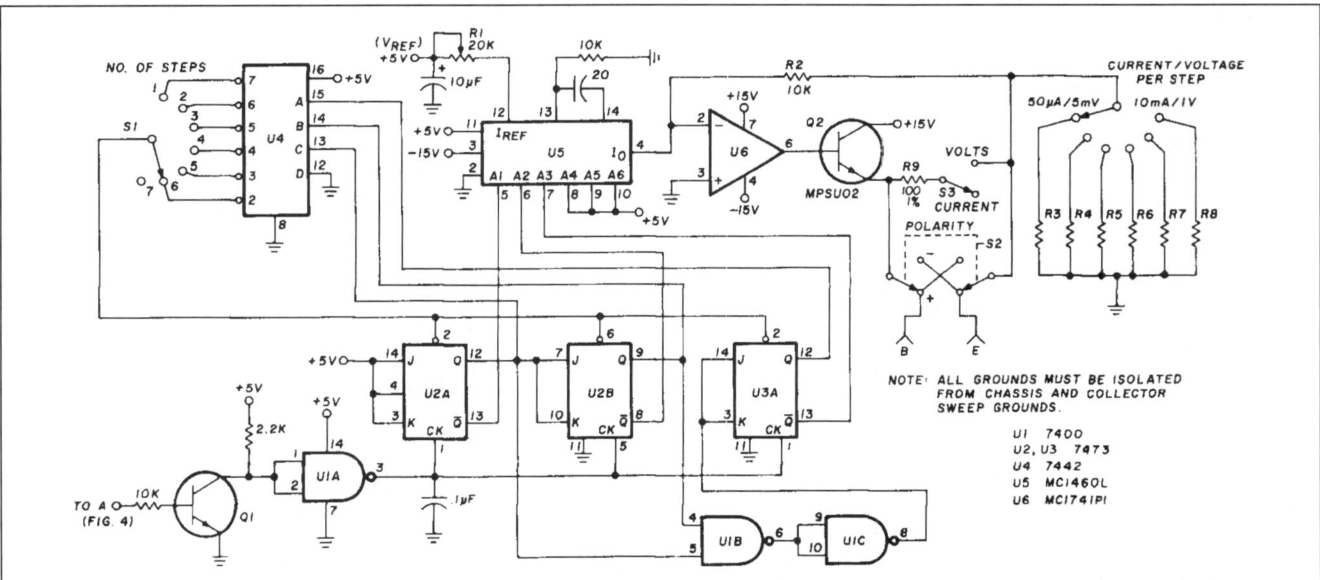

fig. 1. The base-step generator using the MC1406L six-bit, digital-to-analog converter. The values for resistors R2 through R8 are listed in table 1. R1 is a 10-turn 20k potentiometer. Q1 is a general-purpose npn with dc current gain of about 30.

1. Be extremely careful *not* to apply voltage steps to the base of bipolar transistors, or excessive base current will result.

2. The ground for the step generator circuit *must* be isolated from the collector sweep circuit and chassis grounds for proper operation.

construction

Normal construction practices are applicable. However, it is important that the power supply leads be properly bypassed. For the digital portions, a 0.01 μF disc capacitor for every five IC packages is satisfactory. All linear device supply voltages should be bypassed as close to the device as is possible with 0.1 μF disc capacitors.

table 1. Values for R3 through R8 are shown in last column as a ratio of R2 for various gains and voltage- and current-step ratios.

current step	voltage step	gain	R
50 μA	5 mV	1	∞
100 μA	10 mV	2	R2
500 μA	50 mV	10	R2/9
1 mA	100 mV	20	R2/19
5 mA	500 mV	100	R2/99
10 mA	1000 mV	200	R2/199

With the exception of the current amplifier's gain determining resistors, circuit components are noncritical and appropriate substitutions can be made. The accuracy of the circuit is directly dependent upon the selection of resistors R2 through R9.

As was previously mentioned, the gain of the current amplifier is determined by the ratio of two resistors. If an accurate digital ohmmeter or resistance bridge is available, it is a simple matter to "bridge" the resistors necessary to obtain the desired gain. For example, if it is desired to obtain a gain of 100, then R2/R = 99. Since R2 equals approximately 10k, R should be approximately 110 ohms. R2 would be accurately measured and its value noted. Then resistors of 100 ohm nominal value would be measured and a resistor selected whose measured value is closest to being 1/99 of R2's measured value. Highly accurate resistor ratios, and therefore gains, can be obtained in this manner.

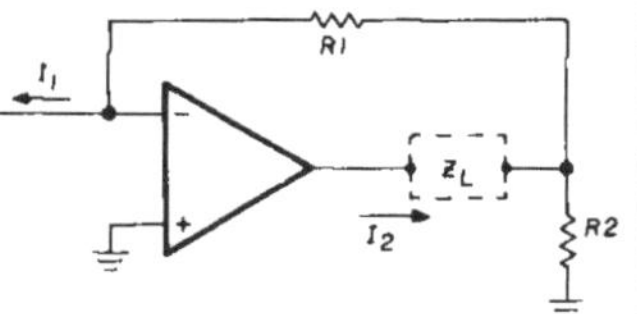

fig. 3. Current amplifier used to buffer the output of the base-step generator follows the configuration shown here. If input current is constant, then output current will also be constant, independent of voltage changes across Z_L.

Of course, this method can not be used if a digital ohmmeter or resistance bridge is not available, or if you don't have a healthy stock of resistors on hand. In this case, precision (1%) resistors must be obtained and used in combinations to obtain the appropriate resistor values.

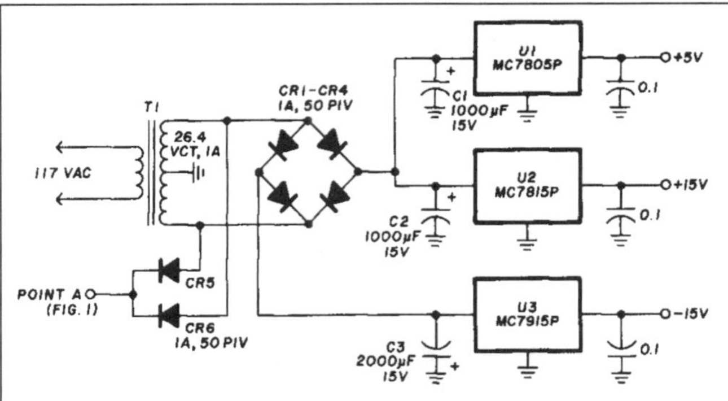

fig. 4. A suitable power supply for the base step generator of fig. 1.

To calibrate the circuit, set switch S1 to step position 7, switch S3 to the VOLTS position, and switch S4 to a midrange position. Connect an oscilloscope to the B and E output terminals and adjust R1 to obtain the correct voltage steps. The circuit is now ready for use.

conclusion

This circuit offers versatility and adaptability. In addition, its accuracy is limited only by the accuracy of the instruments that are used to select the components and to calibrate the circuit. The inclusion of this base-step generator circuit in a curve tracer adapter provides a highly accurate and extremely useful accessory for an oscilloscope.

appendix

The following are some guidelines for proper circuit operation of the MC1406L six-bit, digital-to-analog converter. They are by no means maximum limits, nor are they intended to define all the possible regions of operation. Instead, they are given as an aid for individual design and are appropriate for most circuit configurations. For more detailed information on the MC1406L's capabilities and applications, see the manufacturer's data sheet.

1. Normal operating voltages:

$$V_{CC} = +5 \text{ volts}, V_{EE} = -15 \text{ volts}$$

2. I_{ref} should be equal to 500 μA to 4 mA

3. V_{ref} should be equal or less than +5 volts and well regulated.

4. If V_{ref} is obtained from a logic supply, it should be heavily bypassed close to R12 (fig. 2).

5. R13 should be approximately equal to R12.

6. Voltage on pin 4 (output pin) should never exceed plus or minus 0.4 volt. This may be accomplished by the use of op amp buffering.

references

1. Albert Klappenberger, K3KWX, "An Accurate Solid-State Component Curve Tracer," *CQ*, July, 1974, page 20.
2. Daniel Wright, WA9LCX, "Transistor Curve Tracer," *ham radio*, July, 1973, page 52.

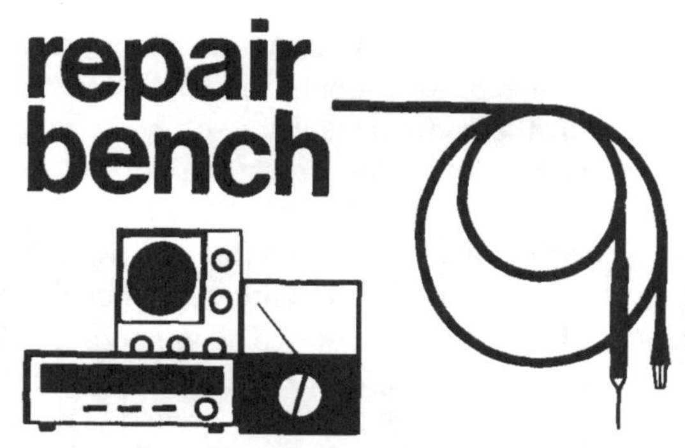

Joe Carr, K4IPV

troubleshooting the power supply

It's probably safe to say that few circuits fail as often, in almost every type of equipment, as the power supply. In fact, a dictum often given younger electronics people by more experienced hands is to look for the power supply problem before attempting to find any others. I know this is a rather large claim, but it's supported by almost two decades of electronic service experience and will probably be corroborated by others who have such experience.

In this month's *repair bench* we'll discuss the troubleshooting techniques appropriate to receivers, transceivers, low-to-medium-power transmitters, and other items of station equipment. It's assumed that the reader is limited as to available test equipment, so the procedures are given in terms of simple voltmeters and ohmmeters, although it's recommended that you also obtain at least a low-cost oscilloscope.

full-wave supply

Consider the rather ordinary full-wave supply of **fig. 1**. This circuit is of a type frequently encountered in amateur radio equipment. We'll not spend any time describing the operation of this type of circuit because this is an article on troubleshooting, not theory. If you want a review, see the *Radio Amateurs Handbook* and the other appropriate amateur literature.

By Joseph J. Carr, K4IPV, 5440 South 8th Road, Arlington, Virginia 22204

The rectified output voltage waveform, without filtering, resembles a series of pulsating dc waves (or inverted parabolas). But with capacitor C3 connected the waveform more nearly resembles that of **fig. 2A**. The output waveform taken across bleeder resistor R1 should be very close to a straight line and be ripple-free.

common problems

Several different types of problems will be found in power-supply circuits. Those most often found are:

1. Hum or ripple on the dc output.

2. No output voltage, but the fuse is okay.

3. Fuse blows.

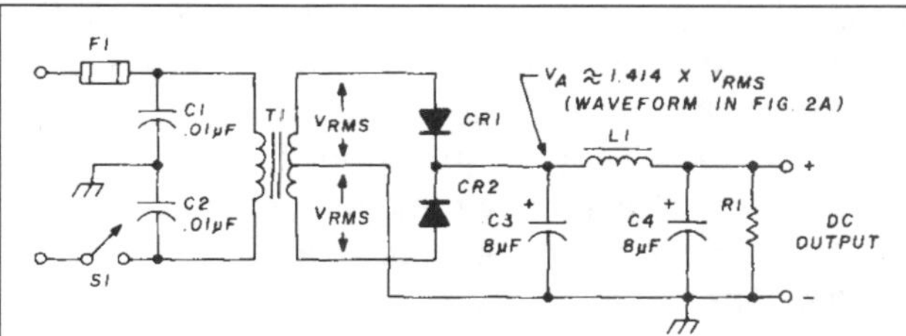

fig. 1. Typical dc power supply using full-wave rectification. The text describes methods for troubleshooting problems that cause ripple or hum in the output.

Ripple appears in any number of ways, depending mostly on the nature of the circuit drawing power from the affected power supply. In a communications or broadcast receiver, for example, the hum will most likely be heard in the speaker or earphones; while in a transmitter, the hum will most likely modulate the output and be heard by others. It doesn't only affect ssb/a-m transmitters, incidentally, so also suspect ripple in many types of problems in CW and fm transmitters. On a television screen, ripple will usually show up as either single (60 Hz) or double (120 Hz) horizontal black bars, sometimes of low contrast, that seem to migrate vertically up the screen.

In almost all cases where ripple or hum originates in the power supply, it's reasonably safe to "pre-indict" the filter capacitors. These components are the most frequent source of trouble. Realize, however, that not all sources of hum or apparent ripple are the fault of the power supply. There are several other sources for such spurious signals.

One of these is a resistive short circuit in a vacuum tube. The heaters, and sometimes the cathode are powered directly from low-voltage ac and this voltage will either appear on the grid or modulate the cathode-plate current directly in some cases. Another case is low-frequency oscillation with a frequency that approximates that of the power line or its second harmonic. Also, in many types of equipment, the possibility of ground loops must be considered as well as common-mode rejection problems or shielding defects as a source of hum.

One differentiating technique is to examine the frequency of the hum. In all equipment that uses any form of *fullwave* rectification, the ripple frequency in the power supply is 120 Hz (or twice the line frequency for those outside the United States and Canada). If the hum is nearer to the line frequency, then you can suspect one of the other causes and temporarily forget the power supply.

These articles on power-supply servicing contain much useful advice for newcomer and old timer alike. Especially appropriate are the author's remarks on "dos and don'ts" in working with this type of circuit. Joe Carr's remarks are *must reading* for anyone thinking about working on power supplies, regardless of the voltages involved. **Editor**

The waveforms in **figs. 2A** and **2B** were taken from the same power supply, on the same oscilloscope, with the same sensitivity setting, during the same session. The actual power supply circuit was not unlike that of **fig. 1**, but the filter capacitors were higher in value. The waveform of **fig. 2A** was taken at point **A** (in **fig. 1**) with capacitor C3 connected and in good condition. The waveform in **fig. 2B**, on the other hand, represents the waveform at the same point but with C3 disconnected. This exercise effectively simulates an open capacitor as might be found in actual equipment. Note the rather dramatic increase in ripple amplitude.

table 1. Steps to follow when troubleshooting a power supply when no dc output is obtained.

measurement across	approximate value	check for open circuit if no voltage at*
1. T1 primary	115 Vac	F1, F1 holder; S1; ac power cord
2. T1 secondary	high-voltage ac	T1 primary; T1 secondary
3. C3	high-voltage dc	T1 center tap; center tap grounding; diodes CR1 and CR2
4. Across C4	high-voltage dc	L1

*All checks are performed with an ohmmeter and with the ac power plug removed from the wall outlet. Wait a few seconds, then connect an alligator cliplead across the filter capacitors. Connect the negative or grounded end first, then attach the hot end. Leave the cliplead in place for 25 seconds, then remove it. The supply should be safe for both you and the *ohmmeter*. In the last two steps it may be necessary to remove the two diodes before making resistance measurements, as they can give erroneous readings.

Always check the associated wiring and solder connections for each component! Many bad solder connections are difficult to eyeball, so you may want to re-do each one if no other possible cause is found.

component substitution

If you don't have an oscilloscope, then you'll have to trouble shoot using the component substitution technique. This technique consists of shunting a known good capacitor across each filter capacitor in its turn. It's important to a) use a filter capacitor with the *same* or *higher* ratings and b) scrupulously observe *polarity markings.*

An aluminum electrolytic capacitor might explode if power is applied in reverse of normal polarity. Even if the danger from shrapnel is reduced, the cleanup afterwards and the pure fright involved should be reason enough to take care!

Capacitor bridging must be done with care, not only as mentioned above, but in the actual manner in which you perform the job. You may well be dealing with potentials that can be lethal, so some safety precautions are necessary. In this one respect I advise you to ignore the professional servicer and listen to some expert advice given to beginners:

1. Turn the equipment off.

2. Using a screwdriver, or preferably an alligator cliplead, ground the filter capacitor for a few seconds.

3. Check with a dc voltmeter to make sure all the charge has been *drained off* the capacitor.

4. Using either alligator clipleads or solder tacking, connect the known good capacitor across the suspect component.

If the substitute capacitor makes the symptoms disappear, then make the substitution permanent — first go through the capacitor discharge procedure given above.

Many people, even (or perhaps especially) those professionally experienced in electronic servicing, are tempted to solder a new replacement directly across the open capacitor which was used originally. This approach to "repair" seems especially convenient if the bad capacitor is a multi-section chassis-mounted type, and an available replacement is of the tubular type . This is an example of popular, but *extremely poor* practice. It's often the case that the old capacitor will short circuit, which will have a spectacularly bad effect on the future life expectancy of your equipment! This is one of those cases where we can show a good reason for demanding good craftmanship on repair jobs. Shunting a bad capacitor with a good one, then, is only a *diagnostic* method and is definitely **NOT** a *repair* technique!

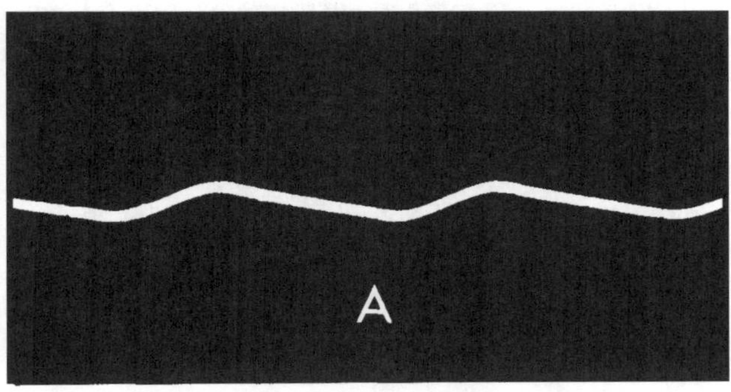

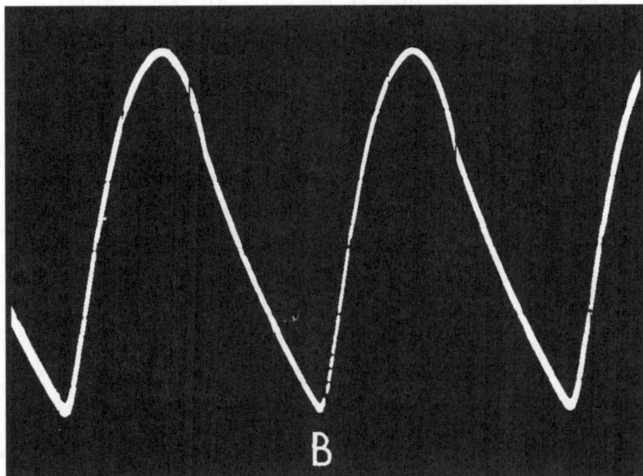

fig. 2. Oscillograms showing the output waveforms obtained from the supply in fig. 1. Picture A shows the supply output waveform taken with capacitor C3 (fig. 1) connected. Picture B shows the pulsating waves obtained at CR1, CR2 junction (fig. 1), without the succeeding filter, C3, L1, C4 connected.

In cases where there's no dc output but the fuse is intact, or where the fuse blows, it's necessary to use a vom or vtvm. These instruments are so low in cost these days that you should make one or the other a part of your amateur-station equipment. For most amateur work the vom is preferred. Reasons: portability, low cost, and it's not susceptible to rf fields.

But let's get back to our next problem type: no dc output, but the fuse is intact. **Table 1** shows the steps to follow in this type of job. Set your voltmeter to its highest ac scale and connect the probes across the primary of transformer T1. Turn on the power and observe the voltage reading. If it's too far down the scale to be easily read, reduce the voltmeter range, posittion-by-position, until a good reading is obtained. If you get past the 150-Vac scale, however, and still have no reading, then give up.

A good reading (105-125 Vac in the U.S.) indicates that you should go o n to the next step. If, on the other hand, there's no voltage across T1 primary turn off the set, unplug it from the wall outlet, then use ohmmeter continuity checks (*R x 1* scale) to find the open component. Check those items indicated in the right-hand column of **table 1** for continuity.

Similarly, you must check the voltage across the T1 secondary. Again set your ac voltmeter to its highest ac scale and work down until a readable deflection is obtained. If there's no voltage across T1 secondary, suspect either the secondary or primary winding as being open. Use the same procedure as indicated in the first step.

For the last steps in the procedure, it's necessary to use the dc scales of your voltmeter. The same method of starting with the highest scales and working down is necessary to avoid an expensive surprise.

the open fuse

A somewhat more spectacular type of power supply defect is the case of the blown fuse. It's reasonably true that "fuses don't cause trouble, they *indicate* trouble." Whenever a fuse blows you must assume that there's a root cause, and that does not usually include a surge on the ac line, as many often believe. To be sure, such activity can occur during a thunderstorm, but in most residential areas few truly destructive surges do occur. Something overloaded that power supply, and it bears at least an attempt at diagnosis!

One of my first employers in electronics was fond of saying that, "All electrical troubleshooting involves the finding of an unwanted path for current, or locating a lost but required path for current," and that "The best tactic is to divide and conquer." Good advice 18 years ago and still good advice today. Such is the approach with the case of the blowing fuse.

We want to find an unwanted path for current (a short circuit) and must follow either (or both) of two schools of thought on how to divide and conquer.

One school maintains that the best method is to obtain a good supply of new fuses and keep replacing them as they blow. In the meantime, between fuse changes, you disconnect first one component then another until one is disconnected that causes the fuses to stop blowing. This technique has a certain validity, but can be expensive, especially if there are a large number of options. A better idea is to mount a television receiver-type ac circuit breaker in a plastic box (well insulated). Solder one end of two alligator clipleads to the breaker. The alligator-clip ends are then used to shunt the blown fuse. This can be a little dangerous if not done correctly, so if you have any doubts, then use up a couple of boxes of fuses.

The other school of thought is to use an ohmmeter to locate the short circuit to ground. This is the method of choice if there is some way of knowing the expected resistance to ground at various critical points. This type of information is often given in amateur equipment service manuals, but in many cases you will have to guess. This technique is the more elegant of the two, provided it works, because one must always be aware that operation of an overloaded circuit, even if the fuse is expected to go up in smoke, can further damage the equipment. Under **NO** circumstances may you use a higher-value fuse!

The actual procedure in "divide-and-conquer" servicing of a power supply circuit depends somewhat upon the type of equipment being serviced. In audio circuits you may have a shorted output tube or transistor, shorted output transformer, and so forth. In a transmitter, on the other hand, a final amplifier operated without protective bias may well blow a fuse if the excitation fails. If an ohmmeter check fails to reveal a short circuit in such a case, then suspect a loss of drive, or a sagging filament, as the cause. The latter problem may occur after a few moments normal operation and it is most frequently seen in transmitters where the final-amplifier tubes are mounted horizontally.

If a short circuit does show up on the ohmmeter, however, it can be conquered by first disconnecting the rectifiers (the most frequent single cause of trouble), and then the load, followed by the filter capacitors. Don't overlook the possibility of a carbon path from an arc or dirt to ground, which could cause the trouble.

power-supply dos and don'ts

1. If at all possible, use an isolation transformer to power all instruments and equipment being used or serviced.

2. Always use *well-insulated* alligator clipleads and meter probes. If they are in disrepair then repair them before use.

3. *Always* unplug the power cord when connecting or disconnecting clipleads or when soldering tacking.

4. Work on a bench with a master power shut-off switch and, if possible, a ground-fault interrupter.

5. Use the "buddy-system" and inform the buddy where the master shut-off is located and how it's operated. Similarly, inform all members of your family of the master-switch location and under what circumstances it is to be operated.

6. If your equipment and instruments don't have three-wire power cords, install them. In this type of power cord a third wire (usually green in color) is grounded to the equipment chassis.

7. *NEVER* work outside or in an area with a concrete or dirt floor* unless the equipment is designed for that purpose by reason of *double-insulation* or *three-wire* power line grounding of the equipment case. Many people have been killed by indoor appliances taken out of doors.

8. Now for a seeming paradox: *NEVER* defeat or trust interlocks.

9. Never service ac/dc equipment unless operated from an isolation transformer. Similarly, never use ac/dc equipment outdoors.

10. Always do quality work, use quality components, and *never* button up a piece of equipment cabinet with a temporary or unorthodox repair inside. A Murphy's Law corrollary states that, "Temporary repairs become permanent if there are more than two screws holding the cabinet together."

11. Switch to safety, think safe, work safe, be safe, and live.

*The editor of this article, W6NIF, after 40 years as an amateur, was careless recently when testing a 4500-Vdc power supply. He was working in a garage with a concrete floor and the +4500-volt lead accidentally fell onto the concrete. W6NIF was lucky — he survived.

understanding and using electronic counters

Only a few short years ago, it was extremely rare to see an electronic counter outside of a laboratory or a specialized service installation. Today counters can be found in ham shacks all over the world. Of course the reason for this proliferation is obvious — the integrated circuit, and in particular, medium- and large-scale integration. In the early sixties, a typical 10-MHz counter weighed nearly 120 pounds (55kg), occupied about 5.75 cubic feet (165,000 cubic centimeters), and dissipated approximately 600 watts as heat. By way of contrast, a 520-MHz counter currently produced by the same manufacturer weighs 4.75 pounds (2.16kg), has a volume of approximately 213 cubic inches (3890 cubic centimeters), and dissipates less than 20 watts.

As size has decreased, so has cost. That antediluvian counter cost $2600 in "1966 dollars;" today you can buy a counter for under $100 if you want a bare-bones instrument, and can get a 250-MHz multifunction counter for less than $400. Because of today's relatively low costs, counters have become a versatile tool in the ham station and on the work bench. If you have one, this article may help you make better use of it. If you are planning to buy one, it may help you to decide what to look for.

You may have noticed that the title of the article uses the term *electronic*, rather than *frequency*, counter. This was not a pedantic choice; electronic describes the type of counter and is inclusive of all functions that the counter may perform, only one of which may be the measurement of frequency. We shall discuss these various functions, although emphasis will be placed on frequency measurement, which is of primary interest to the average ham.

Before discussing the applications and limitations of the frequency counter, it is important to cover the method by which frequency is measured by the counter. Regardless of the type and complexity of the instrument, all counters measure frequency by comparing the frequency of the input signal with a known frequency or time period. **Fig. 1** shows the basic functional blocks of a typical counter. The main function of the signal conditioner is to convert the input signal to one whose amplitude and waveshape are compatible with the internal circuitry or logic of the counter. It generally includes an amplifier to increase the amplitude of the incoming signal, and may also contain an attenuator for input signals of high amplitude, trigger level and slope selection circuits, and so on. No matter how the signal is processed, the output of the signal conditioner is a pulse train in which each pulse corresponds to one cycle or event of the input signal.

The conditioned signal is applied to a gating circuit, which is shown symbolically as a single logic gate, but which is actually a more complex circuit. The gate is opened for a predetermined, accurate time interval, during which the signal passes through to the decade counters. These counters count the number of pulses which are gated through, and transfer the count to the display. The number of decade counters determines the number of digits which are displayed, one counter being required for each digit. The display can utilize any type of visual readout device, such as gas-discharge numeric tubes, light-emitting diode arrays, or liquid-crystal displays.

Since the decade counters count the number of pulses which pass thorugh the gate, it follows that the accuracy of the instrument is a function of the time that the gate is open. This interval is, in turn, a function of the time-base accuracy. The time-base oscillator in the modern counter is invariably a crystal-controlled oscillator operating at a frequency between 1 and 10 MHz, although there have been counters made in the past which used crystal frequencies as low as 100 kHz, or even used the ac line frequency as a time base. Even though the oscillator frequency must be divided, crystals in the 1- to 10-MHz range are used because they are inherently more stable than those which work at lower frequencies; the optimum range for stability is between 4 and 10 MHz for most types of crystals.

The divided time-base frequency drives the gate-

By Robert S. Stein, W6NBI, 1849 Middleton Avenue, Los Altos, California 94022.

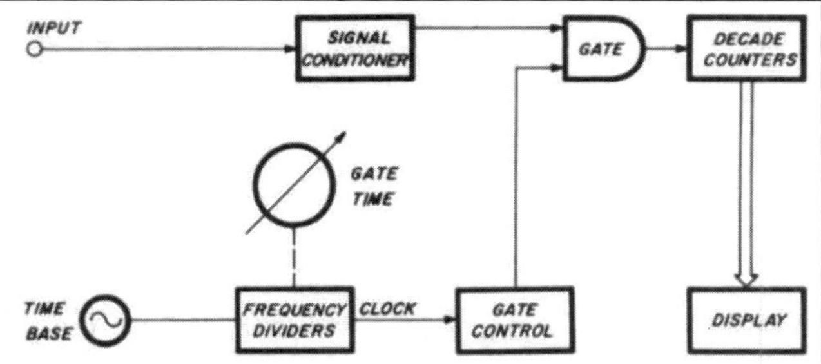

fig. 1. Functional block diagram of the basic frequency counter.

control circuit, which controls the gate-time interval in accordance with the divided time-base frequency. To explain the need for dividing the time-base frequency, we must at this time discuss resolution, or the smallest frequency increment which the counter displays.

Let us assume that the frequency of the time-base oscillator is 10 MHz, and that the gate control holds the gate open for exactly 10 clock pulses (clock being the term used to designate the time-base signal or the signal derived, through the dividers, from the time base). Since each cycle from the 10-MHz oscillator has a period of 0.1 microsecond, the gate will be open for 1 microsecond. If a 1-MHz input signal were being measured, only one pulse would be gated through, and the counter would display a **1**. Frequencies below 1 MHz might or might not produce a reading at all, and those above 1 MHz could be read only to within one digit of the nearest megahertz (more about this later). Thus, the resolution would be 1 MHz at best, an obviously unsatisfactory arrangement.

Suppose, instead, that the time-base frequency were divided down to 10 Hz, or a period of 0.1 second. The gate will now be open for 1 second, and 1 million pulses from a 1-MHz input will be counted. Now the counter will display 1000000, which provides us with a resolution of 1 Hz. Thus, the resolution is the reciprocal of the gate time, and in fact, some counters with selectable gate times have the switch positions designated by the gate time.

The limiting factors governing resolution are the number of digits in the display and the tolerable gate time. Usually 0.1 Hz is the smallest resolution practical, in that it involves a 10-second gate time and a 9-digit display up to 99.9999999 MHz. The gate time can be reduced in a computing counter, but that is outside the scope of this discussion.

Of course, it is not always necessary to read frequency to a tenth of a hertz, nor is it particularly convenient to have to wait for a 10-second count. By selecting the appropriate output from the frequency-divider chain, you can reduce the gate time and resolution to values which may be more appropriate to the measurement. The normal range of gate times is typically between 1 millisecond and 10 seconds, corresponding to resolutions of 1 kHz to 0.1 Hz.

The number of digits in the display can be reduced by switching both the displays and gate times, a technique which is used in many low-priced counters having a 5-digit display. A 2-position switch is used to select gate times of 1 second (1-Hz resolution) and 1 millisecond (1-kHz resolution). When the gate time is 1 second, the five decade counters can produce a display up to 99,999 Hz; when the gate time is 1 millisecond, up to 99,999 kHz or the frequency limit of the counter can be displayed. Thus by switching the clock, the equivalent of eight digits is obtained, with overlap between the two readings. This is an economical, but oftentimes inconvenient, way of obtaining improved resolution.

It should be reiterated at this point that a counter displays a pulse count. Whether the display reads out in Hz, kHz, or MHz is simply one of convenience and the location of the display decimal point. The decimal point is either fixed, or is switch selected with the gate time, and its position is independent of the actual count.

time-base accuracy

It should be apparent from the preceding discussion that the time-base oscillator is the most critical part of the counter, in that it determines the overall accuracy of the instrument. Let us examine its effect, in terms of the specifications usually given for a counter.

The Yaesu YC-500 Frequency Counter has a frequency range of 10 Hz to over 500 MHz. Its six-digit display provides the equivalent of eight digits when the gate time is switched. A room-temperature crystal, TCXO, or ovenized crystal time base may be ordered (*photo courtesy Yaesu Electronic Corporation*).

First of all, the accuracy of the time base, either in per cent or parts per total, is directly translatable to the measurement of frequency, period, interval, or any other function which the counter may be capable of measuring and which utilizes the time base. This holds true, regardless of the magnitude of measurement. For example, if a 1-MHz time-base oscillator is off frequency by 2 Hz, that represents an error of 0.0002 percent. The gate interval, therefore, will also be in error by the same percentage, and the displayed count will have the same error. If the frequency being displayed is 50.000000 MHz, the error will be 100 Hz. If the time-base frequency is high, the displayed count will be low, since the higher the frequency, the shorter the gate time. If the time-base frequency is low, the opposite will hold true.

Time-base accuracy specifications should include the parameters listed in **table 1**, although most lower-priced instruments may omit one or more. Typical values for the various types of oscillators are included as examples. It can be seen that temperature change has the greatest effect on frequency. In the examples listed, the specification for temperature stability can be improved by one order of magnitude by using a TCXO (temperature-compensated crystal oscillator) instead of a room-temperature crystal. In the real world, however, a good room-temperature crystal may be better than a poor TCXO; you must compare the specifications.

The oscillator aging rate is not as important, since this will manifest itself as a gradual change in frequency, and is predicated on the oscillator running continuously. If the counter is designed so that the oscillator circuit is powered as long as the instrument is connected to the primary power source, the specified aging rate is valid. If the

The Fluke 1910-A Multi-Counter is one of a series which provides frequency, period, period-average, ratio, and totalize functions. The 1910A is rated to 125 MHz; the 1911A and 1912A are similar in appearance and will measure frequencies to 250 and 520 MHz, respectively (*photo courtesy John Fluke Manufacturing Company*).

table 1. typical specifications for time-base oscillators.

	room temperature crystal	TXCO*	oven oscillator
Aging rate (long term stability	5×10^{-7}/mo	3×10^{-7}/mo	5×10^{-10}†
Temperature, 0-50°C	5×10^{-6}	5×10^{-7}	7×10^{-9}
Line voltage, ± 10%	1×10^{-7}	5×10^{-8}	5×10^{-9}
Short-term stability per day			1×10^{-10}

*Temperature-compensated crystal oscillator
†After 24-hour warm-up
‡rms/sec

oscillator is deenergized along with the rest of the counter when the instrument is turned off, however, the aging specification means little or nothing.

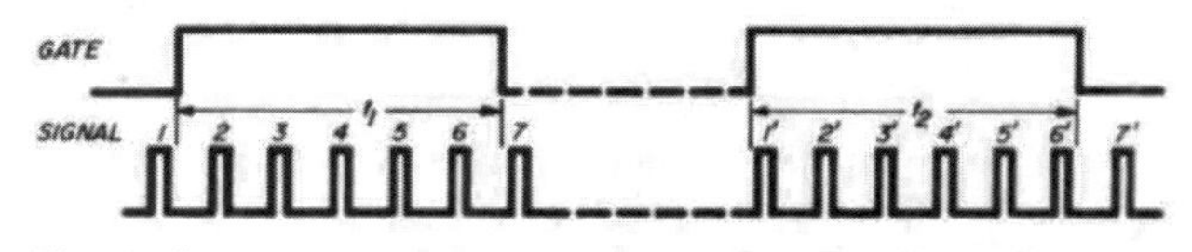

fig. 2. Gate time and signal pulse train, showing ±1 count ambiguity.

Short-term stability is generally specified only for very stable, ovenized oscillators and is pertinent only to laboratory-type measurements.

Time-base errors can be corrected by recalibrating the oscillator against a known standard or against WWV. Virtually all counters incorporate an adjustment control for this purpose. The techniques used in recalibrating the oscillator will be covered later in this article.

frequency-measurement accuracy

Although the preceding discussion of time-base

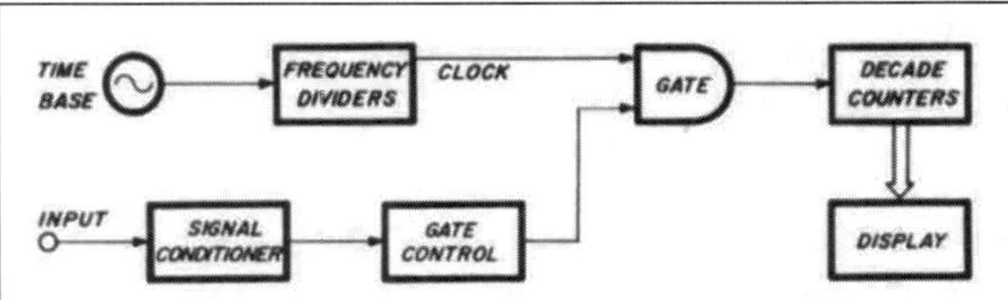

fig. 3. Functional block diagram of an electronic counter configured for period measurement.

accuracy would appear to account for any inaccuracies in the measurement of frequency, such is not the case. The specification for the frequency-measurement accuracy of all electronic counters is invariably stated as ± time-base accuracy ± 1 digit. The last term of that statement is known as the 1-count ambiguity — but what does this mean?

Fig. 2 shows the signal under measurement and its relationship to the gate. Although the successive gate times, t_1 and t_2, are equal in duration, the gate is not synchronized with the signal.

Therefore it is possible, during gate time t_1, for five signal pulses (numbered 2 through 6) to be gated, while during time t_2, six signal pulses (numbered 1′ through 6′) may be gated. Thus there is always an irreducible ±1-count ambiguity in the least significant digit of the display.

The per cent of error due to the 1-count ambiguity is reduced as the measured frequency increases, since it becomes increasingly less significant compared to the total count. The maximum error is an inverse function of the frequency being measured and the number of pulses being counted, *i.e.* the gate time, and is expressed as

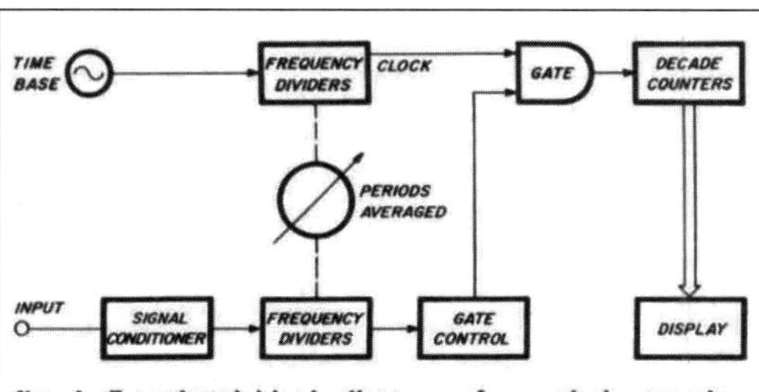

fig. 4. Functional block diagram of a period-averaging counter.

$$\%\ Error = \pm\ \frac{100}{f \cdot t}$$

where f is the frequency in Hz, and t is the gate time in seconds.

From the above equation, it can be seen that measuring a 10-MHz signal with a 1-second gate time will be subject to a ±0.001 per cent error. However, measuring a 20-Hz signal with the same gate time may result in a counter display between 19 and 21 Hz, a ±5 per cent error. This would not be very satisfactory if you were attempting to calibrate the low-frequency end of an audio oscillator, and must be taken into account.

period and period-averaging measurements

One of the ways in which accurate low-frequency measurements may be made is to measure the period of the signal, rather than the frequency. Since the period of a signal is the reciprocal of its frequency, the frequency can be calculated accordingly. It might also be expected that a simple reciprocal arrangement of the functional blocks of an electronic counter would provide a measurement of period, which turns out to be true.

In **fig. 3**, the time-base and signal inputs have been interchanged. If the gate-control circuit is configured so that the gate is open for one period of the input signal, and the 1-MHz clock is applied to the gate input, a series of pulses having a period of 1 microsecond will be gated through to the decade counters. Therefore the counter will indicate the period of the input signal in microseconds. If the clock frequency were reduced to 1 kHz, the counter would display the signal period in milliseconds.

Let us now reconsider the frequency measurement of the aforementioned 20-Hz signal to see how we can improve the possible ±5 per cent error which can occur with a 1-second gate time. The 20-Hz signal period is 0.05 second, so that if the counter were configured to measure period, it would display 50,000 microseconds. Thus, the number of significant digits in the display has been increased from two to five, and the gate time reduced from 1.0 to 0.05 second. Now if you were calibrating the audio oscillator and wanted a 1 per cent dial accuracy, you could accept any reading between 49,500 and 50,500 microseconds, subject to the correction for period-measurement accuracy.

Period measurements are inherently less accurate than frequency measurements because it is the signal, rather than the time base, which controls the gate time. Noise on the input signal, regardless of the measurement mode, causes an uncertainty in the point at which the trigger circuit in the signal conditioner switches. (It is the trigger circuit which converts the input signal to a waveform which is compatible with the counter's circuitry.) If the noise is not great enough to cause false triggering which would result in more or less output pulses than correspond to the input, no significant error is introduced in a frequency measurement.

For period measurements, however, this uncertainty results in an error in the gate time, since the in-

Ballantine's model 5720A Frequency Counter covers the range from 10 Hz to more than 80 MHz and provides frequency and ratio measurements. This counter also includes an audio multiplier circuit for input frequencies from 50 Hz to 1 kHz which provides resolution of 0.01 Hz with only 1-second measurement time (*photo courtesy Ballantine Laboratories*).

Heath's model IM-4130 is capable of measuring period, period average, events (totalizing), and frequency over a 5-Hz to 1-GHz range. Since it has provisions for connecting an external time base, ratio measurements can also be made, as explained in the text (*photo courtesy Heath Company*).

put signal controls the gate time. This error is known as trigger error, and is part of the instrument specification for period measurement, usually expressed as ±time-base error ±trigger error ±1 count. Notice that the trigger error has been added to the previously discussed expression for frequency-measurement error. For low-frequency noise on a sine-wave input, the approximate worst-case errors are ±3 per cent for a 20-dB signal-to-noise ratio, ±0.3 per cent for a 40 dB signal-to-noise ratio, and ±0.03 per cent for a 60-dB signal-to-noise ratio.

In addition to the trigger error caused by noise, the stability of the input signal may be such that successive gate times are of differing durations. Even though the differences may be minute, they will manifest themselves as a continuously changing display on the counter, especially at high resolutions. This is not to be considered a counter error, since it does not occur with a stable input signal.

Period errors may be minimized by averaging the readings over several periods of the input signal. If the input-signal frequency is divided to a lower frequency, the gate will remain open for a multiple of the input-signal period, so that the counter will display the number of clock pulses for 10, 100, 1000, or more periods. A typical counter configuration for the period-averaging mode appears in **fig. 4**. The frequency-divider chain is split so that both the time-base oscillator and/or signal frequencies are divided to obtain the desired resolution and number of periods which are to be averaged. The counter will display the period measurement, regardless of the number of periods averaged, simply by having the

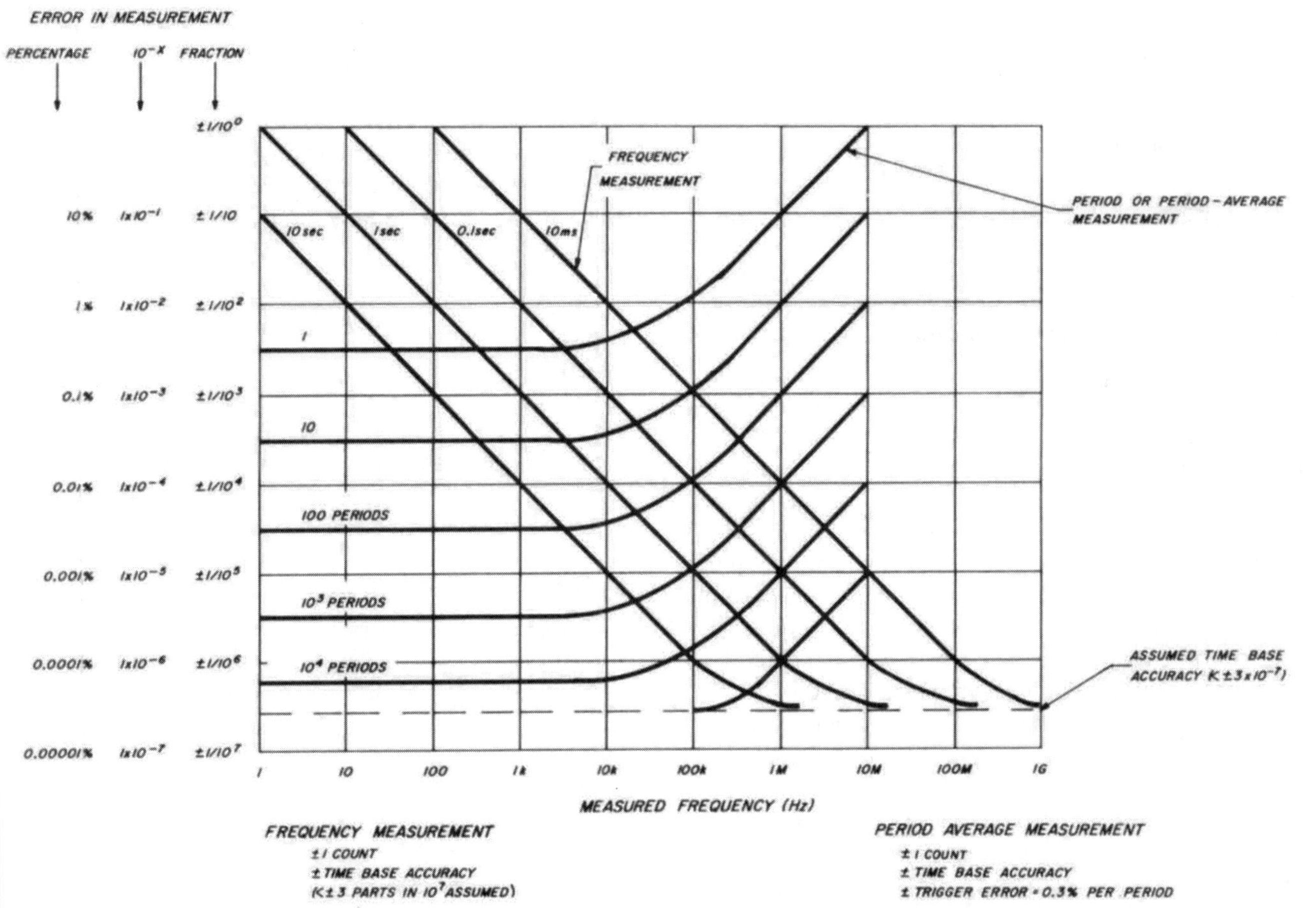

ERROR IN MEASUREMENT

PERCENTAGE | 10^{-X} | FRACTION

±1/10⁰ → $\pm 1/10^0$
10% 1×10^{-1} $\pm 1/10$
1% 1×10^{-2} $\pm 1/10^2$
0.1% 1×10^{-3} $\pm 1/10^3$
0.01% 1×10^{-4} $\pm 1/10^4$
0.001% 1×10^{-5} $\pm 1/10^5$
0.0001% 1×10^{-6} $\pm 1/10^6$
0.00001% 1×10^{-7} $\pm 1/10^7$

FREQUENCY MEASUREMENT

10sec 1sec 0.1sec 10ms

PERIOD OR PERIOD-AVERAGE MEASUREMENT

1

10

100 PERIODS

10^3 PERIODS

10^4 PERIODS

ASSUMED TIME BASE ACCURACY ($K \pm 3 \times 10^{-7}$)

1 10 100 1k 10k 100k 1M 10M 100M 1G

MEASURED FREQUENCY (Hz)

FREQUENCY MEASUREMENT
±1 COUNT
±TIME BASE ACCURACY
($K\pm3$ PARTS IN 10^7 ASSUMED)

PERIOD AVERAGE MEASUREMENT
±1 COUNT
±TIME BASE ACCURACY
±TRIGGER ERROR = 0.3% PER PERIOD

fig. 5. Measurement accuracy of a counter having a 10-MHz time base with an assumed accuracy of better than 3×10^{-7}.

display decimal point moved as the periods-averaged switch is changed.

Period averaging reduces the possible trigger error by a factor equal to N, the number of periods averaged, so that the error for this mode is $\pm$ time-base error $\pm$ (trigger error)/$N \pm 1$ count. If enough periods are averaged, the trigger error can be reduced to a value which may be of little significance. It must be remembered, however, that the gate time increases by the same factor, which may make the measurement time quite long. For example, a 20-Hz signal has a period of 0.05 second; averaging 100 cycles results in a gate time of 5 seconds; averaging 1000 cycles entails a 50-second gate time, which is normally too long for convenient measurements.

From the preceding discussion, we can deduce that there is a point below which period or period-averaging measurements provide a more accurate reading than a correspondng frequency measurement. This can be calculated, taking into account the 1-count ambiguity, time-base error, trigger error, and gate time. More conveniently, it can be plotted, as shown in **fig. 5**. These curves apply to a counter having a 10-MHz time base of the accuracy specified, and indicate which measurement mode should be used for the desired measurement accuracy.

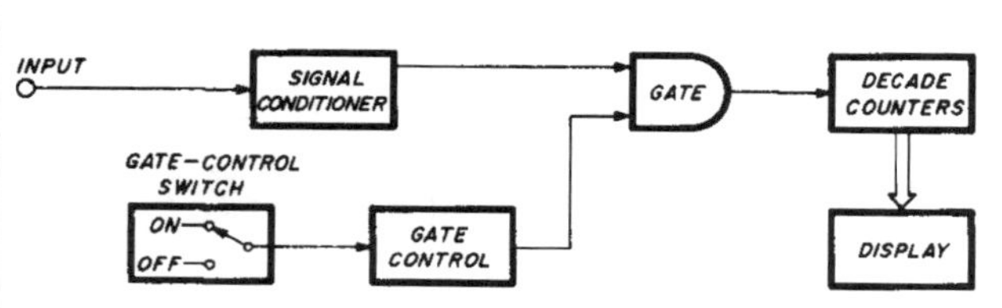

fig. 6. Functional block diagram of a totalizing counter. The gate-control switch can be either a manual switch or an internal switching circuit.

In many instances, measuring the ratio of two frequencies is a time-saving procedure. A typical case might involve designing or troubleshooting a phase-locked loop, where the output frequency is a discrete multiple of a reference oscillator. Since the output frequency may be divided by a factor of up to several thousand within the loop, an error or glitch causing a one-count error in this division may not be readily apparent unless a ratio measurement is made.

In conjunction with **fig. 1**, we discussed the method by which frequency is measured. Another way of defining this measurement is to state that the counter displays the ratio of the input frequency to the clock frequency. By using an internal clock whose frequency is known, the ratio can be displayed in megahertz, kilohertz, or hertz. If an external signal were used in place of the time-base oscillator, the counter would still display the ratio of the two frequencies, except that it would no longer be in hertz or a multiple thereof (unless the external frequency were the same as the time-base frequency). Counters which provide specifically for ratio measurements incorporate provisions for changing the display to a dimensionless number, and position the decimal point accordingly.

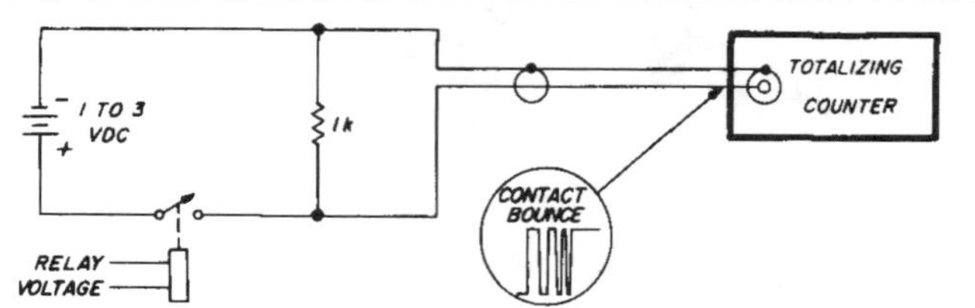

fig. 7. Using a totalizing counter to measure contact bounce. When relay voltage is applied, the counter will display the number of contact bounces.

Many counters do not have an apparent capability of measuring ratio, but can actually be used in this mode. If the counter has provisions for using an external time-base oscillator, the reference signal against which ratio is to be measured can be introduced into the external time-base connector. It is necessary that the amplitude of this external reference signal be as specified for the counter being used, that its frequency be within the range that the counter will accept as an external time base, and that the internal time-base oscillator frequency be known.

The ratio of the input signal frequency to the external reference frequency is determined from the expression

$$\frac{f_{sig}}{f_{ref}} = \frac{f_{ctr}}{f_{int}}$$

where

f_{sig} is the input frequency
f_{ref} is the external reference frequency
f_{ctr} is the frequency displayed on the counter
f_{int} is the internal time-base oscillator frequency.

totalizing

Perhaps the simplest function of which an electronic counter is capable is that of totalizing, or accumulating, a count of input events. Because this mode does not require a time base, as indicated in **fig. 6**, it probably should have been covered previously as the most basic counter circuit. However, totalizing is not usually a function of low-priced counters, nor does it have major applications in amateur work; therefore I have delayed discussing it until the modes of greater interest were described.

The gate-control switch shown in **fig. 6** can be

either a manual switch or an internal switching circuit actuated by the input signal. Switching the gate-control switch to *on* resets the decade counters to zero and allows the processed input signal to pass through the gate for the length of time that the switch is held on. When the switch is turned *off*, the count stops and the number of input events which has occurred is displayed on the counter.

An application which is of interest in these days of digital logic circuits is that of measuring contact bounce. **Fig. 7** shows a simple circuit which permits such a measurement for either a relay or a manually actuated switch. When voltage is applied to a relay coil (or a manual switch is operated), the contacts will usually open one or more times after the initial

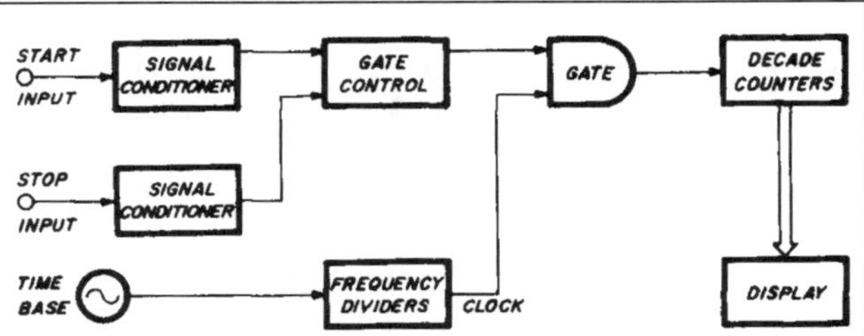

fig. 8. Functional block diagram of a time-interval counter, which counts the number of clock pulses between the time the gate is opened by the START input and closed by the STOP input.

closure because of the elasticity of the switch materials. This results in the waveform shown in the illustration, which is applied to the counter. The counter will then totalize and display the number of bounces.

time interval measurements

A counter may be used to measure the time interval between two input events, but this mode of operation requires two input-signal-conditioning circuits and a more complicated gate-control circuit; it is therefore found only in the more expensive professional instruments. As shown in **fig. 8**, the gate control has two inputs, one from each of the signal conditioners. The gate is opened by the processed *start* input, allowing the accurate clock pulses to pass through to the decade counters until the *stop* input closes the gate. Thus the counter will display the time interval between the two input signals.

The start and stop points are determined by the triggering levels and slopes selected by circuits in the signal conditioners. The time-interval resolution is limited by the clock frequency, and is subject to the same ±1-count ambiguity as all other measurements. As with period averaging, this ambiguity can be reduced for repetetive signals by averaging the time-interval measurements. When averaging, the

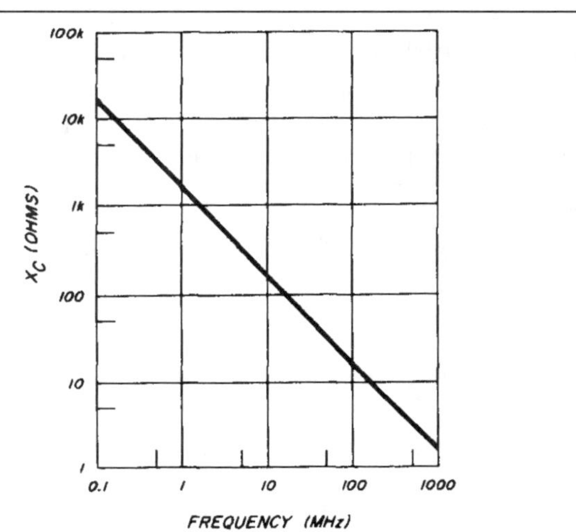

fig. 9. Reactance of 120 pF (typical for a high-impedance counter input with a 3-foot or 1-meter cable) plotted against frequency.

ambiguity becomes $\pm 1\ count \div \sqrt{N}$, where N is the number of time intervals averaged.

An important application of time-interval measurement is the accurate determination of pulse width. The signal under measurement is applied to both the *start* and *stop* inputs. If the *start* channel is set to trigger on the positive slope, and the *stop* channel on the negative slope (or vice versa), the counter will indicate the time interval between the leading and trailing edges of the input signal. Adjustment of the triggering levels will permit the measurement to be made between the desired points on the edges.

The upper frequency limit of the modern basic counter is dependent on the type of digital logic devices used in the signal conditioner and the first decade counter. This frequency limit may be as high as 50 MHz for conventional TTL, 120 MHz for Schottky TTL, and 250 MHz for ECL. Above those frequencies, prescaling is generally used to increase the frequency range, up to about 1300 MHz.

Prescaling simply means that the input frequency is scaled, or divided, down to one which is within the basic range of, and is measured by, the basic counter. The divisor may be any integral number. If the prescaler is external to the counter, it will usually divide by 10 or 100, so that the frequency can be read directly from the counter after you have mentally multiplied the counter reading by 10 or 100, as applicable. If the prescaler is built into the counter, it may scale by any integral factor.

The advantage of using an external prescaler is obvious — it permits extending the frequency range of an existing counter at relatively low cost. Its disadvantages become equally obvious after it has been used. First, there is the necessity of mentally moving the decimal point, since the counter is actually displaying the divided input frequency. Second, one digit of resolution is lost for every decade of scaling. For example, a 145,600.0-kHz signal measured with a scale-by-ten prescaler will read 14560.0 kHz on a counter having a 0.01-second gate time (0.1-kHz resolution). Multiplying by ten yields a frequency of 145,600 kHz; the 0.1-kHz resolution is lost by scaling. It can be re-established only by increasing the gate time by a factor of ten, provided the counter has that capability.

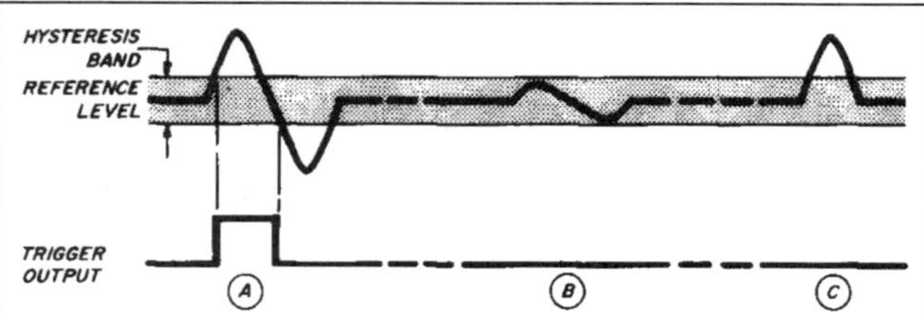

fig. 10. The effect of triggering hysteresis. The waveform at (A) will result in output from the trigger circuit, while those at (B) and (C) will not because neither crosses both limits of the hysteresis band.

If the prescaler is an integral part of the counter, mentally scaling and moving the decimal point is eliminated, since this will be accomplished in the counter when the mode is changed from direct count to prescaled count. Nevertheless, the loss of resolution remains. It can be minimized, however, by scaling by a factor less than ten, and simultaneously increasing the gate time by the same factor.

Suppose that the internal prescaler divides the input frequency by four. If the gate time is increased by the same factor, there will be no change in the number of signal pulses gated through to the decade counters, and the display will read out the correct frequency. Consequently, prescaling is accomplished with only a fourfold increase in gate time, which is generally acceptable.

Switching from direct to scaled operation may be carried out in one of three ways. If a single input connector is used, the counter mode is generally switched manually. If two separate input connectors are employed, one for low-frequency signals and the other for high-frequency inputs, the counter mode may be switched manually or automatically when the input signal is present at the high-frequency input.

input impedance

Counters which measure frequencies below 250 MHz or so usually present a high input impedance — typically 1 megohm shunted by 30 to 40 picofarads. Above that frequency, the input impedance is generally a nominal 50 ohms, although the vswr may be as high as 2.5:1. At audio and low radio frequencies, a high input impedance is normally desirable, since it minimizes the load on the circuit under test. But just how high in frequency is this true?

Consider a counter with an input impedance of 1 megohm shunted by 32 pF, which is used with a three-foot (91cm) cable made from RG-58C/U coax. The capacitance of RG-58C/U is 29.5 pF per foot (96.8 pF per meter), so that the total shunt capacitance presented to the circuit under test is approximately 120 pF. The reactance of this shunt capacitance, plotted against frequency, is shown in **fig. 9.** It can be seen that the reactance drops to approximately 1300 ohms at 1 MHz, and is only about 130 ohms at 10 MHz. So the input impedance can no longer be considered high. On the other hand, if the counter had a nominal 50-ohm input, you would know the loading effect, within the limits defined by the specified vswr.

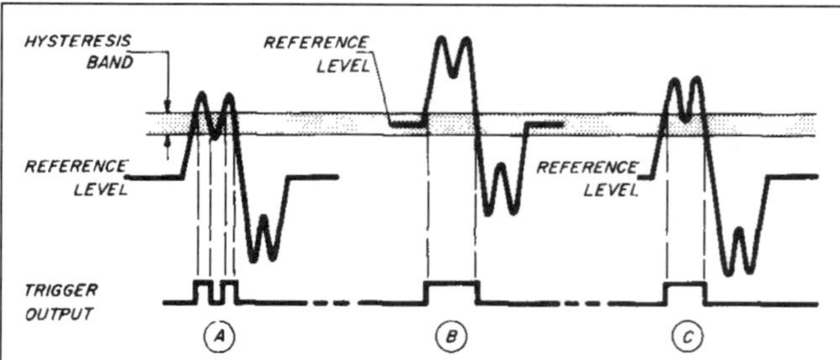

fig. 11. Erroneous counting caused by harmonic distortion is shown at (A). The false count can be eliminated by adjusting the level control, as indicated at (B), or by increasing the signal amplitude, as shown at (C).

Suppose that you had to check the frequency of a 70-MHz crystal oscillator which was designed to feed a 50-ohm load. If your counter has a 50-ohm input, all is well. However, if it has only a high impedance input, the shunt capacitance of the counter plus a cable will more than likely load down the oscillator and change the frequency, if it continues to oscillate at all. Fortunately, a relatively inexpensive accessory will solve the problem. By using a 50-ohm feedthrough termination* at the counter connector, a 50-ohm interconnecting cable will be reasonably well terminated, and will present a load close to 50 ohms at the oscillator.

The same thing may be accomplished by using a 20-dB loss pad at the input connector of the counter, provided that there is enough signal to trigger the counter after having been attenuated by the pad.

Even at low frequencies, the shunt capacitance may be too high for certain applications, such as checking filters. Capacitive loading can be reduced by using a 10X oscilloscope probe. Such probes typi-

*Such as the Heath SU-511-50, Hewlett-Packard 10100C, Tektronix 011-0049-01, Systron-Donner 454, and other similar types.

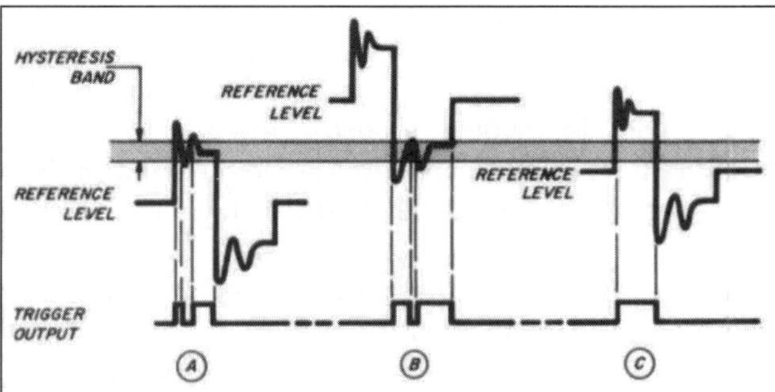

fig. 12. False counting caused by ringing is shown at (A) and (B). Proper adjustment of the reference level and/or amplitude, as shown at (C), corrects the fault.

cally present a 10-megohm resistive load shunted by 5 to 15 pF, but of course attenuate the signal by a factor of 10.

input signal levels

One of the parameters invariably specified for a counter is sensitivity, generally in millivolts, but often in dBm for 50-ohm inputs. This indicates the minimum signal needed at the *counter input* to ensure reliable triggering. Of equal, and possibly more importance, however, is the maximum signal which may be applied to the input without damaging the instrument.

For high-impedance inputs, the maximum signal voltage is usually specified as the sum of a dc value plus a peak ac value. The peak ac value may vary with frequency, going down as the frequency increases. The sum of ac plus dc is limited by the input blocking capacitor; the limiting ac value alone is a function of the input device in the signal conditioner. To be safe, when measuring at any point in a circuit where dc is present, always use an external blocking capacitor of the smallest value which will permit reliable triggering. And if there is any possibility of the ac signal exceeding the specified maximum for the counter, use an external attenuator or dividing probe.

B&K Precision 1820 Universal Frequency Counter will measure frequency from 5 to 520 MHz, and permits high-resolution period measurements from 5 Hz to 1 MHz. Decimal point position and unit-of-measure display is selected automatically for best resolution (*photo courtesy B&K Precision*).

For low (50-ohm) impedance inputs, the maximum signal level is limited by the input circuit of the signal conditioner. This level is generally much lower than for high-impedance inputs, and is typically between +19 and +27 dBm (2 to 5 volts rms across 50 ohms). Because of the relatively high cost of the high-frequency input device and the possibility of applying excessive power from a transmitter, the 50-ohm inputs are fuse-protected in many counters which can function at 500 MHz and higher.

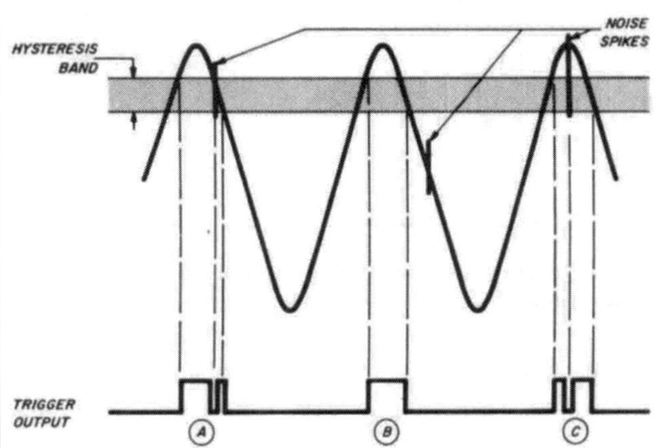

fig. 13. Spurious counts can result from noise on a signal when the noise is of sufficient amplitude to cross the hysteresis band.

Although it should be obvious, the following warning must be included: *Never connect a counter directly to a transmitter or any other high-power signal source*! Use a short length of unshielded wire as an antenna at the counter input connector, an inductive coupling loop at the end of a shielded cable, and/or an attenuator of sufficient power rating. The counter you save may be your own!

If the counter is battery-powered, and there is no direct connection between it and the circuit or generator under test, the counter should be grounded. This will reduce noise pick-up, especially when using a counter with a high input impedance.

triggering

The signal-conditioning circuits in all counters include a trigger circuit which, as previously stated, provides output pulses whose amplitude and waveshape are compatible with the counter circuitry which follows. The sensitivity of the counter depends on the threshold level of the trigger input and the amplification between it and the input of the counter. If the amplified input signal has insufficient amplitude to reach the threshold level, the instrument will not count or will perform erratically.

All trigger circuits have a hysteresis band, through the limits of which the input signal must pass in order

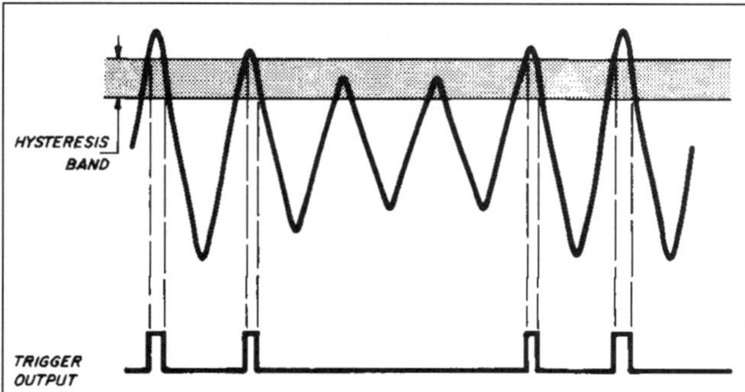

fig. 14. Amplitude modulation of the input signal can cause missing counts when the signal amplitude is too low.

to result in an output pulse. **Fig. 10** shows three input signals in relationship to the hysteresis band. Sine wave **A** crosses both the upper and lower limits of the band, and will actuate the trigger circuit; the amplitude of sine wave **B** is too small, so triggering will not occur; and waveform **C** crosses the upper threshold, but not the lower, so again no output will be produced by the trigger circuit.

It is the action of this hysteresis effect which can result in erroneous counting which is so confusing to a relatively inexperienced operator. Suppose that the input to the counter were a sine wave with considerable second-harmonic distortion, a not uncommon situation. In **fig. 11A**, the amplitude of the signal is such that the positive half-cycle crosses the hysteresis-band limits twice, instead of once. The trigger circuit will generate two output pulses for each input cycle, and the counter will display twice the fundamental frequency of the signal. If the counter has a level control, which adjusts the reference level at the input of the trigger circuit, it can be adjusted to eliminate the false count, as shown in **fig. 11B**. If there is no level control, as is the case with most low-priced counters, the problem can be eliminated by increasing the amplitude of the input signal, as depicted in **fig. 11C**.

A similar problem may arise when measuring the frequency or period of a signal comprised of fast pulses. If the interconnecting cable is not terminated in its characteristic impedance, or if other impedance discontinuities exist, ringing will occur on the pulses. If the ringing traverses the hysteresis band, as shown in **fig. 12A** and **12B** for two different reference levels, a false count will result. Proper adjustment of the signal amplitude and reference level, indicated in **fig. 12C**, will provide the correct count.

Another way of solving the ringing problem, which is useful when the reference level and/or amplitude cannot be changed, is to use a low-value resistor (100 to 1000 ohms) between the circuit point under test and the counter cable. This resistor, in conjunction with the cable and counter input capacitance, integrates the pulse and minimizes the pulse aberrations which reach the counter.

Figs. 13 and **14** illustrate two other conditions which can result in false counts. The noise transients on the signal shown in **fig. 13** will cause additional counts, while amplitude modulation may result in missing counts, as shown in **fig. 14**, if the amplitude of the input signal is too small. In either case, the solution is the same as previously prescribed — change the reference level and/or the signal amplitude.

In our earlier discussion of period and period-averaging measurements, it was stated that the trigger error resulting from noise on the input signal contributed to the measurement error. This is shown in **fig. 15**, in which a sawtooth wave is used to demonstrate the effect of slope, or slew rate, on the trigger error. It can be seen that the noise voltage on the relatively slow rise-time can cause a much greater trigger error than that which occurs on the fast fall-time. Thus we can see that the trigger error can be minimized by triggering on the steepest portion of the input signal to the counter. For a sine wave, this will be that part of the waveform at the zero axis, leading to the conclusion that a signal of the maximum possible amplitude should be used.

It should be apparent from the preceding discussion that an input attenuator on the counter can be of considerable help in establishing the correct input level to the trigger circuit. In many of the lower-priced counters an attenuator has been omitted because of cost and because it was felt that limiting diodes at the input of the signal conditioner would protect the input device. The latter reason is valid

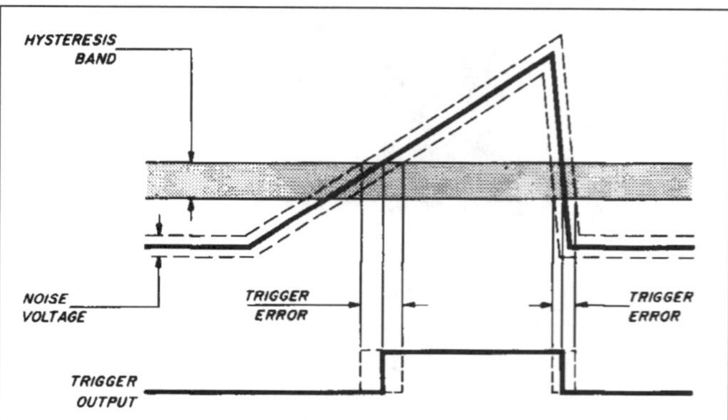

fig. 15. Trigger error in period and period-averaging measurements, caused by noise in the input signal. The error is minimized by a fast slew rate through the hysteresis band.

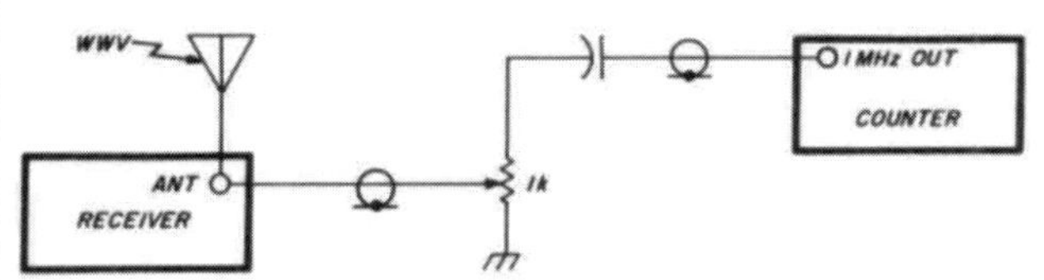

fig. 16. Test setup for calibrating the time-base frequency against WWV.

only where overload is considered, for even a two-position attenuator can be extremely valuable in eliminating false counting.

time-base calibration

Unless an oven oscillator or TCXO is used as the time base in a counter whose oscillator circuit is energized continuously, the oscillator frequency should be checked, and adjusted if necessary, whenever accurate measurements are to be made. Be sure, however, that the counter is fully warmed up before checking or recalibrating the oscillator.

In order to calibrate the time-base frequency, either a standard of known accuracy or a communications receiver capable of receiving WWV is required. If the standard is at least five times more accurate than the best resolution of the counter at the time-base frequency, it can be applied directly to the input of the counter. Then adjust the time-base oscillator frequency control for the correct frequency read-out on the counter.

A more accurate adjustment may be made if the counter has an output connector from which a 1-MHz time-base signal can be obtained. Connect this output to the vertical input of an oscilloscope, and connect the output of the frequency standard to the horizontal input.* The scope will display a Lissajous pattern, which will probably be moving. Adjust the time-base frequency control until a stationary pattern is obtained.

If the counter is to be calibrated against WWV, a time must be chosen during which transmissions are received with an absolute minimum of fading. Select the highest receiving frequency possible (*e.g.* 15 MHz) to achieve the greatest calibration accuracy. The calibration technique involves obtaining a visual beat indication on the receiver S-meter, and adjusting the time-base oscillator frequency for as close to a zero-beat as possible.

In order to obtain a good beat null, the time-base signal which is applied to the receiver must be of the correct amplitude relative to the signal level from WWV. Since we cannot control the latter, we have to be able to vary the signal level from the counter. If the counter has a 1-MHz output from the time-base circuit, make the connections shown in **fig. 16**. If a time-base signal is not brought out to a connector on the counter, substitute an insulated wire for the coax shown connected to the receiver antenna terminal and place it near the counter time-base oscillator or frequency-divider chain. In either case, the harmonic of the 1-MHz signal should result in a low-frequency

Leader LDC-822 Frequency Counter measures frequency to 80 MHz and features selectable gate time and input attenuation (*photo courtesy Leader Instruments*).

beat with WWV. (This will not be an audible beat unless the time-base oscillator is very far off frequency; more likely it will be observed as a rythmic variation in the S-meter reading.) Adjust the potentiometer shown in **fig. 16**, or change the position of the insulated wire, to obtain the deepest beat null on the receiver S-meter.

It should be possible to adjust the time-base frequency so that the beat-frequency period is several seconds, which corresponds to a remarkably accurate short-term frequency setting. To demonstrate this, assume that eight beats are observed on the S-meter in a 60-second period. The beat frequency is therefore equal to 8/60, or 0.133 Hz. If the beat is measured at 15 MHz, the error is $0.133/15 \times 10^6$, or 8.9×10^{-9}. Of course, this degree of accuracy may hold only for a short period of time, because the stability of the counter time-base oscillator, unless it is an oven type, is nowhere near that good. Nevertheless, highly accurate measurements may be made until the counter is turned off or a temperature change affects the time-base oscillator.

*These connections are based on the assumption that both the horizontal and vertical amplifiers in the oscilloscope will pass a 1-MHz signal. If the counter has a time-base output of higher frequency, it can also be used, provided that it is within the frequency range of the scope. The lower of the two frequencies (the standard and the counter time base) should be connected to the horizontal input, since the horizontal frequency limit is usually lower than the vertical.

capacitance measurements with a frequency counter

A digital capacitance meter has been on my shopping list for some time, but it's been difficult to justify the expense of the single-function instruments available. An article in *QST*[1] followed by investigation of the literature on the NE555 led to the design of a frequency counter attachment that allows capacitance to be read from the counter display. Using a 7-digit counter, capacitance values from 1 pF to 1 μF may be read with an accuracy of about 2 per cent ± 2 pF. A range switch isn't required.

design

The NE555, when used as a one-shot, produces a pulse of width

$$T = kRC \qquad (1)$$

where k is inherent in the 555, R is the charging resistor, and C is the capacitance being measured. By ANDing an oscillator output with this pulse, a burst of pulses is produced each time the 555 is triggered. The pulse frequency in the burst is that of the oscillator, while the number of pulses in the burst is determined by the length of the pulse produced by the 555. The value of R may be adjusted so that when a 1-pF capacitor is measured, the 555 output causes exactly one oscillator pulse to appear in the burst. Increasing the capacitor to 100 pF will make the 555 output 100 times longer and allow 100 oscillator pulses in the burst. Counting the pulses in one burst will then indicate the capacitance in pF.

To ensure that exactly one burst occurs per frequency-counter gate time, the 555 is triggered by the opening of the frequency counter gate as shown in the block diagram, **fig. 1**.

The effects of stray capacitance are compensated for by providing a third input to the AND gate which forms the burst, nulling the first part of the pulse from the 555. Typical waveforms are shown in **fig. 2**.

circuit description

The design is implemented as shown in **fig. 3** using a 4009A and a NE555. The functions are shown in the same relative positions in the block diagram and schematic for clarity.

The crystal oscillator is standard and provides a stable square-wave output. The crystal frequency isn't critical but should be greater than 1 MHz to allow measuring 1-μF capacitors using a 1-second

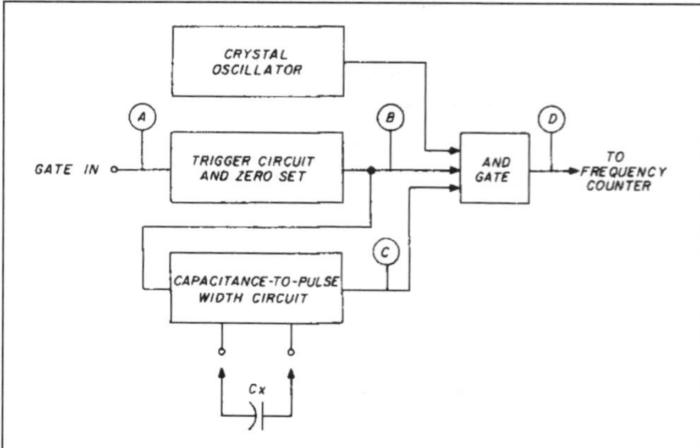

fig. 1. Block diagram of the capacitance measurement attachment for a frequency counter.

By John Moran, 8 Newfield Lane, Newtown, Connecticut 06470

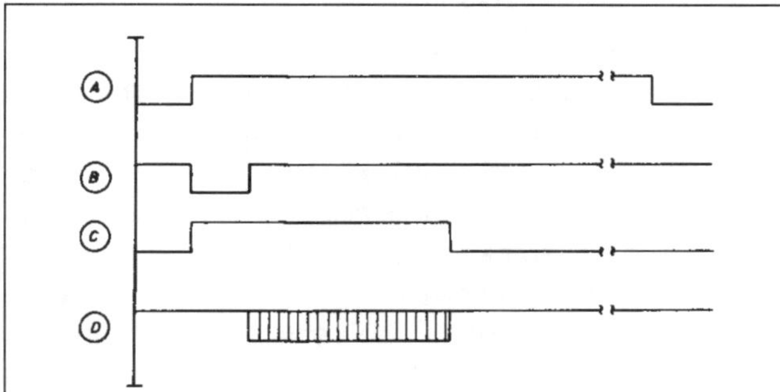

fig. 2. Typical waveforms. Letters at left key the waveforms to both the block diagram and the schematic.

gate time. CMOS speed is restricted when operating from 5 volts, so frequencies above 2 MHz may not work too well.

The trigger circuit consists of a one-shot started by the leading edge of the gate pulse. The output pulse width is adjustable to allow compensation for the effects of stray capacitance, as described earlier. The one-shot is followed by an inverter, which cleans up the pulse shape and provides the proper polarity to trigger the 555 and drive the AND gate. The capacitance-to-pulse width conversion is done by the trusty NE555 in a standard one-shot configuration.

The AND gate is home-grown CMOS/diode logic, which works well and avoids the need for a third IC. The voltage divider on the output reduces amplitude and, more important, the output impedance, thus avoiding stray pickup from the oscillator.

construction and calibration

Perf board construction is adequate if the ICs are bypassed with 0.01-μF capacitors directly at the V_{CC} pins and if the output voltage divider is located a short distance from the crystal.

A metal box should be used to house the unit, since it's sensitive to 60 Hz pickup. This phenomenon is evidenced by a slow variation of about one-half per cent in the reading as the counter gate beats with the line frequency. A variation of over 2 per cent occurs without an enclosure.

The 100-pF capacitor at the oscillator output affects starting and should be adjusted for reliable

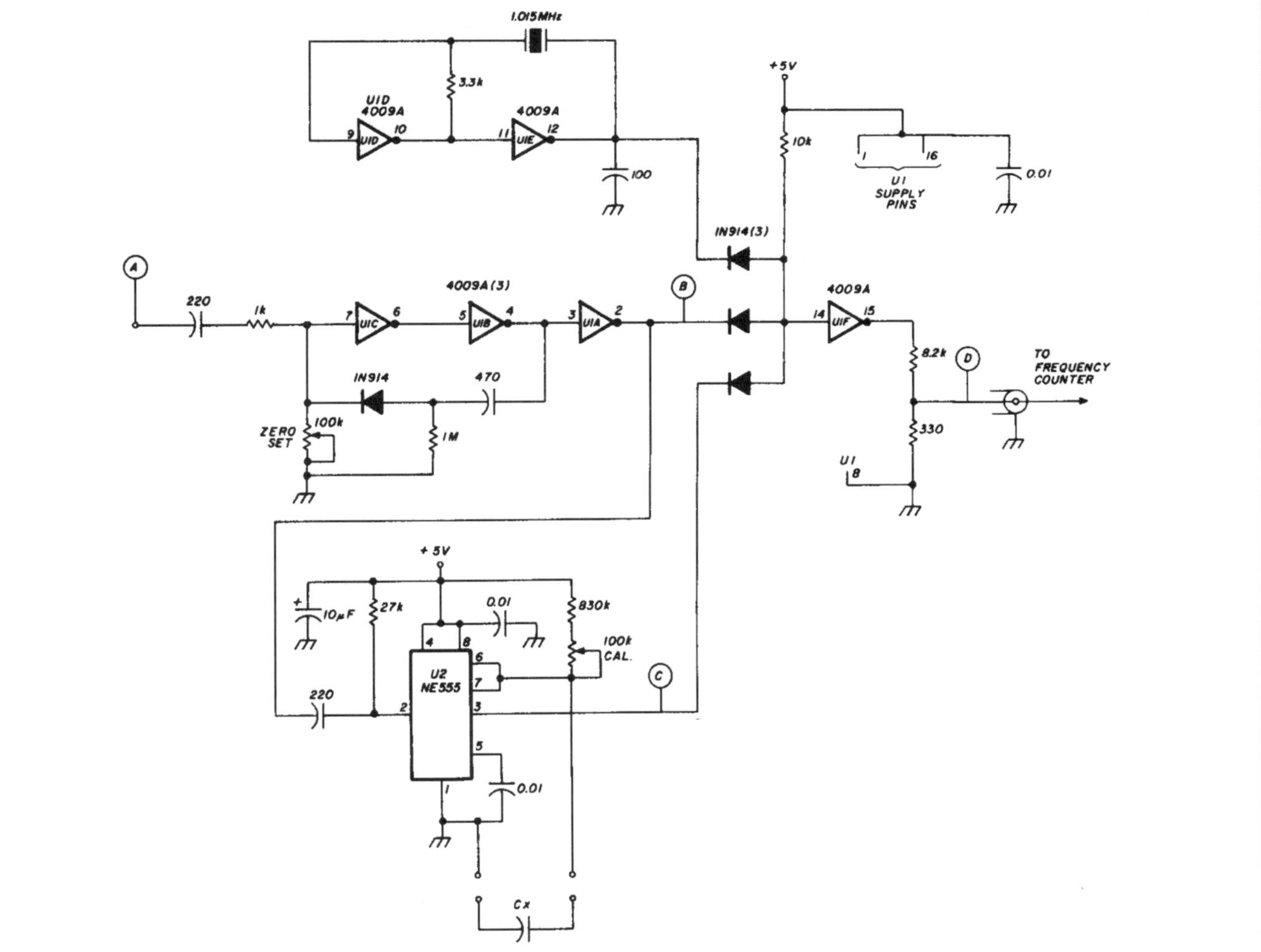

fig. 3. Schematic diagram. Relative position of functional sections matches the block diagram.

operation. The 830k resistor in series with the calibration pot is actually two resistors in series to get a nonstandard value. Using a higher-frequency crystal may require this value to be reduced for calibration.

The frequency-counter gate signal must be made accessible to allow triggering the unit. I installed two pin jacks in the back of my counter to provide both the gate and +5 volts. I used a 200-pF series capacitor to bring the gate line out, to prevent damage to the counter in the event that the line is inadvertently shorted.

Calibration is best done using a one per cent capacitor with a value of several thousand pF as a standard. Lacking this, several silver-mica capacitors in parallel will probably be adequate, since the deviations from the marked values should average out.

Begin calibration by adjusting the ZERO SET control to the maximum resistance setting. Then, without a capacitor in the measurement socket, adjust ZERO SET until a reading is obtained on the counter. Back off slowly until a zero reading is obtained again. Connect your standard capacitor and adjust the CAL control to obtain the correct reading. Then repeat the procedure to correct for adjustment interaction.

If your counter has 0.1-second and 1-second gate times, be sure to use the 1-second gate when measuring capacitors above 0.01 μF.

variations on the theme

It's possible to add a second range by paralleling the charging resistance (830k + CAL) with values that are a factor of 1000 smaller to allow reading capacitors between 0.001 μF and 1000 μF. Note, however, that capacitor leakage causes artificially high readings, so electrolytics with an internal resistance of 1 megohm or less can't be read accurately.

Much of the circuit is also applicable to using the 555 as a direct-reading ohmmeter covering 1 ohm to 1 megohm in a single range.

concluding remarks

A frequency counter attachment has been described that measures capacitance. My hope is that frequency counters will soon be replaced by multifunction instruments incorporating voltage, current, resistance, capacitance, and frequency-measurement capability, much as the voltmeter was replaced by the VOM. Until that time, the versatility of your frequency counter may be increased with devices such as that described here.

reference

1. D.A. Blakeslee, "An Inexpensive Capacitance Meter," *QST*, December, 1978.

digital capacitance meter

Easy to build
digital capacitance meter
for the home shop
features ranges from
1000 pF to 100 μF

Amateurs who build or service electronic equipment sooner or later encounter the situation where replacing a capacitor with a "larger" one produces the wrong results: power supply ripple worsens or the time constant of a timing circuit decreases when it should increase. Highpass or lowpass audio might have their actual 3-dB rolloff points at 200 Hz instead of the intended 300-Hz point. Such differences often occur because the actual value of the capacitor used is different from its marked value. The best performance of narrow bandpass filters and notch filters is obtained when matched capacitors of exactly the same value are used. There are many good "100-for-a-dollar" capacitor buys available, but they often included unmarked or house-numbered units. Those 25-cent, 68-μF capacitors I bought at a hamfest were actually 6.8 μF — the reason, no doubt, they were only 25 cents!

Capacitors are among the most common components used in electronics. Most users assume that the value marked on the capacitor is its actual value; specifications simply guarantee a minimum value. Most electrolytics, for example, are specified to be within +80 to −20 per cent of their indicated value. There are a few that are within ±10 per cent of their marked value; some small capacitors are available with 1 per cent and 5 per cent tolerances. The true value of a capacitor is not important in some cases, such as audio bypass applications, while in other applications the capacitance must be accurately known to produce the desired results.

The digital-capacitor meter presented in this article was built to preclude the type of problems described above. It measures capacitors from 0.001 μF to 999 μF in six ranges, with accuracy of about 1 per cent. The three-digit display has the decimal point correctly positioned as the ranges are switched. The circuit uses low-cost components which are readily available. It requires no difficult adjustments for reliable operation and is easy to duplicate with the printed circuit board layout shown. The meter requires about 100 mA from a 5-volt regulated source, so it lends itself to battery operation if desired. The circuit includes a flashing overflow indicator.

circuit description

The circuit is based upon a digital counter that counts a reference oscillator. The input to the counter is gated by the C_x monostable which has its period determined by the capacitor to be measured.

By Marion D. Kitchens, K4GOK, 7100 Mercury Avenue, Haymarket, Virginia 22069

Construction of the short lead adapter.

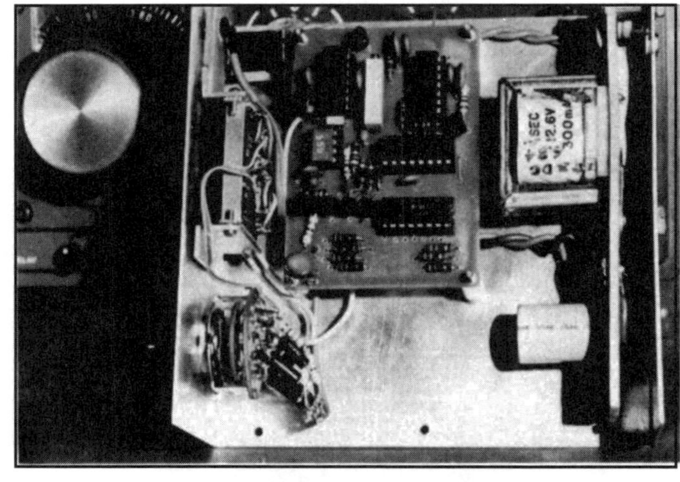

An interior view showing arrangement of the display, circuit board, power supply, and range switch. This was a prototype circuit board which has its overflow circuit mounted below the main board.

Digi Cap

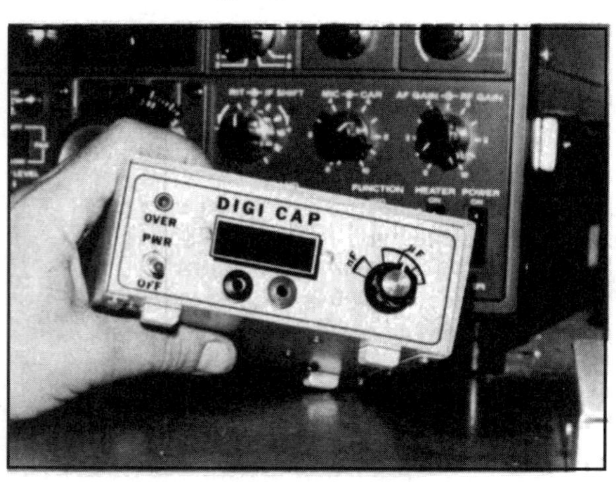

Two different meters showing suggested range switch labeling for right-hand decimal displays per the text (on top) and left-hand decimal displays. The unit with the small display (top) was used to develop the circuit. The bottom unit was built by WA4RVN to verify the circuit reproducibility and performance consistency.

Closeup view of the point-to-point wired power supply. The 7805 voltage regulator is snugged beneath the 1000-μF filter. Yes, the capacitor was measured before use. Would you believe 998 μF?

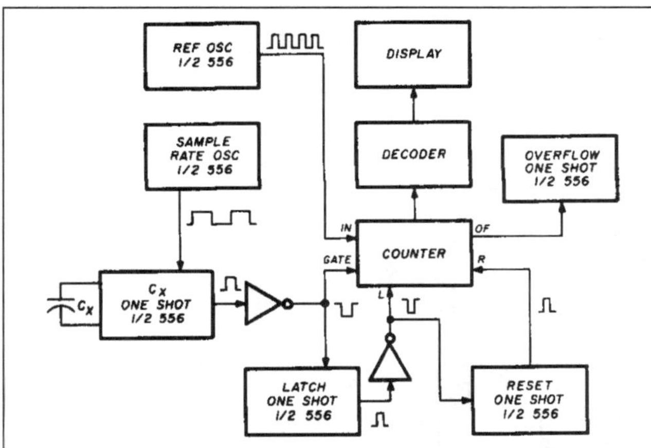

fig. 1. Functional block diagram of the digital capacitance meter. The meter is based upon the 14553 counter. The other ICs provide the necessary gating for the oscillators and display functions.

The functional block diagram is shown in **fig. 1**. About twice a second, the sample rate oscillator triggers the C_x monostable circuit. This monostable output is inverted and applied to the counter control gate. The duration of this control gate input is directly dependent upon the value of the capacitor being measured. If the reference oscillator input to the 14533 IC counter is at the proper frequency, the resulting display will indicate the value of the capacitor. One half of a 556 dual timer serves as the sample rate oscillator, while another 556 dual timer is used as the C_x monostable and reference oscillator.

The 14553 counter chip contains all the circuitry to count and multiplex three digits. It has built-in latch and reset functions and an input control gate. The counter chip's BCD output is applied to a single seven-segment decoder which drives the multiplexed LED displays. The required latch and reset functions are provided by another 556 dual timer with each of its sections operating in the monostable mode. The latch signal is applied to the 14553 at the end of the input gate enable period to store and display the accumulated count. Immediately thereafter the reset signal is applied. The 14553 holds the outputs for the displays, even though the internal counters have been reset, until the latch signal is again low. The latch signal goes low only after the capacitor value has been measured again. This produces a constant or steady display that does not flicker or count up to the final value.

The circuit timing diagram is shown in **fig. 2**. The overflow signal from the 14553 is applied to one half of a 556 dual timer to provide an overflow indication. The timer is run as a monostable to produce a flashing LED overflow indicator. **Fig. 1** shows wave forms at significant locations and indicates the direction of information flow in the circuit. The complete schematic diagram is shown in **fig. 3**.

Construction is uncomplicated when using the printed circuit board. **Fig. 4** shows the location of components on the board, while **fig. 5** shows the circuit board foil pattern. Careful examination of **fig. 4** will reveal the location of the numbered and lettered points to be wired to the display and the range switch. These points are shown on the schematic for easy reference. Switch wiring is shown in **fig. 6**. Points **X**, **Y**, and **Z** are not used.

The circuit uses a common-anode multiplexed display. The seven 82-ohm resistors near the 7446 decoder are the recommended value for displays that require around 10 mA per segment. The suggested value for displays rated at 5 mA per segment is 150 ohms. These values can be varied to achieve the desired display brightness. One unit was built without the seven current limiting resistors (to achieve the maximum brightness) and has worked without any LED burnout problems.

None of the circuit component values are critical, but best performance can be obtained with a good quality capacitor, preferably plastic, for the reference oscillator. This particular capacitor is the 0.001-μF capacitor located near the 100k pot and connected to pins 2 and 6 of U2. Q1 is used to boost the current-handling capability of the C_x monostable (U2) and should have low capacitance and a power rating of ½ to 1 watt. A 2N3906 will work with good results. Transistors Q1, Q4, Q5, and Q6 are PNP transistors, while Q2, Q3, and Q7 are NPN transistors; 2N3906s and 2N3904s can be used, respectively. Q4, Q5, and Q6 should be installed so that their emitters go to the 5-volt land, bases go to the 1 kilohm resistors, and their collectors to the anodes of the display. The overflow LED is connected with its anode to point **F** on the circuit board and the cathode to ground.

A well-regulated, 5-volt power supply capable of 100 to 150 mA is required. **Fig. 7** shows a schematic for a suitable supply. Point-to-point wiring on a insulated board is an easy way to build the supply.

Care should be taken to keep the wiring between Q1, the range switch, and the C_x input jacks as short as possible and away from the 60-Hz ac line.

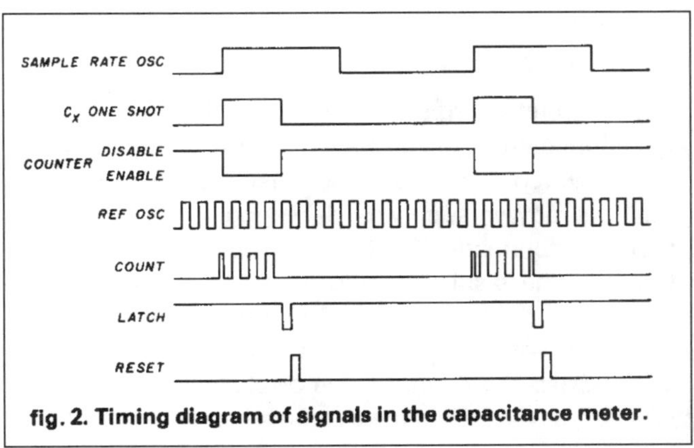

fig. 2. Timing diagram of signals in the capacitance meter.

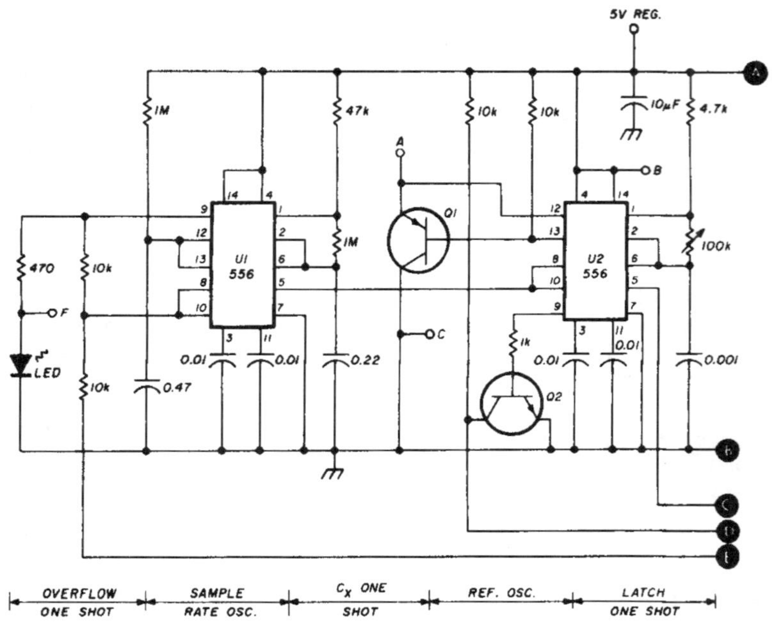

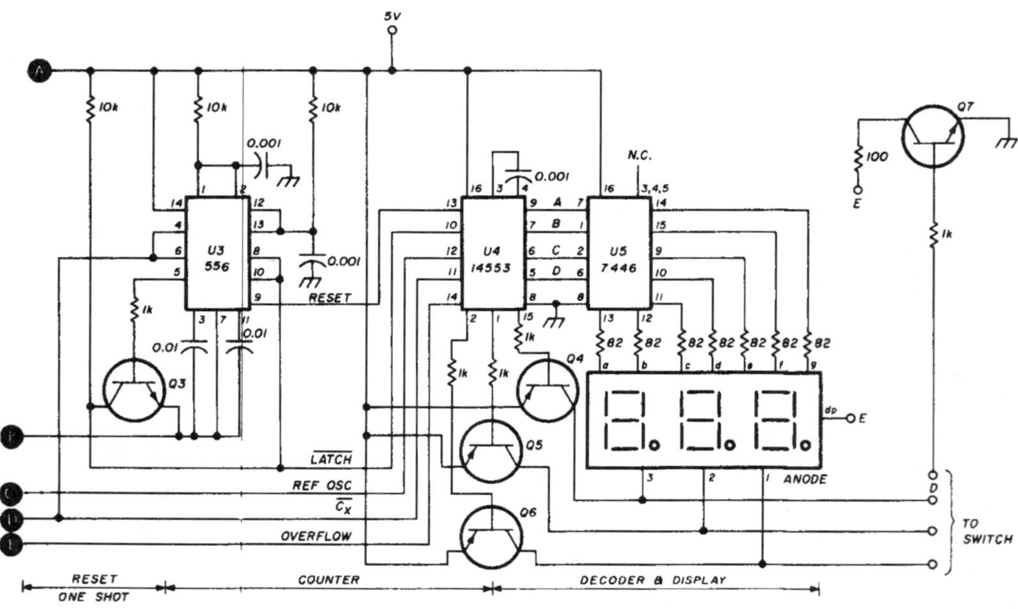

fig. 3. Complete schematic diagram of the digital capacitance meter. Suggested types for the transistors are given in the text. The current requirement of the meter is approximately 100 mA, small enough that a battery supply can be used for field use.

checkout and calibration

The circuit board should be completed and all wiring connected to the display, overflow indicator, and range switch before starting checkout. Make sure that the power supply is delivering 5 volts and is properly connected to the circuit board. At power turn on, the display should light and the overflow indicator should flash once. The display should show 000 or 001 with no connection at the C_x input. With a short across the C_x input, the display should show a number, say 433, and the overflow indicator will flash continuously. This number should not change when the range switch is moved to other positions. The display should show a number of 000 to 002 with the range switch in position 1 (see **fig. 6**) and no connection at the C_x input. An unsteady count ranging from 000 to about 060 indicates that the meter is picking up stray 60 Hz. If this happens, try redressing or rerouting the wiring between the circuit board, range switch, and C_x input jacks. K4ZKU found that reversing the ac line cord at the wall outlet would help with such a situation. A simple test of U5, the display, and

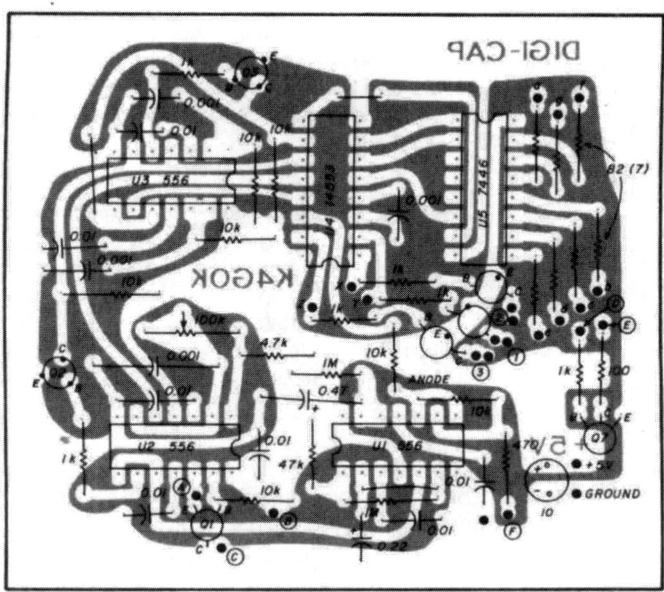

fig. 4. Parts placement diagram for the printed circuit board.

the wiring between can be made by temporarily grounding pin 3 of U5; the display should show 888.

The unit must be calibrated before use. Capacitors of known value are required. Surplus computer and audio boards are a good source for precision capacitors. I found 1 per cent capacitors from 0.001 to 2.5 μF at local hamfests. The meter should be allowed to warm up for about 20 minutes before calibration. If precision measurements in the 10s and 100s of microfarads ranges are not required, the 2000- and 200-ohm pots at positions 5 and 6 of the range switch can be replaced with 1000- and 100-ohm fixed resistors. To calibrate the meter, connect a 0.1- to 0.3-μF capacitor of known value, and with the range switch in position 3, adjust the 100-kilohm reference oscillator pot on the circuit board so that the display indicates the correct capacitor value. This calibrates the 100k-pF range (switch position 3) as well as the 10k-pF (position 2) and 1-μF (position 4) ranges. The 1k-pF is range calibrated by the 1-megohm pot at switch position 1; the 10-μF and 100-μF ranges are calibrated by the 2000- and 200-ohm pots at positions 5 and 6.

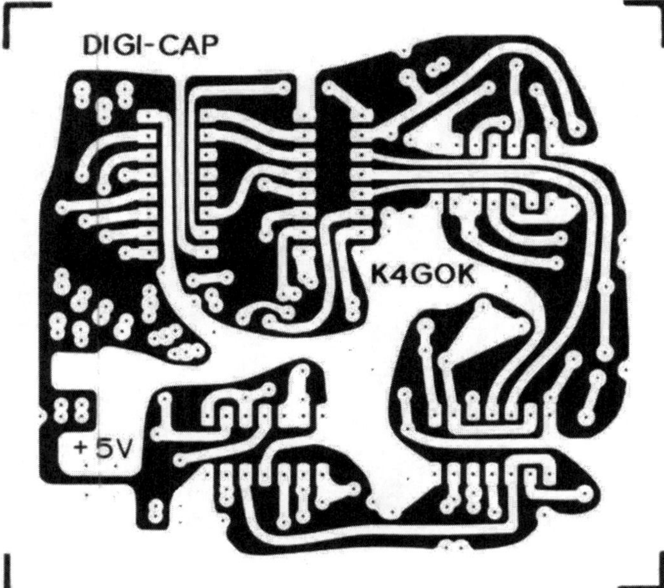

fig. 5. Foil layout pattern for the digital capacitance meter.

using the meter

Operation of the meter is simple. Observing proper polarity, connect the capacitor to be measured, select the largest range that does not cause an overflow, and read the capacitor value shown on the display. **Table 1** shows examples of how the display indicates various capacitor values for each of the range switch positions. The first three ranges measure in thousands of pF and the last three ranges measure in μF. The decimal point is properly positioned. Note that if a 22-μF capacitor is being measured the range switch should be in position 5 and the display will show 22.0. A 0.047-μF capacitor is 47k-pF, and it will be measured with the range switch in position 2. The display will show 47.0. Labeling the first three positions of the range switch as kpF (or nF for nanoFarads if preferred), and the last three positions as μF will make the meter very easy to read.

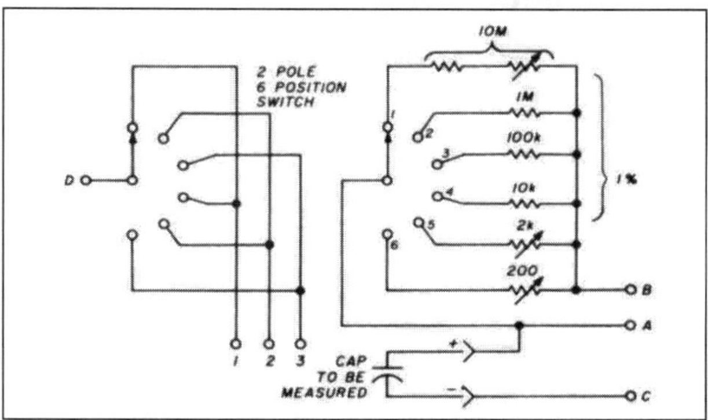

fig. 6. Switch connections for the range switch of the capacitance meter. The points specified are connected to the appropriate location on the circuit board (see fig. 4).

An open capacitor will cause a 000 to 001 to be displayed. A shorted capacitor will cause the overflow indicator to flash and the display to indicate a fixed number that is independent of the range switch position.

Lead lengths should be kept short when measuring small value capacitors. The photographs show a plug-in device made from banana plugs, a small piece of copper clad board, and sheet brass.

conclusion

The digital capacitor meter has been a fun project

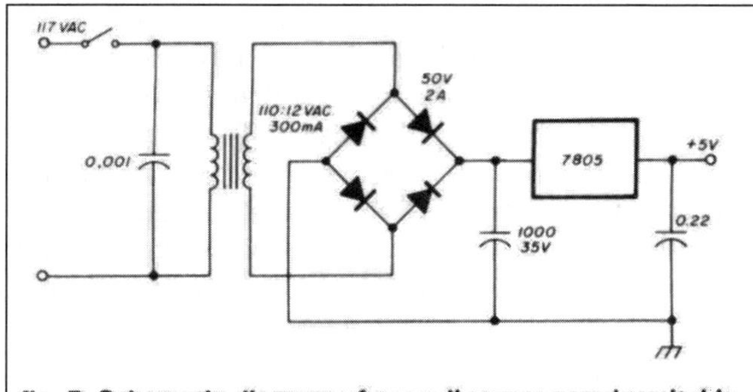

fig. 7. Schematic diagram of a small power supply suitable for home station use of the capacitance meter.

table 1. Switch positions for various measurement ranges showing display and associated capacitance value. In switch position 1, a display of 1.50 indicates a capacitance of 0.015 μF (1500 pF), a reading of 2.20 indicates a capacitance of 0.002 μF (2200 pF), etc.

switch position	display	capacitance	range
1	1.00	0.001	1000 pF (1 nF)
2	10.00	0.010	10k pF (10 nF)
3	100.00	0.100	100k pF (100 nF)
4	1.00	1.000	1 μF
5	10.00	10.000	10 μF
6	100.00	100.000	100 μF

to build and it has been a time- (and agony-) saver around the ham shack. I hope that others who enjoy building and experimenting will find it to be the same. I will offer film negatives (or positives) so that builders can make their own circuit boards. Correspondence regarding the meter will be answered if an SASE is included.

acknowledgments

Several hams have been of great assistance in developing the digital capacitor meter, in particular WA4RVN, K4ZKU, and W4PVA. K4ZKU provided valuable information on driving the display to full brightness, and W4PVA helped with the information on the 14553 counter chip without which the project could not have been undertaken. WA4RVN built his meter according to this article to verify the construction and checkout notes.

digital logic probe

A discussion and several designs for TTL and CMOS logic probes, featuring short pulse type memories

When electronic equipment consisted primarily of analog circuitry, most maintenance and troubleshooting could be handled with a simple volt-ohmmeter and some common sense. The VOM was the one instrument that could always be found on the bench of any ham or electronic experimenter. In addition to being generally useful, the VOM had much going for it. It was relatively small and easily handled. It was generally affordable even by Radio Amateurs of very modest means. Although it wasn't a precision instrument, if one knew how to use it, very good results could be obtained. By and large, it is still a most useful tool, but it's one whose relative importance has considerably diminished.

Complementing the VOM today, as the general purpose test instrument of digital circuitry, is the logic probe. Like the VOM, the logic probe is easy to handle, convenient, inexpensive, and, when used intelligently, capable of furnishing the needed *troubleshooting information*.

There isn't much to a logic probe. It is simply a device that will indicate the "state" of an accessible point in a digital circuit. Using some sort of quick response display, like an LED for example, the probe will indicate whether the voltage at the test point is high, low, or, perhaps, alternating. Does every ham and electronic experimenter need one now? Well, that's pretty much for the individual to decide, a decision to be based on what other equipment is around the shack and whether he does his own maintenance or pays someone else to do it.

There is no getting around the fact that each new piece of electronic gear hitting the market contains more digital circuitry than did its predecessor. Not too long ago the digital circuitry in the typical ham shack might have been limited to that in the electronic keyer or, if there was a new and expensive oscilloscope, in the trigger circuit of the horizontal sweep. Today, digital circuitry is the heart of the fully synthesized, VFO, of frequency counters, modern capacitance meters, fashionable readouts and displays, to name a few places, and it's becoming ever more commonplace.

For troubleshooting these digital circuits, it is pretty tough to find better instrumentation than the simple logic probe. Coupling that observation with the fact that a generally adequate logic probe can be quickly and easily assembled at a cost of anywhere between a few quarters and a few dollars, depending upon the status of your "junk" box, it's hard to justify not having one around the shop bench or shack. Just how elegant or sophisticated a probe is needed for any given shack is easily determined and, using one or more of the following circuit ideas, built.

If all of the digital circuitry currently in your shack operates on +5 Vdc and you have no particular interest in detecting very short pulses (10 μs or less) occurring at very low frequencies, then the very simplest of TTL (transistor transistor logic) probes should suffice. If your equipment is all in the 5-volt category, but there is a reasonable chance that you'll be looking for fleeting pulses, troubleshooting the triggered sweep of a modern oscilloscope is a prime example, then the logic probe should be slightly more elegant. The probe will still be based on a TTL integrated circuit, but some technique for capturing those elusive pulses need be added.

Some of the newest frequency synthesizers and frequency counters are built around LSI, or large-scale integrated circuits. Many of these are made up of mosfet rather than bipolar transistors. They may be operating on any voltage between about +4 and

By Raymond S. Isenson, N6UE, 4168 Glenview Drive, Santa Maria, California 93454

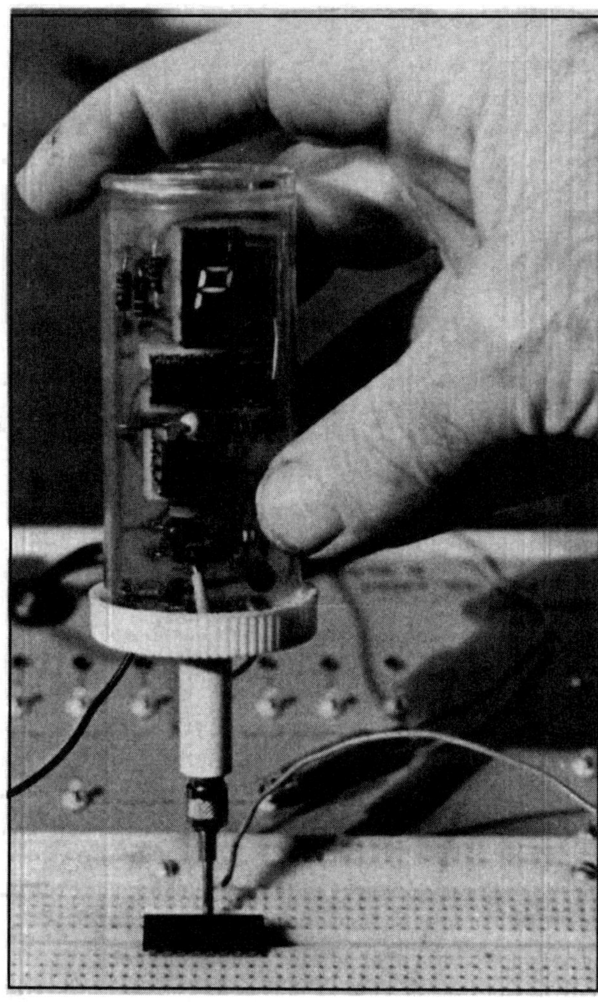

View of the digital logic probe showing the mounting and connections within the pill bottle.

+15 Vdc. A logic probe suitable for working on these circuits must be able to cope with the entire span of voltages. Such a probe is going to entail more than the basic probe discussed above. If the capability to detect very short pulses or noise spikes is desired, it will be still more elegant. However, it will be only slightly more complex. In fact, a probe capable of handling the full spectrum is so simple and little more complex than the simplest one that the only reasons for building the lesser one is that it's all you need for the foreseeable future, every part needed for it is in hand, and you have time to build it right now! Well, this very simple probe, the circuit of which is shown in **fig. 1**, has done yeoman service for me over five years and has on only the fewest of occasions been inadequate to a task at hand. It's not to be sold short.

simple logic probe

I previously pointed out that a logic probe is simply a device that will indicate the "state" of an accessible point in a digital circuit. Thus, the probe must have some sort of a readout, and its readout should be unambiguous. The probe ought to be convenient to use, and it must not upset or influence the circuit being observed. **Fig. 2** is a circuit diagram for a very simple logic probe. It should be adequate for most purposes, just as long as its use is limited to 5-Vdc circuits. With a minor exception, it's the circuit of a kit that was widely available a few years ago.

Isolation between the logic probe and the circuit under test is provided by the 10k resistor in the base of Q1, which can be any NPN switching transistor with a low-current beta equal to or greater than about 50. A 2N2222 would be just fine. All of the logic and the display drive is provided by an inexpensive 7404 hex inverter. The two LEDs make up the readout. LED A is turned on when the probe sees a high, LED B a low. If the test point has an alternating voltage, the two LEDs will blink alternately or appear to be both on depending upon the pulse frequency. Power to operate the probe is taken from the circuit under test. As each inverter of the 7404 can source about 15 to 20 mA, there's adequate current to drive the LEDs to a reasonable brightness. The current limiting resistors protect the 7404 and the LEDs. The whole thing is assembled in a salvaged plastic pill bottle or the like.

If the two LEDs are of different colors this logic probe, exactly as described, could be satisfactory. However, if only LEDs of the same color are available, the resultant readout might be something less than desirable. The logic probe is small, hand-held, and is frequently used to probe densely packed circuitry. It isn't advisable to let one's eyes stray too far from the tip of the probe if the risk of accidentally shorting a couple of pins together is to be avoided. This means that it is most advantageous if the readout is such as to permit reading "out of the corner of the eye." A distinctive difference in the readout for high or low is most desirable. An elegant and inexpensive way of accomplishing this is shown in the

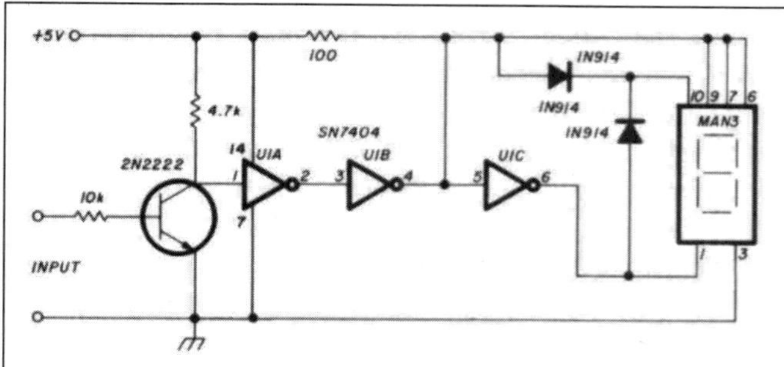

fig. 1. Schematic diagram of a simple logic probe for use with strictly 5-volt circuitry.

circuit of **fig. 1**. MAN 3A or MAN 3M LED readouts, or their equivalents, are available for as little as $1.00 (or less) on the surplus market. One of these devices coupled to the output of the SN7404 hex inverter, as shown in **fig. 1**, makes for a most ingenious little display. A high at the probe tip causes the hex inverter to drive the numeric display so that a *1* appears. A low at the probe tip results in a *0* being displayed. If there is a pulsating signal at the test point you will see a *P*.

I did not originate this circuit; it is merely one of several that were available as inexpensive kits over the past several years. The output transistors of the hex inverter source the current to drive the LED segments. As this circuit uses no current-limiting transistors between the 7404 and the display, a short in the latter will likely destroy the inverter. That's the primary weakness of an otherwise very clever circuit.

The MAN 3A was used because it was available. A common-anode display could be used with, of course, appropriate interchange of connections. The significant point is that the output transistor is capable of sinking up to 30 mA, so that you can use either a common-cathode or common-anode display. This is not true, as will be discusses later, if the TTL device is replaced with an MOS device. Referring again to the probe shown in **fig. 1**, the two leads are connected to the V_{CC} and ground of the circuit under test, and the probe is held against the test point. **Fig. 3** shows the printed circuit board layout as seen from the foil side. The overall size of the board is tailored to fit snuggly into the pill bottle so it may be necessary to make some slight changes to the printed-circuit board layout to fit any given plastic bottle.

pulse memory

When the instrumentation target is a very fleeting positive going pulse, such as in the previously suggested example of the trigger circuit of a modern oscilloscope, or if you are trying to ferret out some suspect random noise pulses in that new desk top computer, the logic probe must see and retain the high long enough to produce a visible signal on the

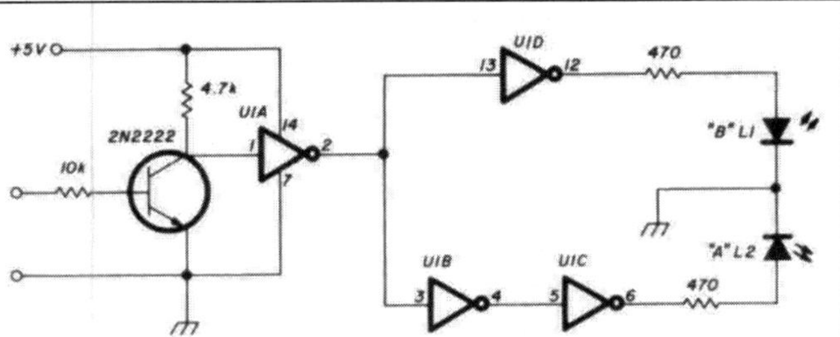

fig. 2. Schematic of the basic logic probe which used single LEDs to indicate either a high or low logic level. A cyclic signal will cause the LEDs to flash or both appear to be on due to the repetition rate.

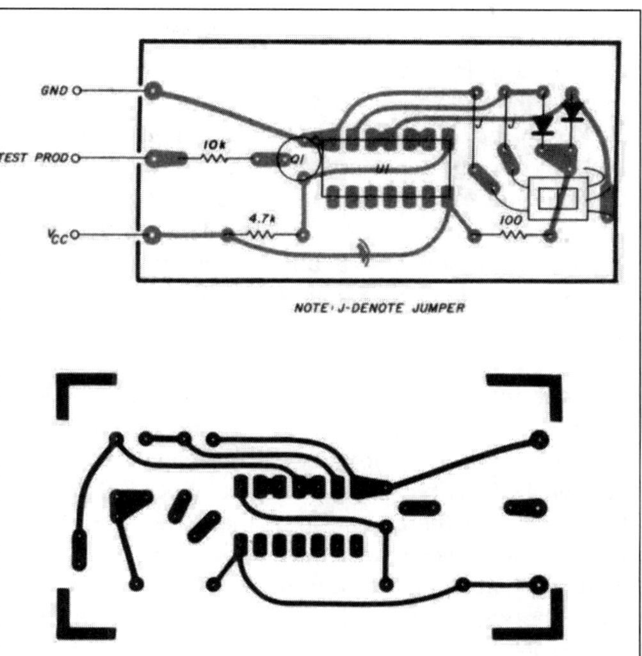

fig. 3. Etching pattern and parts placement diagram for the simple TTL logic probe.

LED display. An acceptable way of accomplishing this pulse "stretching" is through recourse to a one-shot multivibrator as a memory circuit. A very short incoming pulse, too short to be seen directly on the LED, is fed by the input amplifier of the probe to the input of the one shot, triggering it on. The output of the multivibrator is placed in parallel with one of the normal outputs of the hex inverter. The pulse duration at the output of the multivibrator is tailored by the time constants of the circuit so as to ensure a visible signal. Where the logic probe is to be used only on a +5 Vdc supply, and if there is an SN74121 in the junk box, a possible probe circuit is shown in **fig. 4**. The length of the output pulse duration needed to yield an acceptable display can be determined experimentally by varying R1, C1, or both.

versatile logic probe

An alternative short term memory uses the ubiquitous 555 timer. The connections for the 555 as a pulse stretcher is shown with a CMOS rather than a TTL hex inverter. Nevertheless, it can be used with the latter in exactly the same way. This, the most flexible and elegant logic probe, is presented in **fig. 5**. It can be used with any logic circuitry operating between approximately 4 and 15 Vdc. This means the logic probe can be used for RTL, DTL, TTL, or CMOS; in fact, for about any existing digital circuitry except for I^2L. On the lower side, the voltage limitation is the efficiency of the LED display. With prime LEDs, it might be possible to work down to about 3 volts. On the high side, the limitation is the upper limit on V_{DD} for the 4049 CMOS hex inverting buffer used as the logic chip and display driver. If a Fairchild F4049 is used, V_{DD} could safely go as high as 18 Vdc without damaging the probe. At any rate, it is most unlikely that the user will be confronted by voltages exceeding the range of 4.5 to about 13.8 volts so any 4049 will be acceptable and almost any surplus common-anode, seven-segment readout will work satisfactorily.

Being aware that the 74C04 and CD4069 CMOS hex inverters are pin compatible with the TTL 7404, one might well ask why go to the trouble of redesigning the circuit. Why not just replace the 7404 of the previously described logic probe with its CMOS counterpart and let the circuit go at that? There are two reasons why this cannot be done. The first has to do with the nature of the output mosfet of the CMOS chip, the second with the current limitation of the LED display.

In discussing the logic probe built around an SN7404 TTL hex inverter, note was made that the output transistors could each source about 15 mA or sink 30 mA. This permits the designer to select either a common-anode or common-cathode seven-segment LED display with the full confidence that there will be adequate current for safe direct drive of the LEDs. The 74C04 or 4069, on the other hand, are specified as being able to sink or source considerably less than 1 mA for +5 Vdc V_{DD} operation and 1.5 to 2 mA for 15-volt operation. The chip might be used with a common-cathode display, but the light intensity would be low. Used with a common-anode display, the CMOS output stage would quickly fail if it were forced to sink enough current for the LED to be acceptably visible, a function of the voltage applied to the common anode of the display and the size of the current-limiting resistors.

The problem is circumvented by turning to the CMOS 4049, a hex inverting buffer. These CMOS buffers provide both the necessary logic for the probe and a high current output capable of safely driving the LED load. It is not, however, as flexible as the TTL 7404. The CMOS buffer will typically sink about 5 mA with a V_{DD} of 5 volts and about 20 mA for a 15V V_{DD}. Under the same operating voltages, it will source only 1 to 3 mA. Thus, the TTL design option of using either a common-anode or a common-cathode configured display is closed; only a common-anode device can be used. How this is done is shown in the circuit in **fig. 5**.

The other major concern when designing the logic probe for this very wide range of operating voltages is the current limitations of the LEDs themselves. The generally useful current range of most LEDs is about 2 or 3 to 1. That is, starting with no current through the LED, current is gradually increased until first, the light output is barely adequate to be seen in a lighted room and then second until the LED fails. The current at failure will be about 2 to 3 times that at "visible." By the way, this isn't offered as a "scientific truth," but rather as an observation based on experience and generally supported by pertinent specification sheets. Q2 and ZD1 in **fig. 5** provide a voltage regulator whose output is applied to the common anode of the display. As the applied voltage at the V_{DD} lead of the probe is varied between +5 and +15 volts, the voltage at the output of the regulator varies between 4.4 and 6.2 Vdc. In the path between the output of the regulator and ground there is the 1N914, across which there will be about a 0.6-volt drop, the LED itself, which will account for a drop of 1.7 volts, and the current-limiting resistor which must make up the rest of the drop. The variation in voltage across the resistor, for the 4.4 to 6.2-volt swing, will thus be 2.1 to 3.9 volts, considerably less than 2:1 range. The LED current will be limited to the same range, one that is quite safe.

Three 1N914 diodes are shown between the 4049 and the LED readout in **fig. 5**. These diodes perform several functions so, unlike the diodes in **fig. 1**, cannot be replaced by slightly larger current-limiting resistors. This probe is designed to be used with operating voltages as high as 15 volts. Under this condition, and when the output of the buffer is in the high state, the output will approach 15 volts. Meanwhile, because of the voltage regulator, the anodes of the LEDs are close to 5 volts. The 1N914s protect the LEDs from what otherwise would be about a 10-volt reverse voltage, some 4 to 7 volts more than the maximum permitted according to the manufacturer's specifications. A second function of the diodes, or at least two of them, is to isolate the output mosfets of

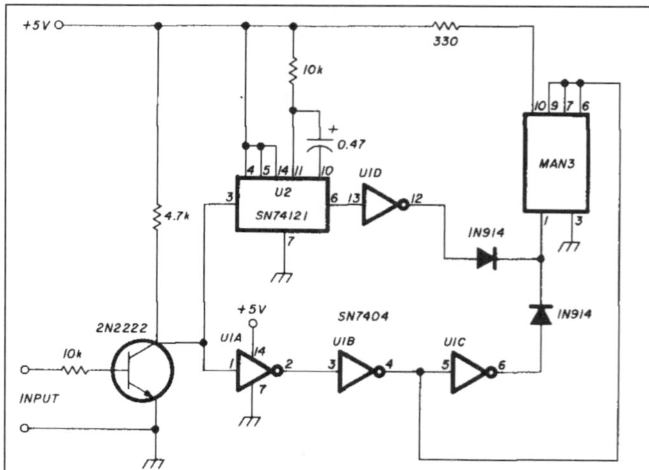

fig. 4. Diagram of the TTL probe with a short pulse memory. The monostable is used to capture any short-duration pulses for display on the LED display.

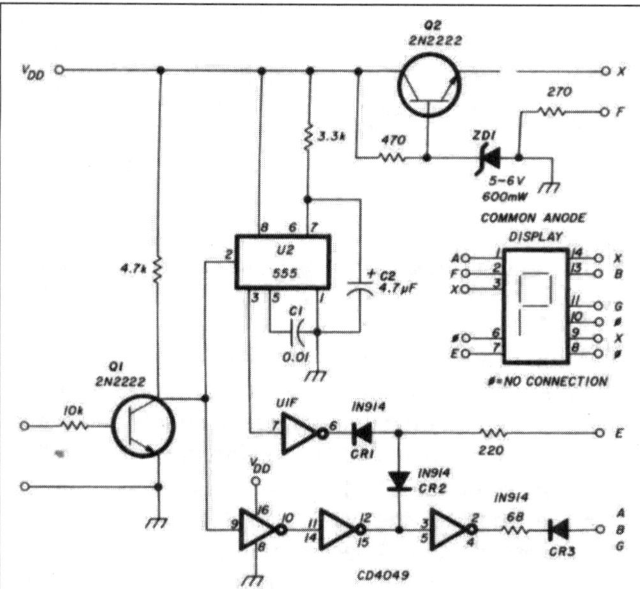

fig. 5. Schematic of the CMOS type logic probe with a short pulse memory. A 555 timer is used as the memory element; a common-anode display must be used in this version.

the inverter buffers from each other. Either of two inverters may go low to turn on segment E of the LED while the other is high. The diode isolation permits this to occur without risk to the 4049.

Unused inputs of CMOS ICs are never allowed to float. They are tied high, low, or to a used input. In the design of the circuit of **fig. 5** the inverters were simply paralleled as necessary so that no inputs were allowed to float.

Just as the pulse stretcher for the TTL-based logic probe design could have been a 74121, this CMOS-based design could as well be a CD4047A monostable/astable multivibrator. The 555 timer was used because it's smaller and was available; it is also less expensive. There's nothing unusual about the employment of the 555; the one-shot configuration is right out of the book for a negative going trigger input and a one-shot stretched output. The output pulse length is given as 1.1 RC where the RC applies to the resistor between V_{DD} and pins 6 and 7 and the capacitor between this point and ground. The component values shown on the schematic were found, experimentally, to give a pulse that was just long enough to barely flash the Litronix readout. For test purposes, 0.25-microsecond pulses were generated at a pulse frequency of one pulse per second. Readout visibility was very acceptable. The test circuit is described briefly at the end of this article.

Obviously, the readability of the output for a stretched pulse can be enhanced simply by increasing the RC time constant in the 555 timer circuit. In designing the probe, however, the duration of the stretched pulse was deliberately kept to the useful minimum; the probe readout differentiates between short pulses or noise pulses at low frequency reoccurrence rates and low frequency "clocking" phenomenon.

With a low-frequency, alternating state signal at the probe point (10 Hz or less), the readout will alternate between *1* and *o*. At a higher frequency and in particular where the duty cycle is between 20 and 80 per cent, the eye of the observer is fooled into seeing a steady *P*. For very short positive-going pulses at low-frequency rates, the display is a brief *P* followed by an extended *o*. At higher frequencies, the display takes the form of a fairly bright *o* with a dim staff to form the *P*.

None of the described logic probes will indicate the presence of a brief negative-going pulse. This is a design limitation accepted because I have never found need for that capability and because providing for it does cause some additional circuit complication. If the added capability is required, it can be achieved by modifying the CMOS logic probe as follows. Replace the 555 timer with a 556 dual timer. Isolate the two paralleled hex inverters. Connect the input of one of these inverters to the collector of Q1 and its output to the input of the added timer. Connect the output of the added timer to the input of the other freed inverter and its subsequent output to the cathode of an additional 1N914. The anode of the 1N914 connects to the anode of the existing 1N914 in the circuit coupled to the three LED segments that

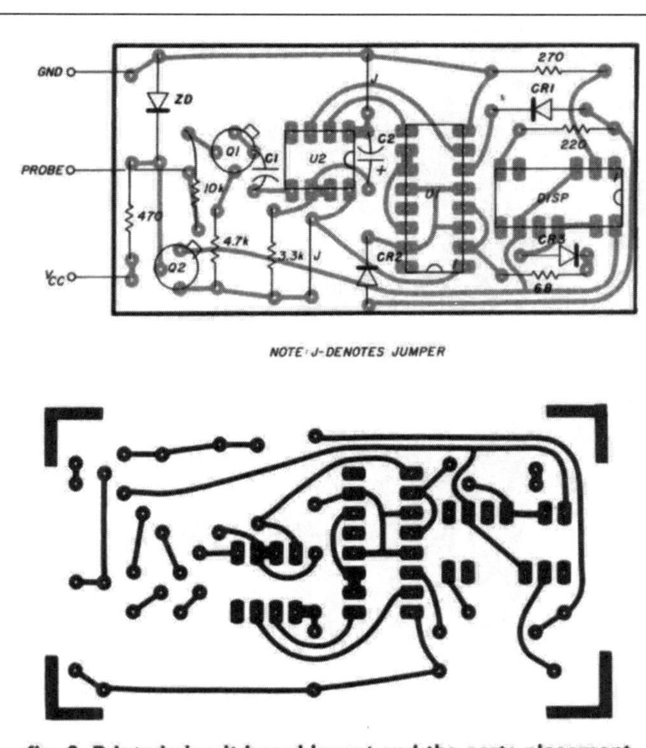

fig. 6. Printed circuit board layout and the parts placement diagram for the CMOS logic probe.

make up the switched element of the *o*. In the presence of an occasional negative-going pulse, the display should be a *1* changing to a *P* when each pulse appears. The passive components associated with each half of the 556 should be the same as those shown for the probe in **fig. 5**.

Having decided upon the circuit to be implemented, the next step is to collect the pill bottle to be used as the case. This is an important step because the size and shape of the pill bottle will determine the size and layout of the circuit. **Fig. 6** shows the circuit board layout of the CMOS logic probe described in detail above. It will be useful if your pill bottle will take a 1¼ × 2½-inch (3.0 × 6.0-cm) board.

Printed circuit board techniques were used for both probes shown in this article only because it was convenient to do so. Wire wrap techniques or even point-to-point wiring on sockets mounted in perf board would be just as good. The logic probe is fundamentally a low-frequency device. It would be difficult to find a poor construction technique as long as the workmanship is good!

Test of the completed circuit for all but the pulse stretching feature is easily accomplished. Connect the V_{DD} and V_{SS} or V_{CC} and ground wires, depending upon your choice of CMOS or TTL, to an appropriate power source. If everything is working properly the readout will display a *o*. Touch the probe to the positive voltage terminal of the power supply and the *o* should change to *1*. The circuit shown in **fig. 7** will test the pulse capture feature if one has been included. Pulse length of the output pulse of the 74121 is approximately 1400•C1 seconds. If C1 is 180 pF, the pulse at the test point would be $1400 \cdot 180 \cdot 10^{-12}$ second or approximately 0.25 microsecond.

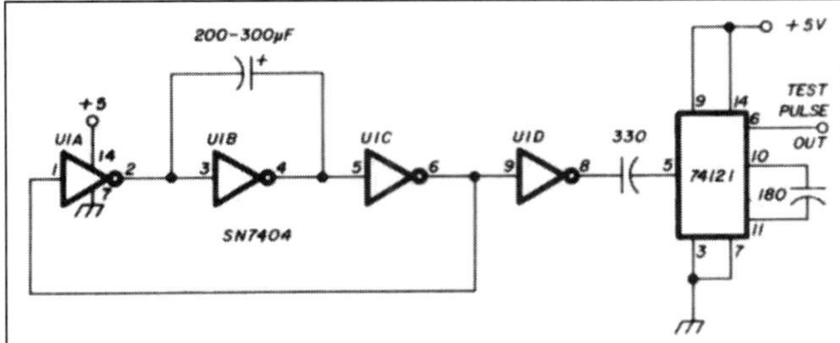

fig. 7. Pulse generator to test the short pulse memory capability of the logic probes.

The logic probe is a very practical instrument to have around the shop or shack. If you don't have one and can squeeze out an evening, try one of the circuits presented here. It won't be long before you begin to wonder how you ever managed to get along without it!

a new look at dip meters

Some updated versions of that most useful tool — the grid-dip meter

One of the most useful accessories in the Amateur station is the grid-dip oscillator, grid-dip meter, or just plain dip meter. It can help solve many problems associated with resonant circuits. With proper calibration the dip meter can be used to determine the frequency of resonant circuits and, with capacitors and inductors of known values, it can be used to measure unknown inductances and capacitances. VHF parasitic oscillations can be detected in rf amplifiers, and antennas and transmission lines can be checked for resonance if proper precautions are used to determine harmonic responses with respect to fundamental resonant frequency.

The original grid-dip meter used a tube-type, class-C oscillator with a meter in its grid circuit. An external resonant circuit coupled to the oscillator's tank caused a decrease, or dip, in grid current; hence the name of the instrument. If this external circuit was a high-*Q* tuned circuit, the dip in grid current was noticed at a rather sharply defined frequency as the oscillator was tuned. It is precisely because the grid-cathode diode of the tube rectifies rf voltage that the oscillator operates in a stable class-C mode; and this rectified rf provides the dc grid current to operate a conventional dc current meter. A typical grid-dip meter is shown in **fig. 1**.

Historically, the oldest Amateur reference to grid-dip oscillators was that made by Bud Bane, W6WB, in the March, 1947, issue of *CQ* magazine. W6WB's meter used a 3A5 for battery operation, and covered 80 through 10 meters with three coils. Still earlier use of the grid-dip oscillator is discussed in the "Zero

By Hank Olson, W6GXN, P. O. Box 339, Menlo Park, California 94025

50
13k
+150V
PLUG-IN COIL
50
955
13k
0.01
SENSITIVITY CONTROL
Except as indicated, decimal values of capacitance are in microfarads (μF); others are in picofarads (pF); resistances are in ohms. k = 1,000 M = 1,000,000
M
200μA

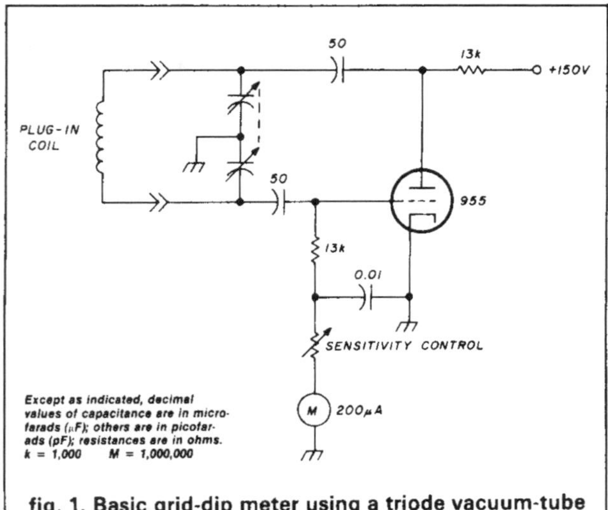

fig. 1. Basic grid-dip meter using a triode vacuum-tube (9002 and 6C4 also commonly used).

5
250k SENSITIVITY
33
RCA 2N1178
3k
PLUG-IN COIL
50
IN34A
39k
3.9k
0.01
0.01
50μA
M
13.5V
Except as indicated, decimal values of capacitance are in microfarads (μF); others are in picofarads (pF); resistances are in ohms. k = 1,000 M = 1,000,000

fig. 2. Typical dip meter using germanium PNP bipolar transistor with separate diode detector.

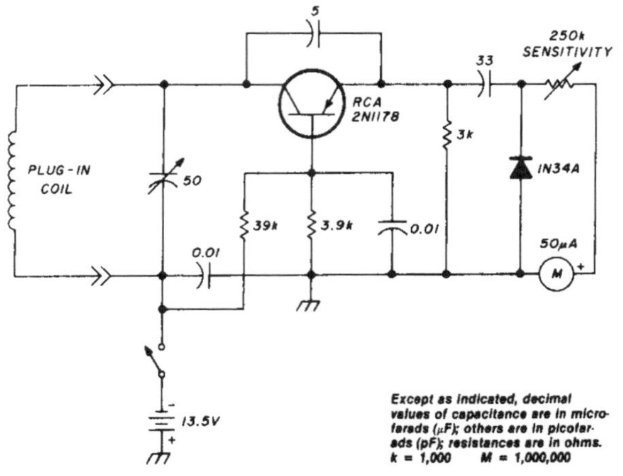

Except as indicated, decimal values of capacitance are in microfarads (μF); others are in picofarads (pF); resistances are in ohms. k = 1,000 M = 1,000,000
560
6.8k
M
0-1mA
3.3
100k
TI363
TI363
2N185
GE STD633
GERMANIUM
PLUG-IN COIL
100k
6.8k
1k
G247 SILICON
100k
SENSITIVITY
0.05
OSC.
DIODE
OFF
47
560
1.5V
0.05

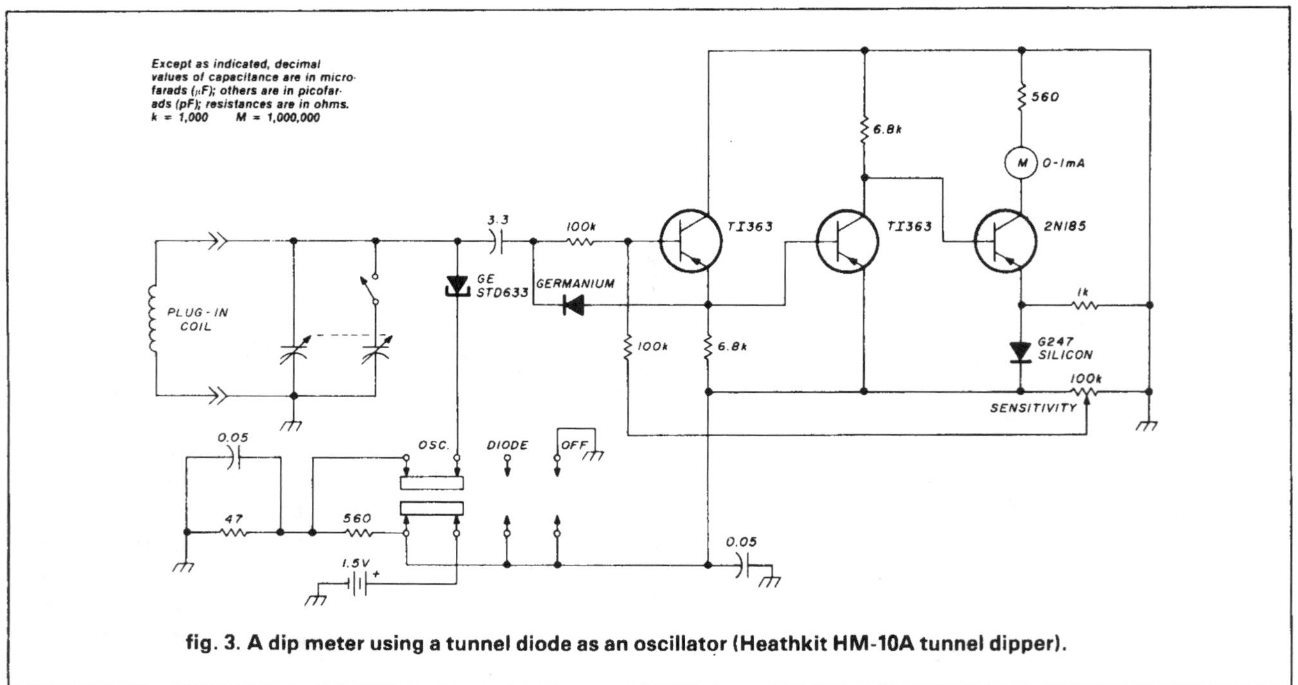

fig. 3. A dip meter using a tunnel diode as an oscillator (Heathkit HM-10A tunnel dipper).

Bias" column of *CQ*, April, 1979, and in "Our Readers Say," *CQ*, October, 1979.

When bipolar transistors became available, they were almost immediately applied as the active elements in rf oscillators — and also applied in dip meters. The term *grid-dip* had to be shortened to *dip* since, of course, there are no grids in bipolar transistors. The bipolar transistor (whether NPN or PNP) is unlike the tube in a number of ways. The bipolar transistor is a current-operated device; that is, it requires base current to control collector current. This generally means that it is a lower impedance active device, and that the base has dc current flowing in it whether oscillation is present or not. Even though the base-emitter junction of a bipolar transistor is a diode, the base current is not a very sensitive indicator of oscillation level because a good portion of the base current is dc bias current.

So we find, that, although a number of dip meters have been built using bipolar transistors, all of them make use of a separate diode detector to provide dc current for the meter.[1,2] Most of these circuits use a 1N34A or some other germanium point-contact diode as the diode detector, and might be updated by use of one of the newer (silicon) Schottky diodes. A typical bipolar dip meter is shown in **fig. 2**.

It is even possible (and one was commercially built by Heathkit for a time) to use a two-terminal form of oscillator: a tunnel-diode. In the Heath HM-10A Tunnel Dipper, a tunnel-diode was used as the oscillator and a standard diode detector was used as the rf detector to provide dc meter current. The Tunnel-Dipper is shown in **fig. 3** (reprinting is by permission of the Heath Company).

junction field effect transistors

The junction field effect transistor (JFET) is the first solid-state device to come along that really behaves in a "dip-meter" circuit the way tubes used to behave. Like a tube, the JFET requires voltage on the gate (grid) to control current in the drain (plate), and the gate-source junction operates like a "grid-leak" diode. So, with a lower supply voltage, an N-channel JFET can be put in an old tube-type dip-meter circuit and it will usually work immediately.[3] Early silicon JFETs had rather limited performance at VHF, and it was not until the arrival of the TIS34/2N3823 (Texas Instruments) that a good, cheap, silicon JFET was available for Amateur dip-meter construction. Subsequently, the Motorola MPF102, Union-Carbide 2N4416, Siliconix 2N5197, and other excellent N-channel, silicon JFETs became available inexpensively; these too fit dip-meter service quite well. An example of a dip-meter using an N-channel JFET is shown in **fig. 4**.[4]

the MOSFET

The MOSFET, unlike the JFET, does not have an inherent gate-source diode, and there are many MOSFETs that operate very well up through the VHF range. RCA and Motorola have a number of N-channel (depletion mode) MOSFETs that were designed for TV tuner service and are therefore inexpensive enough for Amateur dip-meter service. The 3N128 is a favorite in Amateur designs, and has been incorporated into at least one dip-meter design (see **fig. 5**).[5] Since there is no gate-source diode in a MOSFET, rectification of the rf is accomplished by adding a diode. It is even feasible to use one of the dual-gate MOSFETs, using the No. 2 gate as an oscillation level control (see **fig. 6**).[6]

Note that in **figs. 1** through **6** some of the oscillators use the Colpitts configuration and some use the Hartley circuit. Either form of oscillator will work, or any other form will do, as demonstrated in **fig. 3**. The Colpitts circuit has the advantage over the Hartley that only two leads must be provided for the plug-in coils.

It is possible to build a dip meter without plug-in coils. This has the advantage that the accessory coils don't get misplaced. One author has written at least two articles on bandswitching dip meters.[7,8] In reference 8, Fred Brown adds two worthwhile innovations to dip-meter technology: 1) a broadband 50-ohm output for driving a frequency counter, and 2) a square-wave modulation system which amplitude modulates at 1 kHz without fm.

With frequency counters now available for less than $200.00 that cover the range 0 to 1000 MHz, the

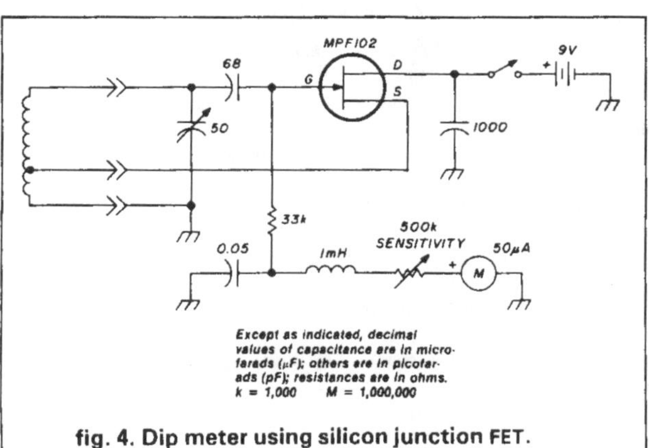

fig. 4. Dip meter using silicon junction FET.

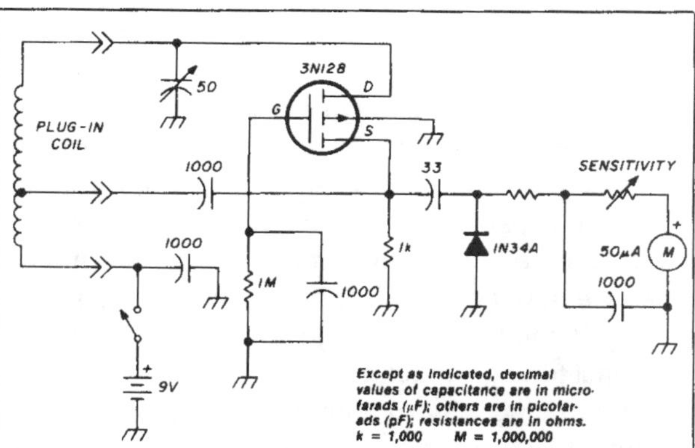

fig. 5. Dip meter using N-channel IG FET (MOSFET) and separate diode detector.

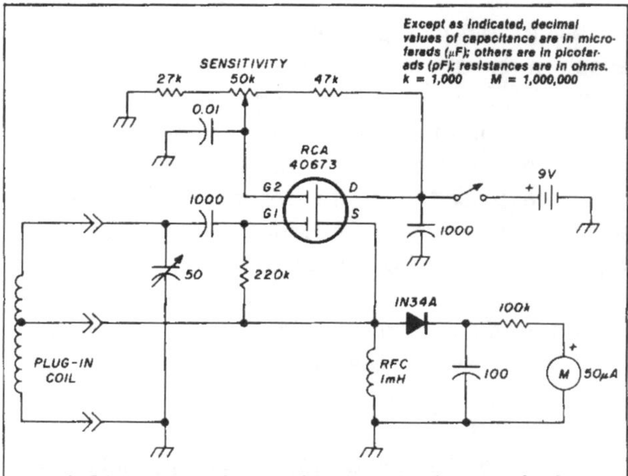

fig. 6. Dip meter using dual-gate IG FET (MOSFET) where No. 2 gate is used to adjust oscillation level.

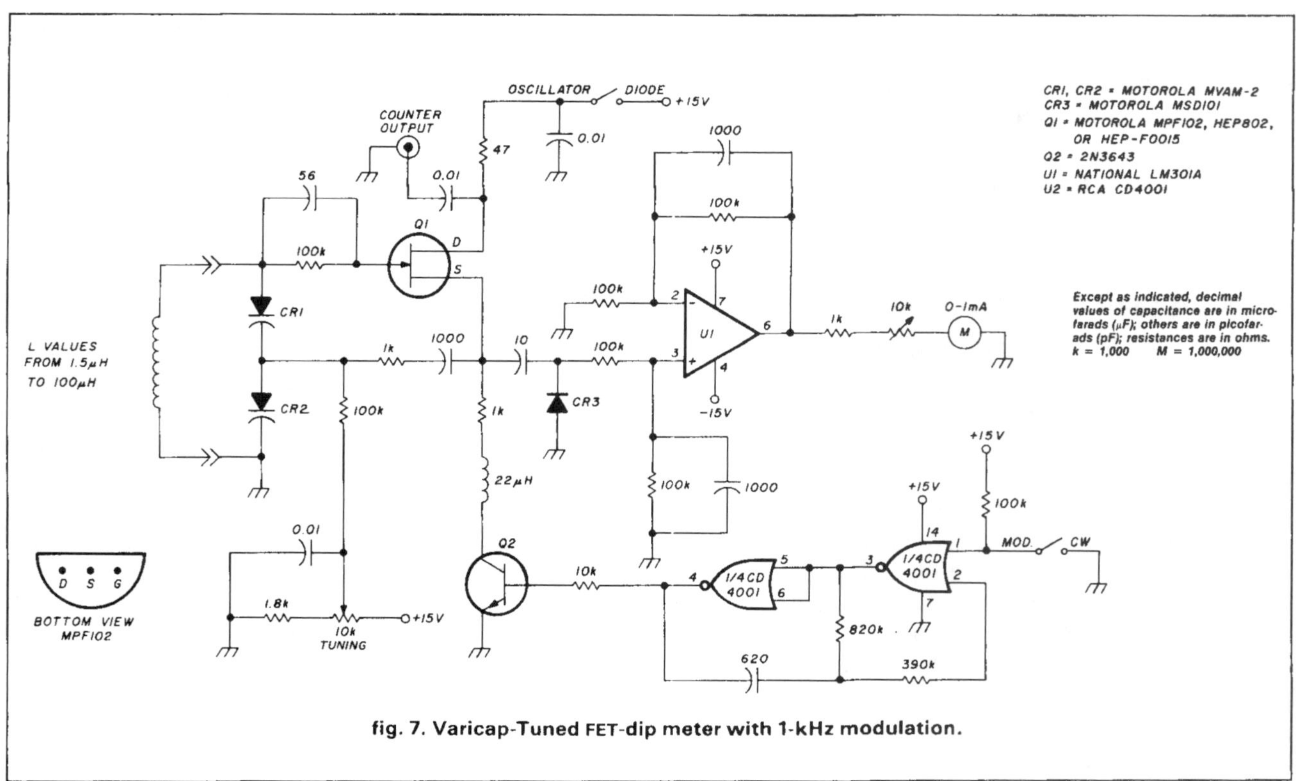

fig. 7. Varicap-Tuned FET-dip meter with 1-kHz modulation.

time is perhaps right to dispense with the calibration dial on the modern dip meter. Once this step has been taken, it is only a logical, short jump to replace the tuning capacitor with varicaps and use a remote dc pot to tune the dip oscillator, since no calibration scales are now needed. If the counter is not desired, the dc pot can be calibrated just as well as could the traditional tuning capacitor dial.

One tempting way to combine voltage-tuning the dip meter and using a frequency counter as a readout would be a microcomputer-controlled dipper. The oscillator varicap input would be a ramp (sawtooth) generated by the microprocessor; the detected dc output of the oscillator would be A to D converted and fed to the microprocessor. The program would sense a point of minimum detection voltage (where the dc slope changes from negative-going to positive-going) and remember the frequency counter digital reading at that point. This system could then simply be placed near the resonant circuit to be measured and the frequency would be read out digitally. The details of oscillator design, programming, integration with a frequency-counter, and provision for multiple dips could be a big development job; but it's certainly a system that seems feasible.

a simple dip meter

As a simple example of how one might build up a voltage-tuned, counter read-out dip meter, the circuit of **fig. 7** is shown. It was built to accept plug-in coils with banana-pins with 3/4 inch spacing; the coils were borrowed for this breadboard circuit from an old tube-type grid-dip meter.

As shown, the circuit operates over the 2-32 MHz range with each coil covering about an octave of frequency. It would probably be worthwhile to have coil units made up with three terminals, so that varicaps are changed with the plug-in coil, to allow for lower minimum capacitance at the higher frequencies, a 2N5197 or other similar VHF FET might also improve things above 32 MHz.

Look at the circuit of **fig. 7**. The 1-kHz square-wave oscillator has been redesigned around a CMOS IC, and the meter's dc amplifier redesigned around an LM301A op amp IC.

references

1. "Grid Dip Meter," *RCA Transistor, Thyristor, and Diode Manual*, SC-15, 1971, section 15-22, page 736.
2. P.A. Lovelock, "Solid State Conversion of the G.D.O.," *ham radio*, June, 1970, page 20.
3. P.A. Lovelock, "Solid State Conversion of the G.D.O.," *ham radio*, June 1970, page 22.
4. C.G. Miller, "Gate Dip Meter That Really Dips," *ham radio*, June, 1977, page 42.
5. L.G. McCoy, "A Field Effect Transistor Dipper," *QST*, February, 1968, page 24.
6. Frans Bruin, "A Dual Gain MOSFET Dip Meter," *QST*, January, 1977, page 16.
7. Fred Brown, "A Band-Switching Grid-Dip Meter," *CQ*, February, 1966, page 60.
8. Fred Brown, "A 1980 Dipper," *QST*, March, 1980, page 11.

Printed-circuit techniques,
solid-state devices,
and some old-time technology —
a project for
the first-time builder

field-strength meter
for the high-frequency Amateur bands

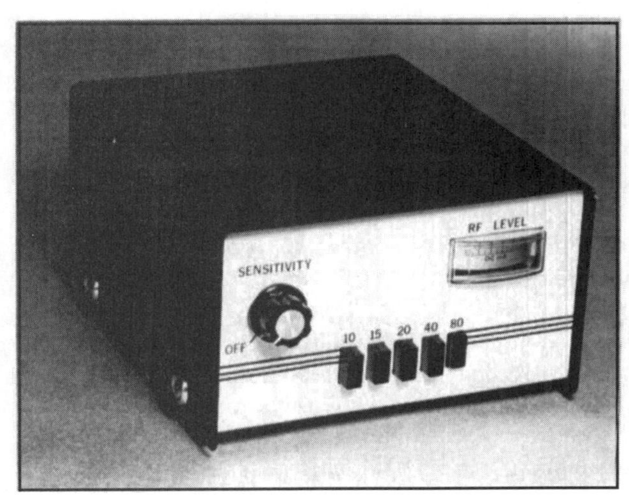

Here's a project for the first-time builder — a band-switching field-strength meter. It's a neat accessory for your station that indicates the relative field strength of a signal. Actually, it's a small, band-switched receiver covering the Amateur 80, 40, 20, 15 and 10 meter bands. In this unit the speaker or headphones have been replaced by a meter that indicates the relative field strength. The receiver is a very basic unit, but it's sensitive enough for critical antenna adjustment and tuning. It's battery powered, allowing you to work at some distance

By Ken Powell, WB6AFT, 6949 Lenwood Way, San Jose, California 95120

from the antenna. And it's sensitive enough so that a short whip antenna, as used on 2-meter portables, will serve as a pickup or receiving antenna.

The field-strength meter is constructed on a single printed circuit (PC) board. Component count is low, overall cost is minimal, and the highest voltage encountered is only 3 volts. The field-strength meter has a minimum number of controls: a push-button bandswitch, an rf-level meter, and a sensitivity control that keeps the instrument within reasonable bounds as you get your antenna and transmitter really humming!

The complete unit is about 5 inches wide, 6 inches deep, and 3 inches high (13 × 15 × 8 cm). It could be squeezed down a bit smaller; but, in the interest of making construction a pleasure rather than a chore, the package was kept on the roomy side. Component values are not overly critical, and no special tools or equipment are required for construction or adjustment.

circuit

The description of the field-strength meter makes it sound a bit more complex than it really is. A glance at the schematic (**fig. 1**) shows the simplicity of the circuit. Band switching sounds complex because we tend to think of large rotary switches and the like; but in this case, band switching is handled by a simple push-button switch mounted on the PC board. This eliminates a lot of wiring and the chance for errors. This type of switch is a real boon to the builder.

The circuit is actually a crystal set, much like those used by the pioneers of early wireless. The antenna is connected to the primary of the rf coil, L1; the secondary of this coil is tapped to provide resonance on each of the high-frequency bands covered by the field-strength meter.

The push-button bandswitch, S1, selects the desired tap on the secondary winding of the coil, L1, and shunts the secondary tap with the required capacitance to provide resonance at each switch position. To make parts procurement easy, the same variable capacitor is used on each band and is shunted by a fixed capacitor where required (see **fig. 1**). This provides a trimmer for setting each band and eliminates the need for a front-panel tuning control.

The voltage developed across L1 is applied to the detector, CR1, a germanium diode. This rectified voltage is filtered by capacitor C10 and applied to the base of Q1. This general-purpose germanium NPN transistor is an amplifier. It allows the use of a low-cost meter movement as an output indicator. (The alternative is a more sensitive meter, which is more expensive and more subject to damage.) The meter is shunted by sensitivity control R1, which allows a selected portion of Q1's collector current to bypass the meter movement. This enables you to maintain the meter needle within a usable portion of the meter scale as varying signals are applied to the antenna terminal. The overall collector current of transistor Q1 is limited by resistor R2. The power OFF/ON switch, S2, is coupled to the sensitivity control. The wiring is such that the full clockwise rotation of the control yields maximum sensitivity.

Power for the field-strength meter is supplied by two AA cells mounted on the PC board. Current consumption of the unit is low: battery life should approximate the shelf life of the batteries. Trimmer capacitors C1 through C5 provide individual tuning for each of the bands, and once the unit is adjusted, it's not likely that any drift will be encountered.

construction

Construction of the field-strength meter is really tailored for the first-time builder, and the parts cost, even in today's wild economy, is low. Components aren't especially sensitive to rough handling, and the package is not crammed or difficult to assemble. Radio-frequency circuits are often difficult to duplicate, but printed-circuit construction and standard coil stock take care of that nicely.

Standard hand tools and a drill are all that are required for construction. The aluminum of the case is very soft, and woodworking tools handle it well. A bit of filing is required to get the rectangular holes for the band switch and meter movement, but it's really not much of a task.

let's get started

The first step in the construction of the field-strength meter is to obtain the components shown in **table 1**. The PC board should be etched and drilled in accordance with the foil layout depicted in **fig. 2**. If you don't have the facilities for fabricating the PC board, an etched and drilled board, complete with the bandswitch assembly installed, is available. The source for this board is shown in **table 1**.

To start construction, mount the bandswitch and solder it in place on the PC board. Next, mount the four spacers on the foil side of the PC board. These spacers elevate the board above the base of the cabinet.

Now, before mounting any other components on the PC board, take the time to work out the mechanical details of the PC board in relation to the cabinet. If you use the same case I did and want to have the same general layout on the front panel, see the layout in the photos of the completed unit.

The front panel will require five rectangular holes for the switch assembly, a rectangular hole for the meter, and a round hole for the sensitivity control.

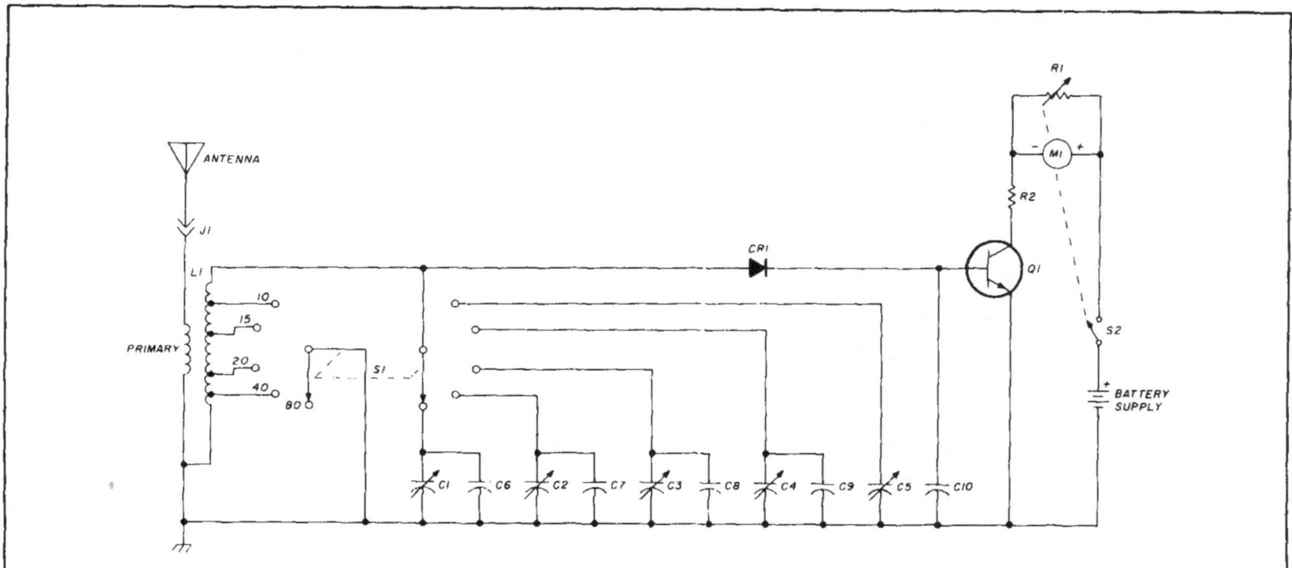

fig. 1. Schematic diagram of the bandswitching field-strength meter. Circuit is kept simple by using simple pushbutton switch mounted directly on the PC board.

The rear panel will require holes for the UHF connector used for the antenna.

After making the cutouts in the case, the PC board can be placed into the case, and the clearance for the push buttons can be checked. If things seem to fit well, mark and drill the four holes in the base for mounting the PC board and the rubber feet furnished with the case. When you're satisfied with the mechanical details, you can go on to complete the assembly of the board.

mounting small parts

The PC-board-mounted components, with the exception of rf coil L1, can now be mounted and soldered in place on the board as shown in **fig. 3**. When mounting components try to develop the habit of installing like components in the same direction so all values can be read from one side or edge of the PC board. In this case, all capacitors would be installed in the same direction. This is a good practice for all your projects and can make service and checkout easier.

Install the four wires that leave the board, using stranded wire to reduce the chance of wire breakage. Leave these leads about 6 inches (15 mm) long and trim them during the final stages of wiring. Color-coded wire is good for this purpose, and if you use color codes, note them on the schematic for reference. The PC board can now be set aside and work can be started on rf coil L1.

the rf coil

Before starting work on L1, study the drawing shown in **fig. 4**. Develop a fairly good idea in your mind of how the coil is to be formulated, as it is a very easy thing to get confused. Initially, the coil stock is 3 inches (76 mm) long and contains 48 turns held in place by four plastic formers.

Starting at the left end of the coil, as shown in **fig. 4**, count off four complete turns then go about one-third of a turn further, just past the second former, and cut the wire. Fold this wire back so it leaves the coil stock parallel to the lead at the start of the coil. This forms the four-turn primary of L1. The remaining end of the wire just cut will be the lead wire for the secondary and is peeled from the coil stock until it is parallel with the primary coil leads. This will leave a single-turn space between primary and secondary windings of L1. From this initial point on the secondary, count off 31 turns and again cut the wire about one-third way around the coil form. Fold this wire back parallel with the other coil leads. Remove the remaining coil stock and you should have a four-turn primary and a 31-turn secondary.

Now to put a few taps on the secondary. Using nail polish or other marking device, put a small mark at the fourth, seventh, ninth, and eighteenth turns of the secondary. These marks should be placed at the former, and the taps will be soldered just past the marks in the direction of the coil winding.

Scrape the coil plating lightly with a sharp hobby knife where each tap will be placed. This will make soldering the taps easier. Tin each of these spots lightly with your iron before soldering the taps in place. Form the four tap wires from resistor leads or solid wire that is well tinned. Don't try to use wire left

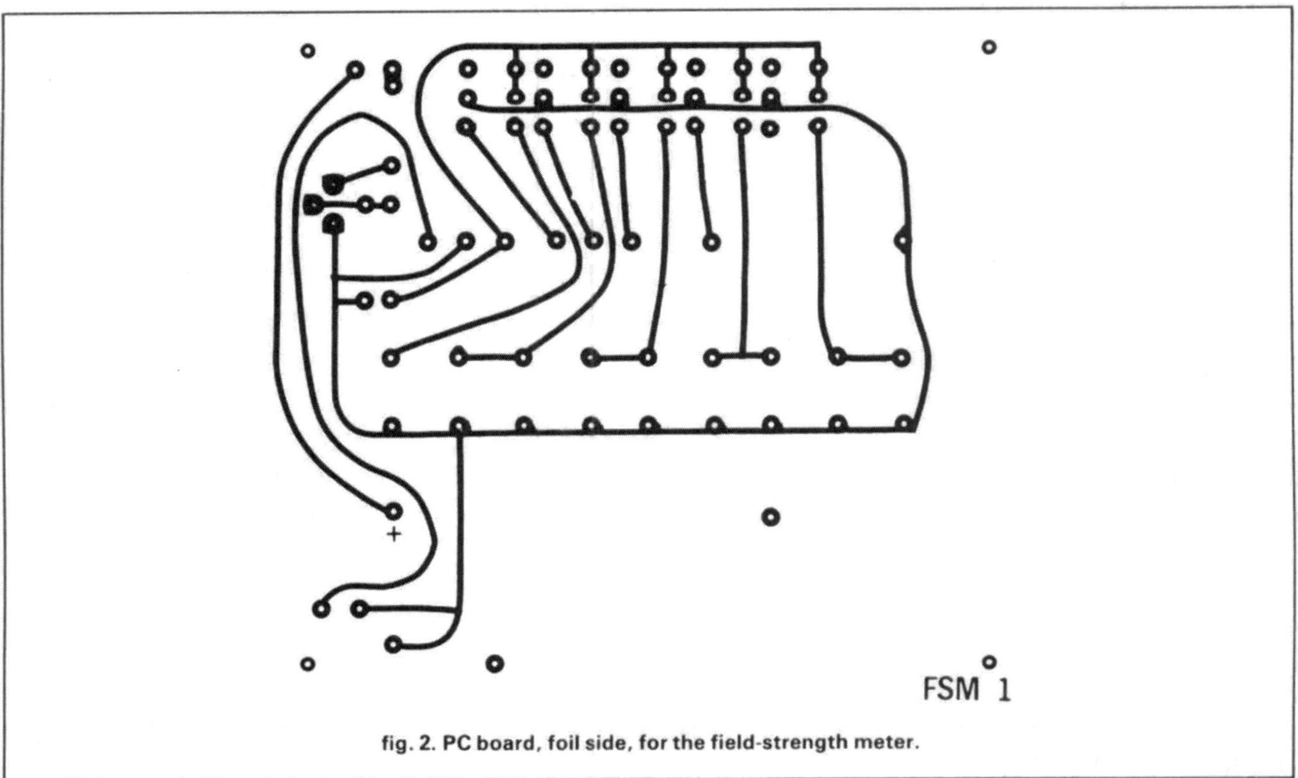

fig. 2. PC board, foil side, for the field-strength meter.

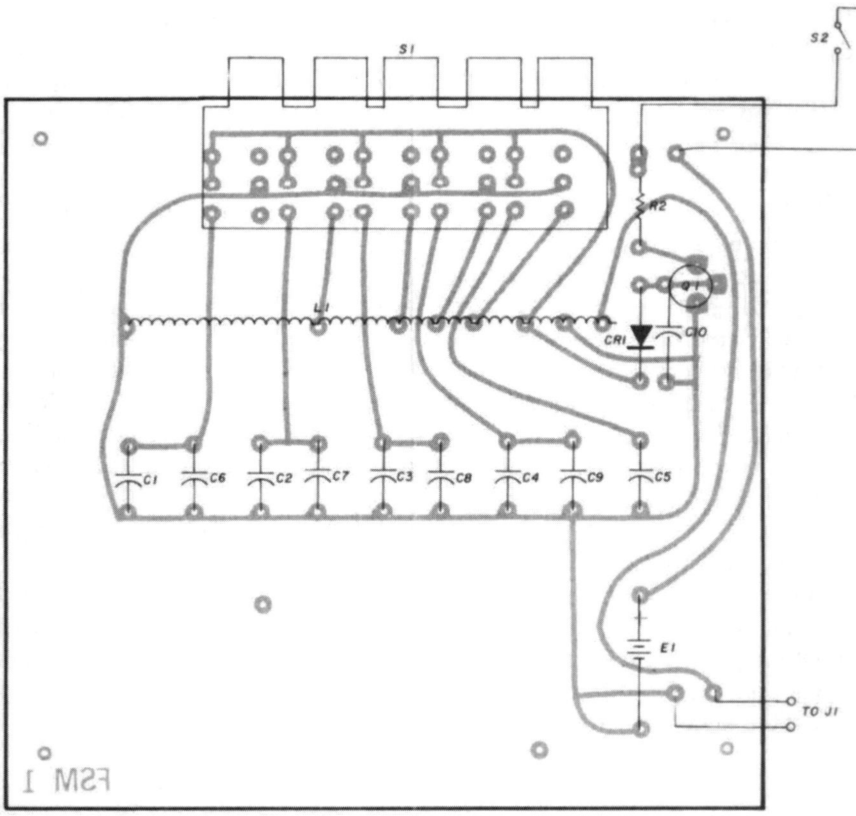

fig. 3. Component layout and associated wiring viewed from component side of PC board.

table 1. Parts list for the field strength meter.

component	description	part number	source*
C1 through C5	5 to 30 pF ceramic trimmer	E.F. Johnson 275-0430-005	CS
C6	150 pF ceramic cap	Sprague 5GA-T15	CS
C7	68 pF ceramic cap	Sprague 5GA-Q68	CS
C8	22 pF ceramic cap	Sprague 5GA-Q22	CS
C9	10 pF ceramic cap	Sprague 5GA-Q10	CS
C10	0.001-μF	Sprague 5GA-D10	CS
CR1	1N34A diode	276-1123	RS
E1	1.5 V "AA" cells (2X)	23-552	RS
E1 Holder	for 2 "AA" cells	12A2016-0	GM
J1	UHF jack, SO-239	278-201	RS
L1	1" × 16 turns/inch	B&W 3015	QE
M1	1 mA meter	Calectro D1-905	CS
Q1	transistor, germanium	276-2002	RS
R1	5 k pot, linear taper	271-1714	RS
R2	270-ohm, 1/2 W resistor	271-016	RS
S1	switch, 5-position pushbutton	18A1731-9	GM
S2	switch for R1	271-1740	RS
case	5-1/4" × 3" × 6"	270-253	RS
knob	0.750" diameter × 1/4" shaft	274-415	RS
spacers	No. 6 screw spacers	64-3024	RS
PC board	with S1 installed	FSM1	JO

CS: Circuit Specialists
1344 N. Scottsdale Road
Tempe, AR 85281
Tel. 800-528-1417

GM: Gravois Merchandisers, Inc.
715 Armour Road
No. Kansas City, MO 64116
Tel. 800-821-3686

QE: Quement Electronics
1000 S. Bascom Avenue
San Jose, CA 95128
Tel. 408-998-5900

JO: Jim Oswald
1436 Gerhardt Avenue
San Jose, CA 95125
PC board with S1
installed $9.75 postpaid
Tel. 408-269-2314

RS: Radio Shack
Local stores

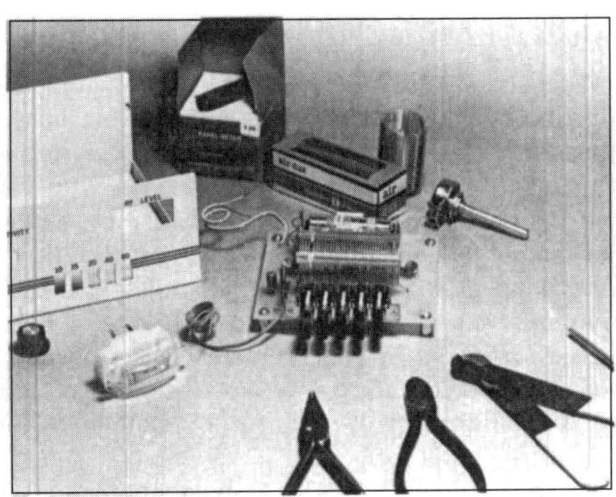

The field-strength meter ready for final assembly. PC-board construction and the use of pushbutton switches make a clean layout and an attractive package. Mounting screws for the meter are avoided by mounting the meter with epoxy cement.

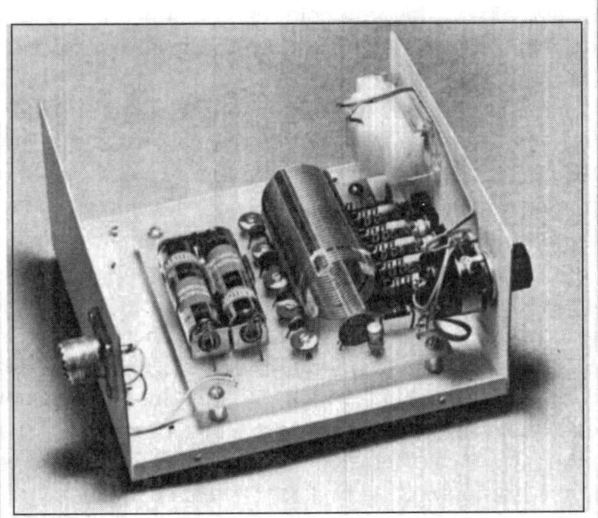

Inside the field-strength meter showing, from left, the battery pack; rf tuned circuit consisting of L1 and C1 through C10; meter, and sensitivity control. Note that component layout pretty much follows the schematic diagram.

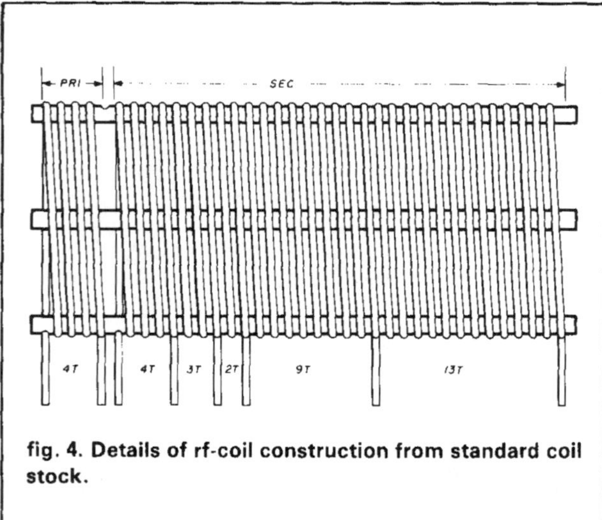

fig. 4. Details of rf-coil construction from standard coil stock.

over from the coil stock, as this is often difficult to solder in place.

Solder the four taps in place as shown in **fig. 4**. This operation is a bit tricky. Should you splash some excess solder on the coil form, it will lift off using solder wick or even a pipe cleaner.

After all the taps are in place, the coil leads can be trimmed to size and the coil can be mounted and soldered to the PC board. Don't get the tap leads too hot during this process or they may pop off the coil!

final details

Now that the PC board is complete, the mechanical details can be finished on the case. If desired, the case can be sanded with 320 or finer paper and a coat or two of your favorite color can be applied. I usually do this, as I invariably scuff up the front panel when making the required cutout.

After the paint is thoroughly dry, the rub-on lettering can be applied. This type of lettering and a few racing stripes can give your home projects a commercial look. A coat of clear lacquer will protect the lettering and level out the paint on the panel.

After all's dry, the PC board can be installed in the case and the sensitivity control, meter, and antenna jack mounted. The meter described in **table 1** has two small mounting holes. However, rather than use these, I mounted the meter with a couple of drops of epoxy cement between the front panel and the mounting tabs. This eliminated the mounting screws in the front panel and makes for a clean package.

The remaining wiring can be completed as shown in **fig. 3**, and the control knob can be installed on the sensitivity control. Give the unit a quick once-over against the schematic and drawings, then install the two AA cells, observing proper polarity. You are just about ready for the smoke test.

Turn the field-strength meter to **ON** and advance the sensitivity control slowly. As you advance clockwise, the meter will move slightly off the zero position in a positive direction. This is normal and indicates that all is well so far. Should the meter move slightly in the negative direction, the meter leads are reversed and must be transposed to correct the problem.

adjustment

The field-strength meter can be tuned with a signal source such as a signal generator, grid-dip meter, or your transmitter. If you use your transmitter, be sure to couple it to a dummy load so you don't radiate signals that will bother other Amateurs. In the absence of a dummy load, turn your transmitter output down and pick a time when the band is dead. Remember to listen on frequency before transmitting and keep each transmission short, ten seconds or less, and identify each transmission as a test transmission.

Choose your favorite frequency on the 80-meter band, and with the field-strength meter in the 80 position, inject a signal from the generator or use a pickup antenna for off-the-air or grid-dip signal source.

Adjust trimmer capacitor C1 for a maximum meter reading. Next, repeat this procedure with your signal source and the field-strength meter set for the 40-meter band. Trimmer capacitor C2 will provide adjustment for this band. Repeat this procedure for the 20-meter band, using C3 for adjustment. Capacitor C4 is for 15 meters, and C5 will be the trimmer for the 10-meter band. This completes the adjustment of the field-strength meter. The cover can be installed and you can sit back and admire your handiwork!

a final word

While the field-strength meter isn't a technological breakthrough or a state-of-the-art device by today's standards, it's a good project for getting started in the fascinating field of home brew. It will help you gain the skills necessary to go on to bigger, better, and more interesting projects. Most first-time builders will be able to complete the project during a weekend once the parts have been procured. And keep in mind that if you do have any problems or areas that are confusing you, ask one of your Amateur buddies for a bit of help. You will both enjoy the project!

The next project? Well, I hope to put an SWR meter together soon, and when we're armed with the field-strength meter *and* an SWR meter, there won't be a tune-up problem we can't handle. So do a good job on the field-strength meter and watch for the SWR meter soon.

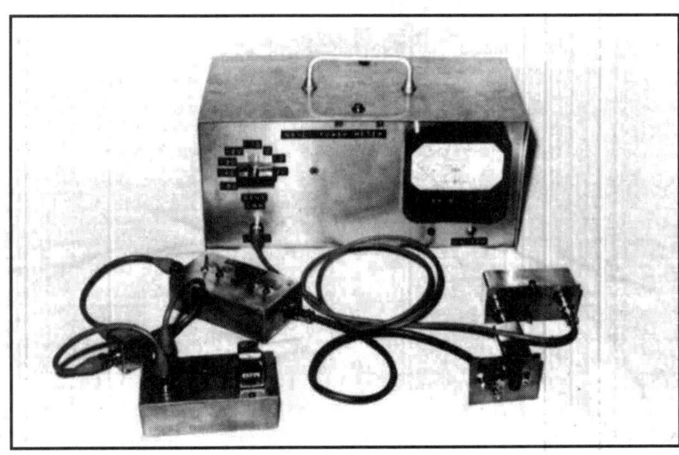

rf power meter

part 1 — instrument description and construction

Homebrewing Amateur gear can be an enormously satisfying experience. There's no reward like being able to say "I built it myself." Homebrew gear generally falls into one of two categories: it's either foolproof in construction and can be assembled with a reasonably good chance of having it work the first time you turn it on, or it's complex and needs test instrumentation for calibration and adjustment, which is simply not available to the average ham. This article is devoted to helping satisfy the need for good, accurate measurement instrumentation. This instrument ranks right along with your VOM, scope, and frequency counter in utility.

types of measurements

Frequency and amplitude of rf signals are, fundamentally, the two measurements made. Others are simply variations. Giant strides have been made with the introduction of simple and relatively inexpensive frequency counters priced within the Amateur's budget. Rf power measurements, however, have traditionally been limited to measuring voltage with an rf probe and VOM or a scope. In many cases this simply is not satisfactory due to the limited sensitivity of the instrumentation, or because the measurement is not easily adaptable to this technique.

This article describes an rf power meter that measures absolute power in dBm, 50 ohms, over a frequency range of 3.5-30 MHz. Measurement range, across a 50-ohm load, is between −60 and 20 dBm with an accuracy of ±1dB.* Armed with a frequency counter and the power meter and accessories described in this article, you can make a variety of scalar network measurements not ordinarily possible without access to sophisticated lab equipment. Best of all, the instrument isn't difficult to build or calibrate. Here are some of the things you can do, just to tickle your imagination, with the power meter and accessories. I have done most of these myself.

1. Evaluate oscillators, QRP transmitters, and small-signal amplifiers with respect to power output, flatness, harmonic distortion, and input return loss (VSWR).

2. Evaluate mixers for flatness, LO and rf suppression and conversion loss or gain.

3. Accurately measure your antenna VSWR down to 1.02 with an uncertainty of ±0.02. It's virtually impossible to get that kind of accuracy and resolution with a simple SWR meter.

*Verified with an HP-8640B signal generator.

By **Ralph H. Fowler, N6YC,** Route 1, Box 254, Pearl River, Louisiana 70452

4. Simplify filter construction. Synthesis of a 50-ohm LC preselector filter or other bandpass filters, for example, can be a ticklish job. Actual component values can depart considerably from theoretical values due to parts substitutions, tolerance unknowns, and stray capacitances. The power meter, combined with a counter and tunable source, enables alignment and evaluation of LC filters with respect to insertion loss, passband shape (ripple), shape factor, and skirt attenuation (up to 80 dB with a 20-dBm tunable source). If your source is stable enough you can even characterize crystal filters.

5. Adjust interstage and output matching networks (50 ohms). Even your transmatch can be adjusted with the power meter and a directional bridge.

6. Make similar measurements on your kilowatt rock-crusher by adding a simple in-line directional coupler (not described here) to your transmission line. You can, naturally, calibrate the coupler with the power meter.

As an academic exercise you can even measure a-m percentages between 25 and 100 per cent to within a few per cent accuracy. Some of these measurements, plus a few provocative ideas, will be discussed in part 2 of this article.

The power meter consists of three basic parts: a biased Schottky diode *broadband* detector (note the broadband emphasis — this becomes extremely important in the measurements described in part 2), a 37-dB broadband preamp, and a relay-switched 0-70 dB attenuator. These three elements are cascaded as shown in **fig. 1**. Each element is described in turn.

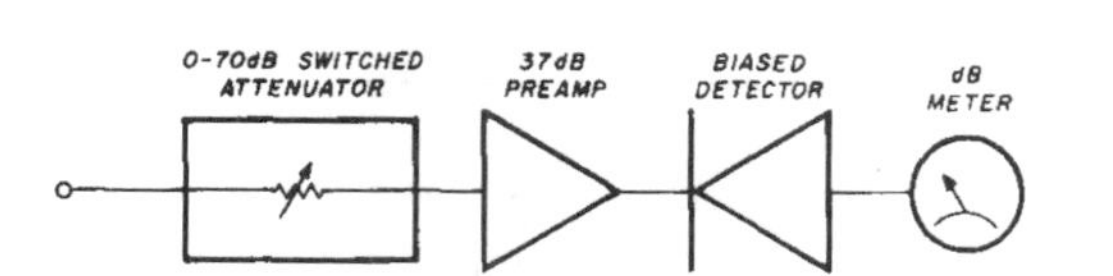

fig. 1. Power meter elements, consisting of broadband detector, broadband preamplifier, and relay-switched attenuator.

Operation is simple. The step attenuator and preamp set the level of the signal being measured to the proper amplitude for the square-law detector. This is essential since the square-law amplitude range of the detector is relatively limited, typically to between −40 and −10 dBm. Square-law, you may recall, means that the output *voltage* of the detector is proportional to the input *power*. Outside of this −40 to −10 dBm range, the detector does not follow the square-law relationship and will be uncalibrated. Even if the square-law range were greater than 30 dB, it would not be useful in this simple scheme since the meter itself can display only 10 dB of range with acceptable resolution (1 dB) without a logging amplifier included in the circuit. With these complicating factors in mind, I chose to accept a 10-dB measurement range per step of the input attenuator. The only inconvenience is that you have to switch the attenuator to bring the reading within the 10-dB display range of the meter. The meter, incidentally, has a 0 to −10 dB calibration scale. If you can't find a meter calibrated in dB (10 $\log_{10}$) it's easy to add your own calibration marks. A calibration table is shown in **fig. 2**.

the detector

The heart of the power meter is the broadband square-law detector. The original circuit design was described by Wes Hayward, W7ZOI, in *Solid State Design for the Radio Amateur*, and the reader is encouraged to consult this excellent reference for more details. This detector is shown in **fig. 3**.

Construction is not particularly critical but should be done on PC or copper-clad Vector board to permit operation above 30 MHz, as described later. Leads up to and including the diodes should be kept short. The detector assembly should then be housed inside a compartment or BUD™ box.

The diodes are Schottkys, which are inherently better matched than conventional types and are essential. In addition, these diodes have a better sensitivity than ordinary silicon diodes, thus providing up to −40 dBm sensittivity with careful biasing.

A variety of op-amps were tried in this circuit with good results, including the 741, LM301A, and LM312. Doubtless there are others which would work equally well.

With this design and a little attention paid to short leads at the front end, my unit had a perfect square-law response between −23 dBm and −13 dBm and was virtually flat up to approximately 500 MHz.* An extra 10 dB of sensitivity could have been achieved by calibrating the detector for the −33 dBm to −23 dBm range, but op-amp drift effects began to show up here. Operation between −23 dBm and −13 dBm is drift free.

With this kind of frequency response, it would pay to provide a jumpered detector input on the back panel of the power meter to allow power measurements at vhf and above using the detector alone.

To access the detector (thus bypassing the switched attenuator and preamp, which are much more frequency limited) you could simply remove the jumper coax and plug right into the detector. I don't operate above 30 MHz so I did not include this in my set.

*Compared to a Hewlett-Packard 8640B signal generator as a reference.

dB	percent full scale
0	100
−1	79
−2	63
−3	50
−4	40
−5	32
−6	25
−7	20
−8	16
−9	13
−10	10
−20	1

fig. 2. Data for calibrating the dB meter.

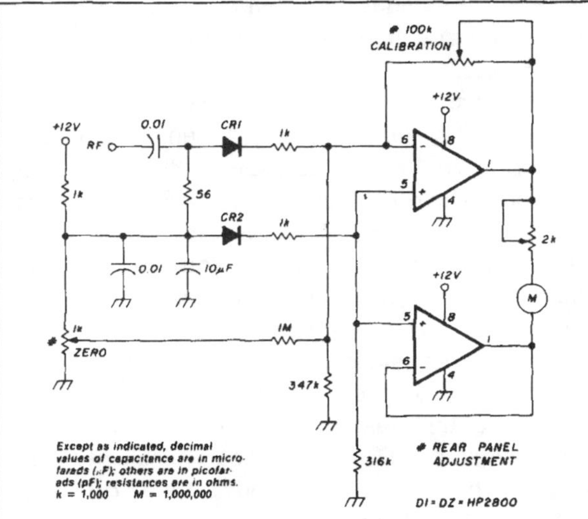

fig. 3. Biased detector schematic. CR1, CR2 are Schottky diodes. Meter is 0-100 microamps or higher. Original circuit was described by Wes Hayward, W7ZOI.

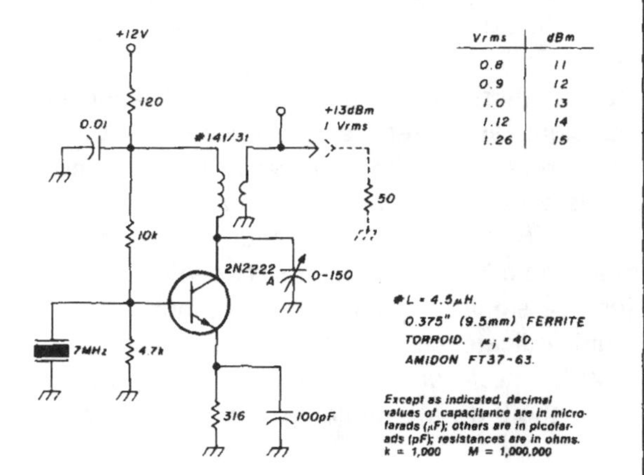

Vrms	dBm
0.8	11
0.9	12
1.0	13
1.12	14
1.26	15

fig. 4. Oscillator for calibrating the detector and preamp. Circuit delivers 13 dBm.

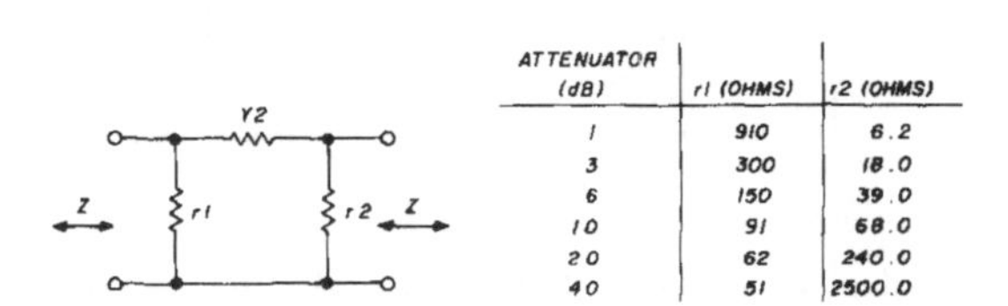

ATTENUATOR (dB)	r1 (OHMS)	r2 (OHMS)
1	910	6.2
3	300	18.0
6	150	39.0
10	91	68.0
20	62	240.0
40	51	2500.0

fig. 5. Symmetrical pi attenuator for bringing the 13-dBm oscillator signal into the range of the detector. *Z = 50 ohms resistive.*

calibration

Calibration of the detector is an intermediate step that should be accomplished to ensure that the detector is working properly and to provide a means of measuring the gain of the preamp once it is built. When the power meter is completed, it will then require only minor readjustment to bring it into absolute calibration.

Since the ultimate accuracy of the power meter depends on the reference used to calibrate it, it is crucial that care be taken in its selection. The ideal reference is an accurately calibrated signal generator that can be tuned over the full operating band. Usually this is impractical, so the next best alternative is to simply build a fixed frequency, high-frequency oscillator with an output of about 13 dBm into 50 ohms, corresponding to 1 Vrms. This level is easily measurable with an rf probe and VOM. This oscillator will be used to calibrate the detector and the preamp. **Fig. 4** shows a simple oscillator that will deliver 13 dBm. The relationship between output voltage and power in dBm (50 ohms) is given so that you should be able to compute the exact power output of your oscillator once the rf voltage is measured. Just remember that you must measure the voltage with the output terminated into 50 ohms (a 50-ohm noninductive resistor will do nicely as a temporary load).

The 13-dBm signal must now be attenuated to bring it into the range of the detector. Adding a total of 26 dB of attenuation to the oscillator output supplies the −13 dBm reference required. This reference is chosen since the detector is preceded by a 37-dB preamp, thus giving the desired −50 dBm (full scale) sensitivity. **Fig. 5** shows component values used for constructing the attenuators. Small adjustments to the oscillator collector voltage can also be made for small corrections to output level.

With no input connected to the detector, the zero adjustment should first be set to bring the meter reading to zero. Next connect the −13 dBm reference signal to the detector and adjust the calibration trimpot to set a full-scale reading. The meter multiplier pot will also have to be adjusted to accommodate your particular meter sensitivity. Disconnect the input signal and reset the zero adjustment. Continue alternating between the zero and calibration pots to complete the initial detector calibration.

To verify square-law operation of the detector, simply add a 3-dB attenuator between the − 13 dBm reference signal and the detector input. The meter should indicate half scale, corresponding to 3 dB below full scale or − 16 dBm absolute power. Replacing the 3-dB attenuator with a 10-dB attenuator should bring the meter indication to 0.1 of full scale, corresponding to 10-dB below full scale, or − 23 dBm. The detector is now absolutely calibrated over the − 23 dBm to − 13 dBm range and will be used subsequently (along with the attenuators) to calibrate the preamp.

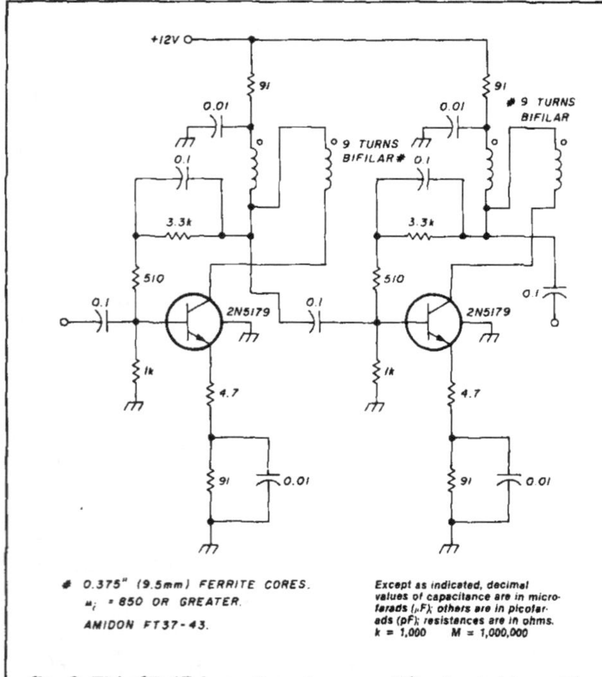

fig. 6. This 37-dB broadband preamplifier is stable, with emitter and shunt feedback. Measured input VSWR is less than 1.2.

preamplifier

The preamp used ahead of the detector supplies 37 dB of amplification using two 2N5179A transistors in a broadband configuration (**fig. 6**). Construction is not critical, and the amplifier should be stable with the emitter and shunt feedback. The feedback also provides a good input impedance match to 50 ohms. Measured input VSWR was better than 1.2. This is important to provide a good match to the step attenuator, which precedes the preamp. If you use another preamp circuit remember not to compromise this parameter.

The broadband transformers should be wound on high-permeability ferrite toroids to ensure the full bandwidth. Using the values shown, the 3-dB bandwidths were 3.0 and 77 MHz, and the 1 dB bandwidths were 3.5 and 37 MHz. Adding a few extra turns should increase the lower 3-dB frequency to allow coverage of the 1.8 MHz band if desired.

Once completed, the preamp should be tested to ensure that 37-dB gain is achieved. More or less gain will require reaccomplishing the detector initial calibration so that the preamp and detector combination have the required − 50 dBm full-scale sensitivity. Assuming that the gain is measured to be 37 dB, ± a few dB, this final adjustment can be deferred until the power meter is completed.

To measure preamp gain, construct a 40-dB fixed attenuator and connect it to the − 13 dBm reference output. This − 53 dBm signal is now applied to the preamp input and the preamp output connected to

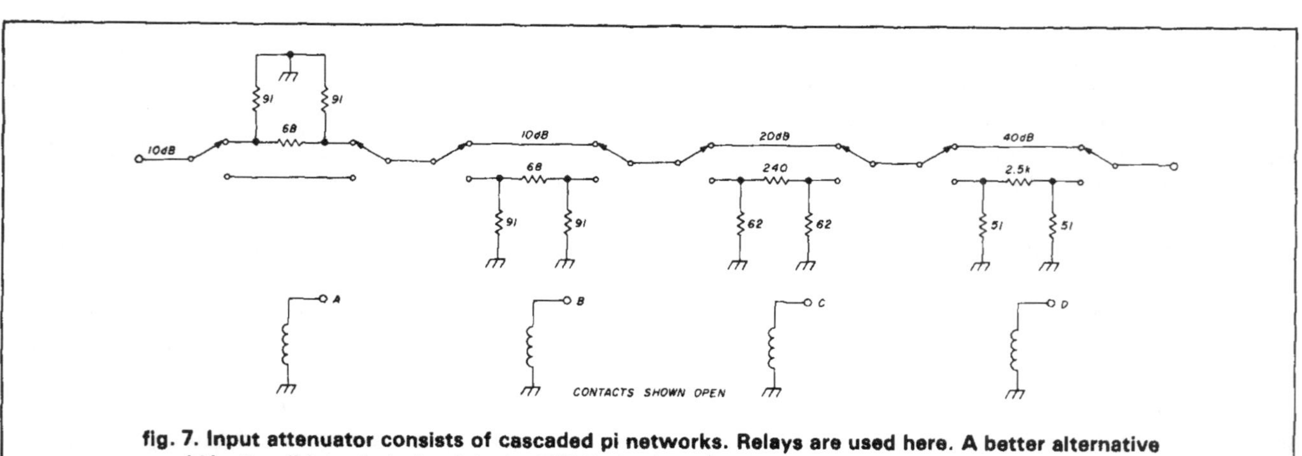

fig. 7. Input attenuator consists of cascaded pi networks. Relays are used here. A better alternative would be the slide-switch circuit in the 1980 *Radio Amateur's Handbook*.

the calibrated detector and meter. The meter should indicate −16 dBm (half scale, assuming full scale is −13 dBm), corresponding to 37 dB of preamp gain.

input attenuator

The switched input attenuator was unquestionably the most difficult part of the project. **Fig. 7** shows the completed unit. A number of prototypes were built with varying degrees of success. The problems encountered most often were (1) flatness variations (a few dB greater than the 1-dB design goal) at the high frequency end of the range, and (2) degraded input-to-output isolation at attenuation levels above 70 dB despite my attempts at shielding. This limits operation of the power meter to 20 dBm (where 70-dB attenuation is used).

As an alternative, you might consider building the attenuator using small slide or toggle switches, perhaps even external to the power meter. Such a unit would have other applications as well as being useful in calibration of the detector and preamp. The 1980 edition of the *Radio Amateur's Handbook* shows construction of a shielded 147-dB step attenuator using simple slide switches; it looks like a good prospect. Using this proven design, you could probably remedy the isolation problems I encountered and increase the measurement range to 30 dBm or higher.

If building the relay-switched attenuator still appeals to you, a couple of construction points are worth remembering:

1. A single-side PC or Vector™ board layout is desirable for minimizing isolation problems and frequency response resulting from impedance mismatches. While the 50-ohm environment of a transmission line is certainly not preserved, I found that by using small PC-mount relays on a Vector board and a physically small layout, an input VSWR less than 1.4 and ±1 dB flatness up to 30 MHz was attainable, even at the 0 dB attenuation setting (all attenuators switched out). Although not shown in my unit, shielding between attenuator sections would still be good insurance against isolation problems despite my experience.

The resistors used to build the attenuator sections are film type, 1 per cent tolerance, 1/4-watt units mounted close to the relay contacts. Physically small, low-wattage resistors reduce the capacitance to ground and help keep the flatness variations reasonable.

2. The attenuator sections consist of 10-, 10-, 20-, and 40-dB sections switched in as required to provide 0-70 dB of attenuation in 10-dB steps. A 10-dB section is used as the input section and remains switched in at all times except at the highest sensitivity setting of the power meter, −50 dBm. This helps establish the good input VSWR and is important when characterizing devices such as 50-ohm LC filters, which require a good 50-ohm termination.

Considering the isolation problems mentioned earlier, measurements at levels above 20 dBm should be done with fixed attenuation ahead of the power meter. Of course, this requires mentally adjusting the readings for the extra attenuation. Remember to construct the external attenuators with higher wattage resistors to dissipate the increased power.

To provide switching the input-attenuator relays, a 4p8t switch or sp8t switch (both make before break) with a diode switching matrix can be used. Since 4p8t switches are not too common, I chose the latter technique. The diode switching matrix is shown in **fig. 8**. The diodes used should be germanium or other low barrier diodes to minimize voltage drops and maintain switching reliability, especially if 5-volt relays are used. If 12-volt or higher relays are used, silicon diodes should be ok.

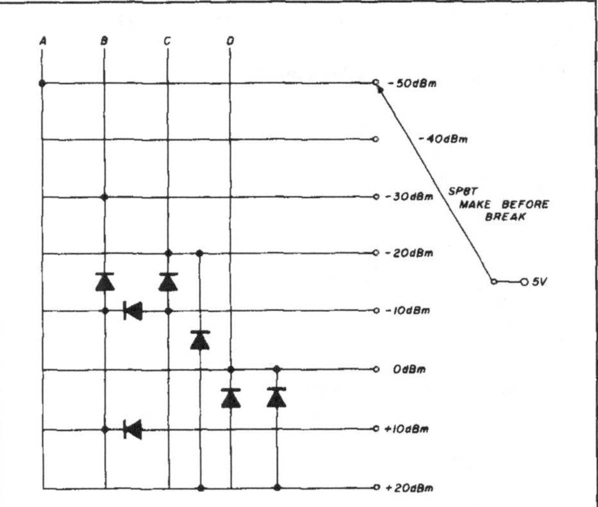

fig. 8. Diode matrix for switching input-attenuator relays. Single pole, 8-throw switch was used.

putting it all together

Once the detector, preamp, and attenuator modules have been constructed, they should be cascaded and the final calibration can be performed. If all goes well, the last step is to mount it all in a suitable box.

Final calibration of the power meter requires simply connecting the −13 dBm reference signal to the input, setting the input attenuator to the −10 dBm full-scale range (corresponding to 40-dB attenuation if an external switched attenuator is used), and tweaking the detector calibration pot to set the meter indication to −13 dBm, or half scale. Power-meter switching accuracy can be verified by using fixed at-

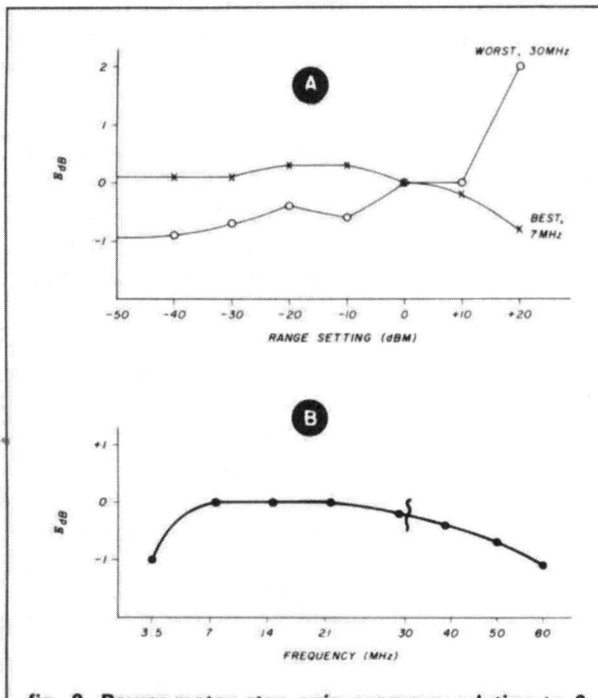

fig. 9. Power-meter step gain accuracy relative to 0 dBm (A) and relative frequency response (B). In-band flatness is less than ± 0.1 dB on all bands.

tenuators to attenuate the reference signal and comparing this known level to the power-meter reading. **Fig. 9** shows the flatness and step accuracy that I achieved with my power meter.

Performance testing was done using a Hewlett-Packard 8640B signal generator and fixed attenuators accurate to better than 0.1 dB. Keep in mind that, while you probably won't be able to perform such quantitative performance tests on your power meter, the *in band* flatness could be expected to be no worse than a few tenths of a dB, and the absolute accuracy could reasonably be less than a few dB, assuming the rf probe and VOM used in the detector calibration were accurate to within 10 per cent. Relative amplitude accuracy over the full −60 dBm to 20 dBm range at a fixed frequency should be even better and depends primarily on the precision of the attenuator sections used in the input attenuator. Not bad at all.

The second part of this article will deal with some of the measurements you can make using the power meter. Details of the construction of a return loss bridge, useful for making accurate measurement of low VSWR, and other measurement accessories will be given. We will look at some important measurement considerations aimed at improving the accuracy of your measurements.

Accessories for use with the rf power meter.

rf power meter

Part 2: Measurements and measurement accessories

Part 1 of this article addressed the construction of an rf power meter capable of measuring amplitudes between −60 dBm and +20 dBm from 3.5 to 30 MHz. This rf power meter is the basis of a measurement system that can be used to make many useful network measurements, as explained in part 1. Some of these measurements are described in detail here.

accessories

Depending upon the measurement, various accessories are required, some of which you may have. Construction of others is covered in other literature sources, such as the *ARRL Handbook*[1] and *Solid-State Design for the Radio Amateur.*[2] I will list only the system components required to make the measurement and will describe the measurement itself. The photo shows accessories that I have built and found useful.

One accessory required for reflection measurements, as in measuring antenna VSWR and amplifier input VSWR, is a *directional bridge*. Hayward[2] briefly mentions the usefulness of such a device, but I haven't found many references that provide details on how to use it or what the limitations are in its use.

One of the advantages in using the bridge described here, as opposed to other "signal-separation" devices, is its simplicity, which results in the fact that it doesn't have to handle large amounts of transmitter power as in other coupler circuits found in SWR meters. This bridge is designed to be used with a low-level signal generator, VFO, or other oscillators delivering less than about +10 dBm (10 mV). Higher power will cause damage or may generate distortion, which could adversely affect the measurement.

directional bridge

The bridge is shown in **fig. 1**. Basically, it's a Wheatstone impedance bridge circuit with an unbalanced output to drive the power meter. Construction is uncomplicated. It should perform well up to and beyond 30 MHz with no adjustment or calibration. If your interests are limited to 30 MHz, no construction points are critical: simply use a reasonably close parts placement, high-quality components (small ¼-watt, film-type, 1 percent or better resistors preferred), and you should have no problems. If you anticipate working above 30 MHz, up to approximately 150 MHz, the bridge can be modified slightly to extend its range.

what's a network measurement?

Don't be concerned. You've probably been making them for a long time already but have been using other names such as VSWR or gain measurements. There are two basic types of network measurements that are made: transmission measurements and reflection measurements.

By Ralph H. Fowler, N6YC, Rt. 1, Box 254, Pearl River, Louisiana 70452

In simple terms, we apply a stimulus signal, generally a low-level sine wave, sometimes swept, to the input of a one-port (antenna) or a two-port (amplifier) device, then we take measurements to see what is reflected from the input (reflection measurement), or what comes out of the output (transmission measurement).

Network measurements are, strictly speaking, linear measurements. That is, we measure the output at frequencies corresponding to the input frequency only. In many cases, however, we can look at nonlinear effects by changing our test procedure slightly and do things such as making an harmonic analysis of an amplifier. **Fig. 2** shows the two types of basic network measurements that can be made with the power meter and accessories. First, let's look at reflection measurements.

reflection measurements

The objective of a reflection measurement is to measure the amount of power reflected from a device, as shown in **fig. 3**. The reflected power is usually referenced to the incident power, and the ratio is expressed as VSWR, σ; reflection coefficient, ρ; or return loss, RL. These three quantities are all different ways of expressing the same thing; that is, how well the device is matched to the transmission line carrying the power. All three are scalar values; that is, magnitude only without phase angle. Any one of the three can be converted to the others using the expressions:

$$\rho = 10^{-\left(\frac{RL}{20}\right)} \quad (1)$$

and

$$\sigma = \frac{1+\rho}{1-\rho} \quad (2)$$

These equations are important since, as you will find, the measurement system described here measures directly in dB of return loss (the advantages will be discussed), which can then be converted to other units as desired. Antenna measurements, for example, are measured in return loss values, dB, which can then be converted into σ values to put them into a form in which they are usually expressed.

Most of us are used to thinking in terms of VSWR and will want to do this in other measurements as well. While you might object to the bother of converting from RL to σ, especially since SWR meters give this information directly, you'll find that the advantages of measuring RL far outweigh the minor inconveniences of conversion. And who knows, once you adapt to thinking in terms of RL, you may want to forget VSWR.

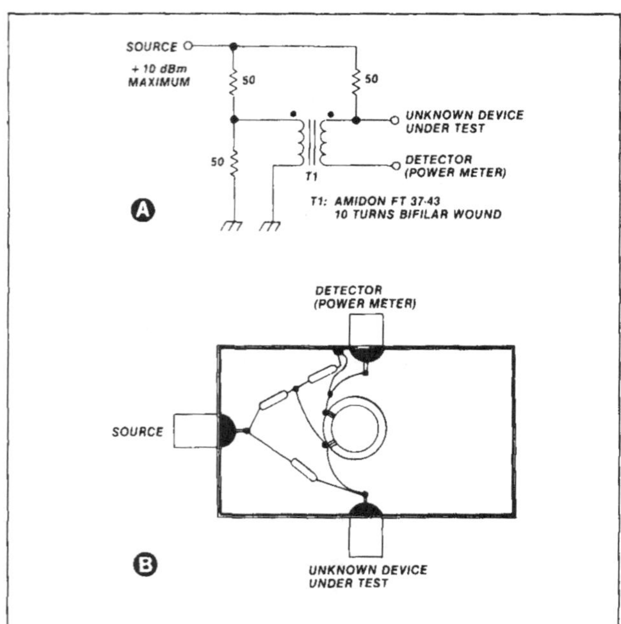

fig. 1. Directional bridge. The circuit (A) is basically a Wheatstone impedance bridge with an unbalanced output to drive the power meter. A suggested parts layout is shown in (B). Construction tips are given in the text.

making a return-loss (RL) measurement

A return-loss (RL) measurement is made in two steps: calibration and measurement. This measurement is analogous to an SWR meter measurement in which the meter sensitivity is adjusted to give a full-scale reading of the forward power (calibration), then switching to read the reverse power as a function of the forward power (the measurement) with a meter readout directly in VSWR.

As a specific example, suppose we want to measure the VSWR and bandwidth at the *VSWR* = *3.0* points of our antenna. This measurement system is shown in **fig. 3**. Calibration is performed by connecting a standard (a known VSWR) to the unknown port of the directional bridge to establish the 0-dB relative reference level on the power meter. The absolute power is unimportant; it serves only as a level with which the subsequent measurement is compared.

The standard used is an open circuit with a VSWR of ∞, corresponding to a perfect reflection where all of the forward going power is reflected. (We would thus say that the open circuit had a 0-dB RL). Assuming that we had 0-dBm forward power, we would measure −6 dBm at the "detector" port of the bridge, corresponding to a 6-dB loss of the reverse power as it passed through the bridge to the "detector" port. A value of −6 dBm would then be our calibration value and should be logged or otherwise remembered.

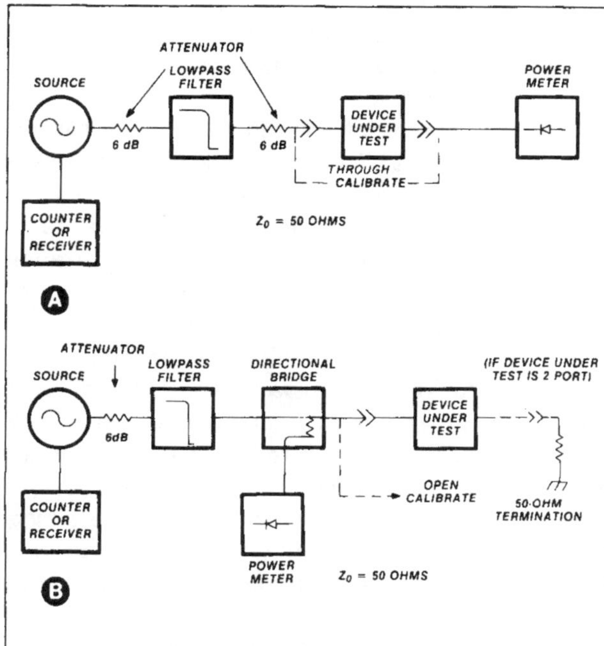

fig. 2. Instrumentation for basic network measurements. Transmission measurement setup is shown in (A); (B) shows instrumentation for reflection measurements.

With the antenna connected to the unknown port, a value is now measured that, when compared to the calibration level, −6 dBm, represents the return loss of the antenna in dB. If the power-meter reading were −15.6 dBm, for example, this would correspond to a 9.6 dB *RL*; 9.6 dB *RL* converts to a VSWR of 2.0, using the conversion equations. By tuning the source across the band you could then plot *RL* versus frequency and determine the *VSWR* = *3.0* frequencies. Assuming the source were reasonably flat as you tuned, the calibration would remain good across the band.

why measure *RL* instead of VSWR directly?

There are three important reasons: resolution, accuracy and sensitivity.

Accuracy is determined to a large extent by how well you can read the meter or indicator, (scale resolution), and how well the directional bridge does its job (there are some other accuracy considerations that will be covered later). SWR meters, like the system described here, use some kind of signal-separation device (bridge, monimatch coupler). Therefore, other things being equal, they perform equally well in this respect. SWR meter resolution, however, is very poor at low values of VSWR below, say, 1.5, where the meter is at 20 percent of full scale. Below 1.5 the meter scale is compressed so much that about the best you can do when adjusting the match is to tune for a null. Dynamic range of the SWR meter (without "expanded" scales) is thus not much greater than about 10 dB with good resolution.

Return loss measurements using the power meter, however, do not suffer from a lack of resolution, even at extremely low values of VSWR (<1.01) because of the increased dynamic range of the power meter, 80 dB. And to make a point, resolution is sufficient so that we could theoretically measure VSWRs of 1.0002 with the system described here if we were not limited by the directional bridge.

Admittedly, while there are few of us who really need to measure antenna VSWRs below, say, 1.5 or so, there are other reflection measurements which can't be made with most SWR meters because of their lack of sensitivity. Input VSWR of small-signal amplifiers, for example, will not admit the power levels necessary to make most SWR meters operate. Due to the increased sensitivity of the power meter, you can drive small-signal amplifiers with as little as −34 dBm and still measure a 1.2 VSWR (20 dB *RL*).

So while the SWR meter is an extremely useful device for continuous monitoring of VSWR under actual transmit conditions (something this system can't do), initial evaluation of antennas, amplifiers, and other devices can best be done with the power meter and bridge.

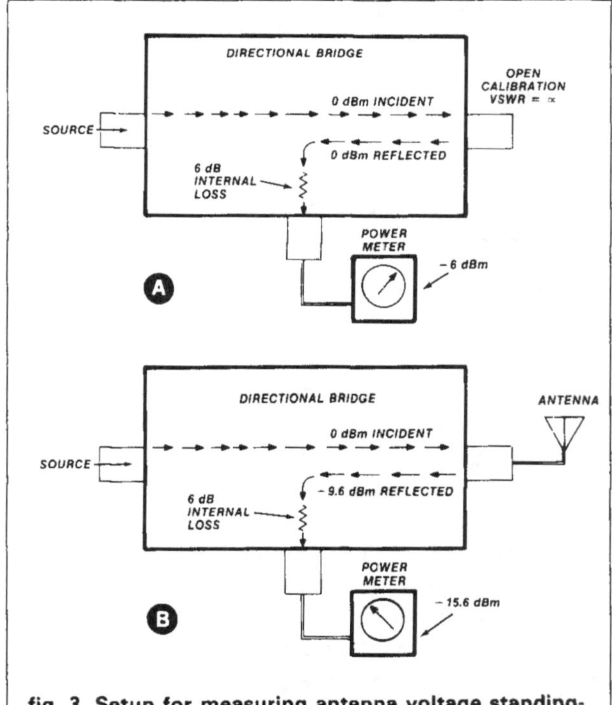

fig. 3. Setup for measuring antenna voltage standing-wave ratio. Reflection calibration is shown in (A); reflection measurement in (B).

how accurate is the measurement?

A detailed analysis of the accuracy of this measurement is not within the scope of this article. However, some general guidelines are given to make sure you make the best measurement possible.

There are two main sources of error in the return loss measurement described: source match and bridge directivity. These errors are limitations on all reflection measurements, whether with an SWR meter or the system described here; see **fig. 4**.

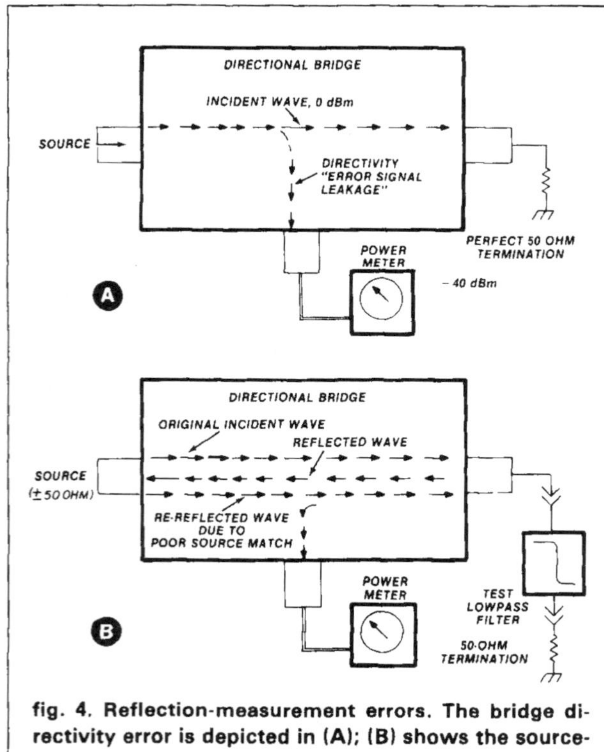

fig. 4. Reflection-measurement errors. The bridge directivity error is depicted in (A); (B) shows the source-match error.

Bridge directivity is a measure of how well the directional bridge separates the forward- and reverse-going waves. It is generally specified in dB. As an example, assume the directivity of the bridge is 40 dB. Practically speaking, this means that if our device under test (the antenna) were a perfect 50-ohm antenna with an infinite RL, we would still measure approximately 40 dB RL, even though there were no reflected waves traveling along the transmission line. Consequently the percentage errors in the measurement become increasingly large as the device RL approaches the bridge directivity. In this example any return loss measurements greater than approximately 35 dB would be unreliable. Fortunately, a 35-dB RL (VSWR 1.03) is usually well beyond the practical requirements of most measurements. The directivity of the bridge described here is greater than 40 dB up to 30 MHz.

The other source of error in reflection measurements is *source-match error*. Source match is essentially a measure of how close the output impedance of the source is to the characteristic impedance of the transmission line (50 ohms). Its importance becomes apparent when you consider the effect of any reflected energy traveling back to the source only to be reflected from the source, becoming part of the new forward-going wave. The forward-going wave now consists of two components (an infinite number if we count the re-reflections) to create an uncertainty in establishing the forward-going power in the calibration step. If the device we are measuring accepts most of the forward-going power (VSWR is low or RL is high), then this source of error becomes negligible.

If, on the other hand, the device has a reasonably high VSWR (a low RL), then source match error becomes appreciable. A source with a VSWR of 2.0, for example, used in the measurement of a device with an actual VSWR of 4.0 would yield a measured VSWR lying between 2.8 and 6.0. The value you would measure depends on the relative phasing of the component waves. When measuring the VSWR of reflective devices such as filters (outside of the passband), source match errors become significant. Ideally, then, the source used to make the measurement should have a 50-ohm (Z_0) output impedance to reduce source-match errors.

Fortunately, there are simple fixes to improve the source match of a given source. Simply adding a 6- or 10-dB attenuator at the output of the source brings the effective output impedance of the source/attenuator combination closer to 50 ohms. In the above example adding a 6-dB attenuator to the 2.0 VSWR source improves the effective source match to 1.4, reducing the uncertain window to 3.4-4.8 in a $VSWR = 4.0$ measurement.

The directional bridge has 6 dB of attenuation built in by design and satisfies this requirement on its own. However, when making transmission measurements remember to add 6 or 10 dB attenuation following the output of your source/filter, particularly when measuring large VSWRs (small RL). More about this later.

other notes on reflection measurements

As mentioned in part 1, the power meter is a broadband detector, one which responds equally well to the stimulus frequency, f_0, coming from the source or to any of the harmonics of the source or the device being tested. Because of this, harmonic filtering of the source is a must, and suitable lowpass or highpass filters must be added to attenuate the source's harmonics adequately. Lowpass filters are recommended because of their relative ease of construction.

The skirt attenuation of the filters outside the passband should be sufficient to reject unwanted harmonic frequencies. The actual value of attenuation required will depend on the dynamic range required in the measurement, although 40 dB or greater harmonic rejection is usually enough, depending on the harmonic content of the source itself. In my system I use seven-pole LC lowpass filters with cutoffs at 5.8, 9.6, 15. 7, 23.1, and 30.4 MHz following my tunable generator to cover the high-frequency bands. These, combined with the inherent harmonic content of the oscillator itself, yield a source with a minimum −50 dBc harmonic content, sufficient for most of my measurements. LC lowpass filter construction is covered in detail in the *ARRL Handbook.*[1]

The attenuators shown surrounding the filters absorb the multiple reflections that result from out-of-band energy reflected from the filters and provide a proper termination for the filters and test amplifier. They should not be omitted from the circuit.

transmission measurements

The other type of measurement that can be made with the power meter is the transmission measurement. As the name implies, it is simply a measurement of the amount of signal power transmitted by a device under test. The measurement can be as simple as measuring the passband ripple and the stopband rejection of a filter, or as complex as meauring the gain, compression level, harmonics, intermodulation distortion (IMD) and third order intercept (TOI) of a small-signal amplifier.

A basic transmission measurement is shown in **fig. 2**. This setup would be suitable for characterizing a filter, for example, and the details of this measurement should be evident. The measurement of a small-signal amplifier's characteristics can be more involved and, therefore, will be discussed as an example of the measurement procedure. This measurement is shown in **fig. 5**. Note the liberal use of attenuators to buffer multiple reflections from the reactive filters, improve the effective source match, and provide a proper termination for the filters and test device.

calibration

Calibration consists of simply bypassing the test device (amplifier) with a through connection and noting the power level on the power meter. This should be done with the 0-40-dB step attenuator adjusted to set the input power level low enough to prevent overdriving the amplifier.

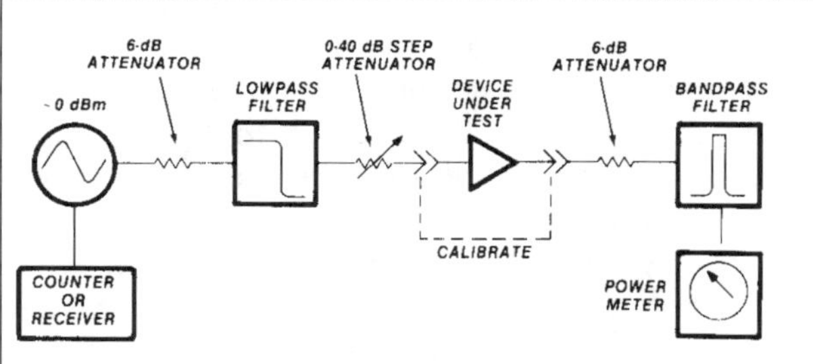

fig. 5. Instrumentation for small-signal amplifier gain, compression, and harmonic-distortion test.

measurement

Gain is measured by connecting the test amplifier and comparing the output power level with the calibration value (the input power level). In cases where the harmonics are known to be more than 10 dB below the fundamental level, the filter following the test amplifier can be omitted without serious errors in the gain measurement. If, for example, the harmonic power were 10 dB below the fundamental level, the error resulting from omission of the filter would be only 0.4 dB in the gain measurement.

The 1-dB gain compression level can be determined by increasing the drive to the test amplifier to a point where the gain, as measured in the preceding step, decreases by 1 dB. At this point the test amplifier is entering the nonlinear region and is generating enough distortion so that by choosing an appropriate bandpass filter we can measure the harmonic distortion. Back off on the drive a bit, however, since the measurement should be made with the amplifier operating in the linear mode and not near gain compression.

Measurement of *nth*-order harmonic distortion required choosing a bandpass filter that will attenuate the fundamental to a level at least 10 dB below that of the harmonic to be measured, as shown in **fig. 6**. We could not expect to be able to measure harmonic

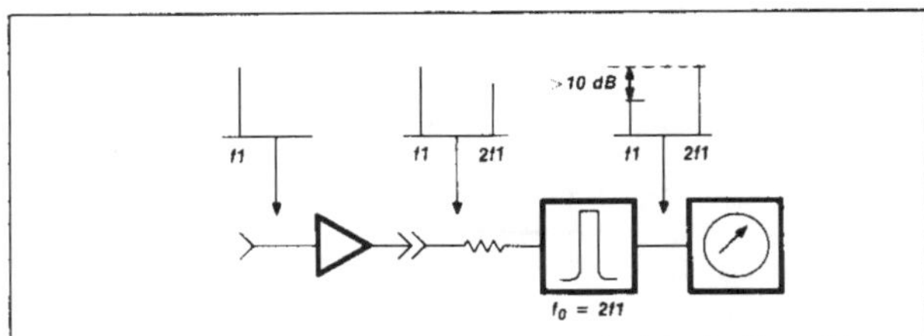

fig. 6. Attenuating the fundamental signal, f1, with a bandpass filter.

distortion 50 dB below the fundamental using a bandpass filter with only 30 dB of rejection of the fundamental, for example. Conventional Butterworth designs using from three to five poles as described in the *Handbook*[1] will provide sufficient attenuation in most cases.

Generally speaking, extreme bandpass-filter selectivity is not required, even in low-distortion measurements, for testing small-signal amplifiers since we can arbitrarily increase the distortion from the test amplifier by increasing the drive level to the input (but staying away from gain compression). **Fig. 7** illustrates this principle.

Note that in **fig. 7A** at an output level of −10 dBm, for example, the second harmonic is at −50 dBm, or relative to the fundamental at f1 we say that it is −40 dBc. If we increase the drive level to the test amplifier by 10 dB, the fundamental output also increases by 10 dB to 0 dBm as shown in **fig. 7B**.

But note that the second harmonic has increased by 20 dB and is now only −30 dBc relative to the fundamental. So given a three-pole bandpass filter with a 40-dB rejection over one octave of bandwidth, it would not be possible to make an accurate second-harmonic measurement with the drive level set as in **fig. 7A**; but by increasing the drive level as in **fig. 7B** we could make the measurement. Once the distortion has been measured at the artificially high drive level, the distortion at any nominal drive level can be computed from the equation:

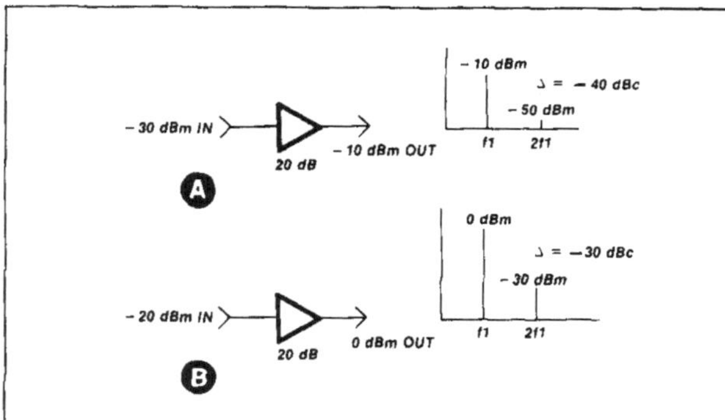

fig. 7. Effect of drive level on amplifier harmonic distortion at 2f1. (A) and (B) respectively show drive at −30 dBm and −20 dBm.

$$HD,_{dBc}\,(nominal\ drive\ level) =$$

$$HD,_{dBc}\,(meas) + (N-1) \times (change\ in\ drive\ level) \quad \textbf{(3)}$$

where N is the order of distortion (that is, second, third, etc.). If, for example, the second harmonic were measured to be −40 dBc at a drive level of 0 dBm, it would compute to be −80 dBc at an input drive level of −40 dBm.

Again, stay away from gain-compression levels by at least a few dB, and remember that you can't measure test-amplifier distortion levels less than that generated by the source used and present at the input of the test amplifier. Also remember to take into account any insertion loss of the bandpass filter (it can be measured separately with the power meter) in your measurement of harmonic distortion.

The relationship between the fundamental and second-order harmonic distortion as the drive level is changed can be generalized to *nth*-order distortion as well: changing the drive level by 1 dB causes the *nth*-order distortion product to change by *nx1* dB. This fact is extremely useful in positive identification of the signal you think you are measuring. For example, when measuring the third harmonic, changing the drive by 1 dB should cause a 3-dB change in the power-meter reading.

IMD and TOI measurements

IMD measurements follow the same general guidelines established in harmonic distortion measurements. One complication arises in the fact that we must now drive the test amplifier with two equal amplitude rf "tones" and somehow provide filtering to reject the fundamentals at f1 and f2 and pass the third-order distortion products at 2f1 ± f2 or 2f2 ± f1. **Fig. 8** shows the measurement setup.

Ideally we would like to present to the test amplifier two rf signals separated by, say, 10 or 20 kHz and measure the IMD product, which would be separated from the test signals by the same 10 or 20 kHz as

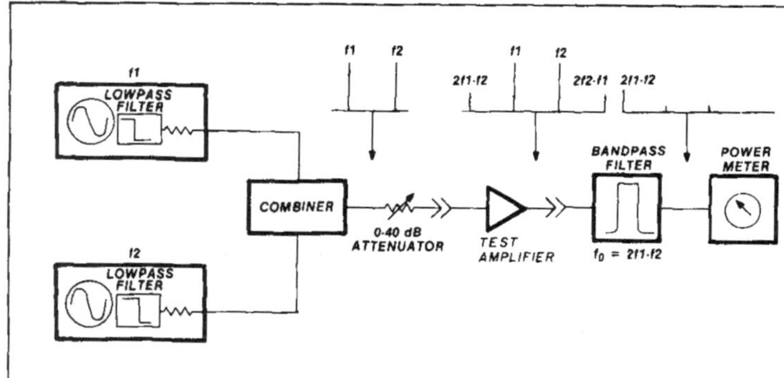

fig. 8. Test setup for measuring intermodulation distortion.

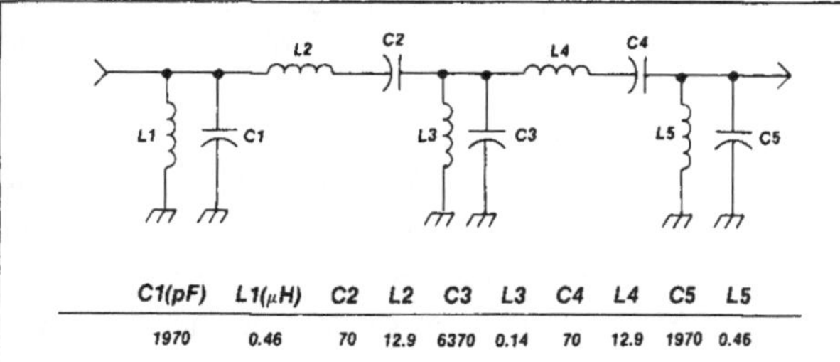

C1(pF)	L1(μH)	C2	L2	C3	L3	C4	L4	C5	L5
1970	0.46	70	12.9	6370	0.14	70	12.9	1970	0.46

fig. 9. A 5.3-MHz Butterworth bandpass filter for testing intermodulation distortion. Filter-component values provide 50-dB attenuation at 7 MHz and above.

shown in **fig. 8**. This tone spacing is desirable since it simulates the situation that exists at the input to a receiver's preselected rf amplifier in crowded band conditions. However, without heterodyning the test amplifier's output to a lower frequency where filtering can be easily accomplished, it becomes necessary to separate the tones by at least a few MHz to allow using LC bandpass filters to select the IMD product and provide attenuation to the fundamentals at f1 and f2. Consequently, the measurement described here is limited to broadband amplifiers or amplifiers with a bandwidth large compared with the tone spacing. Fortunately, tests made with a spectrum analyzer using closely spaced tones show results that agree within a few dB compared with measurements made using tones spaced by a few MHz on broadband test amplifiers.

In my test setup I used two crystal oscillators at 7.0 MHz and 8.7 MHz and measured 2f1 – f2 at 5.3 MHz. **Fig. 9** shows filter component values for a 5.3-MHz bandpass filter that provides 50-dB attenuation at 7.0 MHz and above. As previously stated, the filters and test amplifier must see 50 ohms to provide a good termination. Also remember to verify the test by changing the drive level by a given amount, using the step attenuator, and look for a three times change in the power meter reading.

Once the third-order IMD is measured at the arbitrary drive level, the IMD at a nominal drive level can be computed from:

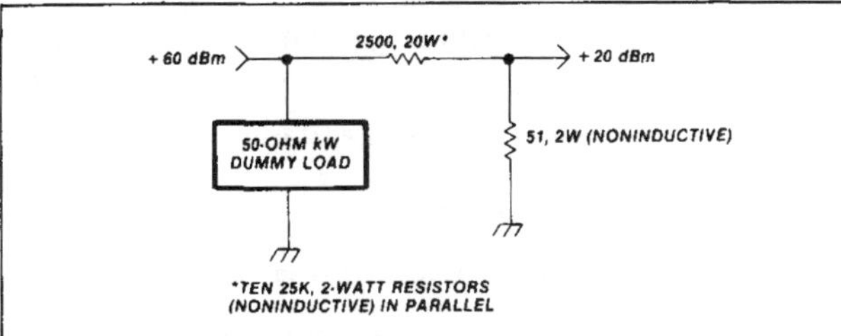

fig. 10. A 40-dB, 1-kW attenuator for making a-m measurements in a high-power transmitter.

$$IMD,_{dBc}\,(nominal\ drive\ level) =$$

$$IMD,_{dBc}\,(meas) + 2(change\ in\ drive\ level,\ dB) \qquad (4)$$

and the *output* TOI can be computed from

$$TOI = \frac{\Delta}{2} + P_{out} \qquad (5)$$

where $\Delta = IMD$ level in dBc relative to the fundamental (one tone), and P_{out} is the *output* level of *one* of the tones. Input TOI is the output TOI minus the gain of the test amplifier.

Measurement accuracy depends to a large extent on maintaining a 50-ohm impedance within the measurement instrumentation as well as the input and output impedance of the test amplifier itself. The measurement is meaningful only to the extent that the test device matches the impedance of the measurement instrumentation. Hence any networks used for input or output matching should be included as part of the measurement device.

One other comment: beware of IMD generated within the power combiner shown in **fig. 8**.[3] IMD caused by saturation of the toroids used in the combiner frustrated me for some time before I figured out what was going on. This combiner had 0.25-inch (6.35-mm) OD toroids that created their own IMD when driven above about + 10 dBm.

Although the foregoing tests were made on small-signal amplifiers such as those used ahead of a receiver, similar tests for harmonic distortion can be made on a high-power transmitter with the aid of an in-line power sampler such as is used in SWR meters. Examples of such samplers can be found in reference 1. Alternatively, a 40-dB pi attenuator capable of handling 1 kW can be made with the aid of a 1-kW dummy load and some 2-watt resistors. **Fig. 10** shows details. This attenuator will reduce the output of a kilowatt rig to a level that can be handled by the power meter.

measuring amplitude modulation

The power meter can also be used to measure amplitude modulation if the need arises and a scope isn't handy. The measurement is based on calculating the modulation percentage after measuring the power difference between the carrier with and without the modulation applied. A tone generator should be used to modulate the transmitter.

With modulation applied, the power meter (or the transmitter output) is adjusted to place the power meter reading near full scale, where scale resolution is best. Remove the modulation and measure the decrease in the power-meter reading. This difference, Δ, when plugged into the equation below, will yield the modulation percentage to within a few percent, depending on how carefully you measure Δ.

$$\text{modulation percentage} = 170\sqrt{10^{(\Delta/10)} - 1} \quad (6)$$

A difference of $\Delta = 1\ dB$, for example, would correspond to an 86.5-percent modulation level. This technique will provide good accuracy to about 30-percent modulation, where the power-meter scale resolution becomes marginal (0.2 dB). Results were verified using an HP-8640B calibrated signal generator. Remember to use an attenuator capable of dissipating the total power of the transmitter, or use a circuit similar to that in **fig. 10**.

extending operating frequency of the directional bridge

As mentioned earlier, the directional bridge has a directivity of about 40 dB to 30 MHz, dropping off to less than 10 dB or so at 150 MHz. By adding a 1-5 pF ceramic trimmer as shown in **fig. 11** and by *careful* parts placement, the balance can be improved to yield about 30-dB directivity at 150 MHz, permitting reflection measurements using the detector assembly (which is usable well beyond 150 MHz) and the modified bridge at 2 meters. The dynamic range of the detector assembly alone is about 20 dB (it departs from square-law operation above − 10 dBm, and sensitivity is about − 35 dBm), so the overall return-loss measurement range is about 20 dB, corresponding to a VSWR of 1.2. The VHF return loss adjustment procedure is shown in **fig. 11B**.

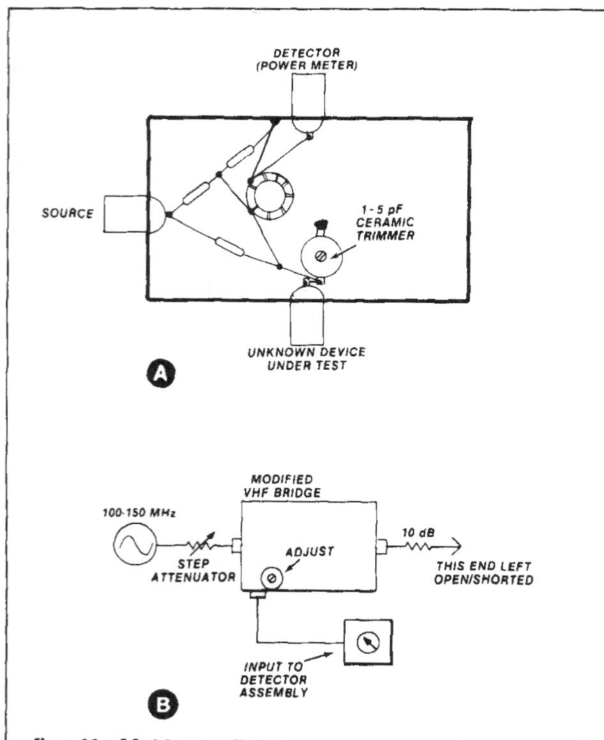

fig. 11. Making reflection measurements at VHF. A modified VHF directional bridge layout is shown in (A); (B) shows how to adjust the modified bridge for best directivity.

Initial adjustment of the modified return loss bridge is rather tedious. For best results it's necessary to adjust the trimmer and tweak the parts placement, particularly the toroid, while observing the return loss of a known device, a 10-dB attenuator connected to the unknown port of the bridge. With the 10-dB attenuator connected, the measured return loss (remember to calibrate with an open circuit first) should be 20 dB, since the signal must make two passes through the attenuator before reaching the bridge. There should be a range of adjustment over which you should measure this value.

A good indication that the bridge has been adjusted for directivity greater than 20 dB is to alternately short and open the open side of the 10-dB attenuator. If the directivity error signal has been nulled satisfactorily, the change in the power meter readings should be small (less than 1 dB). If not, readjust the trimmer capacitor and/or parts placement and try again. At some point the power meter variation as the attenuator is shorted/opened should be small. At this point try the same measurement at a lower frequency, say 50 MHz or 30 MHz, to see that you have not degraded the low-frequency directivity, since adding the capacitance disturbs the circuit balance somewhat.

Once you have adjusted the bridge to your satisfaction, epoxy the toroid in place using a *small* drop of epoxy (avoid coating the windings) to prevent it from moving. The assembly should then be sealed to prevent fingers from touching the insides.

closing remarks

Using this system I've been able to duplicate measurements on many of my homebrew projects that previously required the use of expensive laboratory measurement instrumentation. I can now measure the performance of projects and not wonder what effect a parts substitution had on a particular circuit, or whether adjustments were optimized.

Any questions or comments regarding construction of the power meter or the measurements described are welcome. Please include an SASE. Improvements doubtless can be made.

references

1. *The Radio Amateur's Handbook*, ARRL, Newington, Connecticut 06111, 1980 edition.
2. Wes Hayward, W7ZOI, and Doug DeMaw, W1FB, *Solid State Design for the Radio Amateur*, ARRL, Newington, Connecticut 06111.
3. William R. Hennigan, W3CZ, "Broadband Hybrid Splitters and Summers," *QST*, October, 1979, page 44.

A useful instrument that does a good job on ham-band inductances

easy-to-build inductance meter

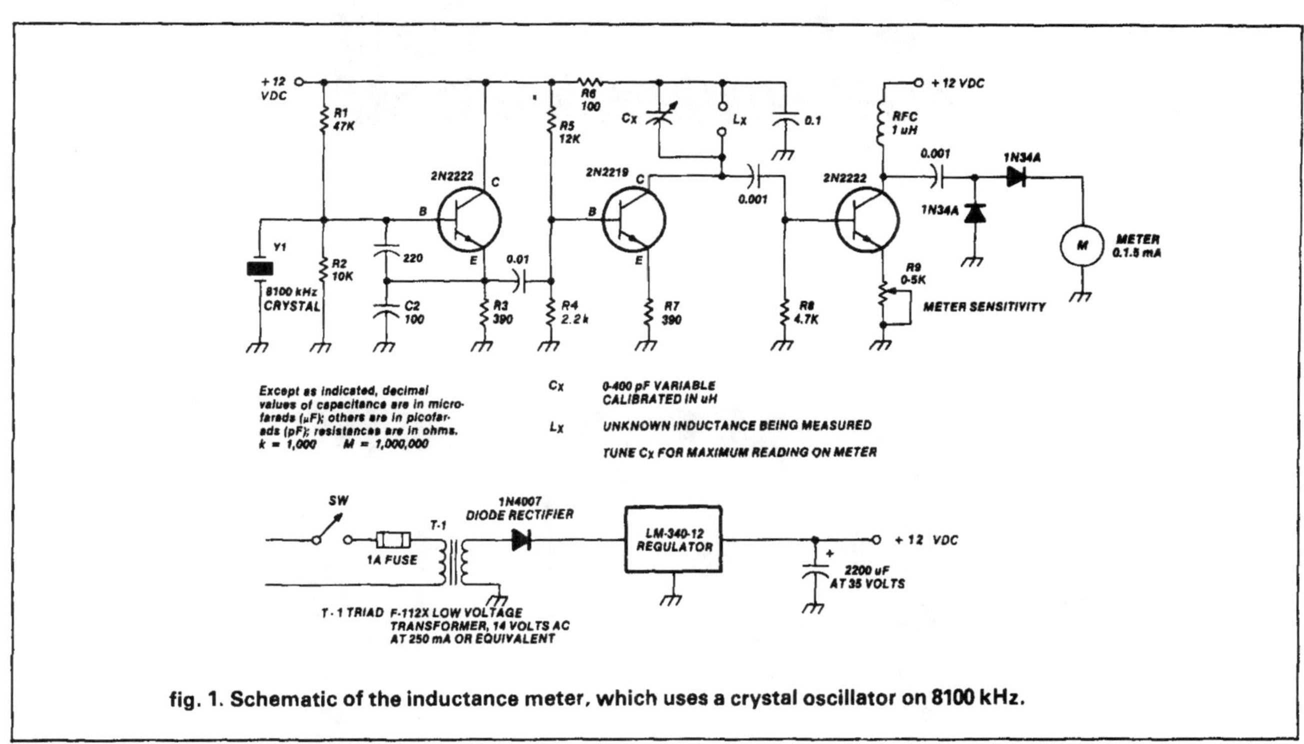

fig. 1. Schematic of the inductance meter, which uses a crystal oscillator on 8100 kHz.

This handy instrument will permit you to measure inductance from 0.5 μH to about 10 μH by reading a capacitor dial. I came up with this idea for my own work bench after searching for a good circuit for measuring inductance. This device does the job for the ham bands, and I hope you will like it as much as I do.

description

This inductance meter uses a crystal oscillator on a frequency of 8100 kHz (see **fig. 1**). The oscillator drives an amplifier. The unknown inductance is placed in the collector lead, and a capacitor across it is tuned for resonance. The dial is calibrated in μH so that the inductance can be read directly.

Another amplifier is used to build up the r-f voltage to operate the 0 to 1.5 mA meter used for tuning to a peak for resonance. A more sensitive meter (such as a 0-50 μA) could be used by shunting a 120-ohm resistor across it, if that is what you happen to have in the junk box.

I suppose the first question that will be asked is, "Why use 8100 kHz for the crystal oscillator?" Well, a surplus 8100-kHz crystal is cheaper to buy than one

By Ed Marriner, W6XM, 528 Colima Street, La Jolla, California 92037

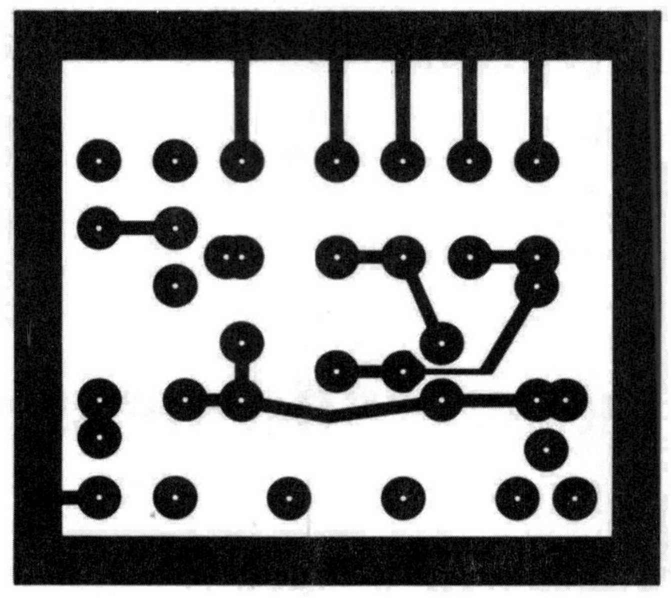

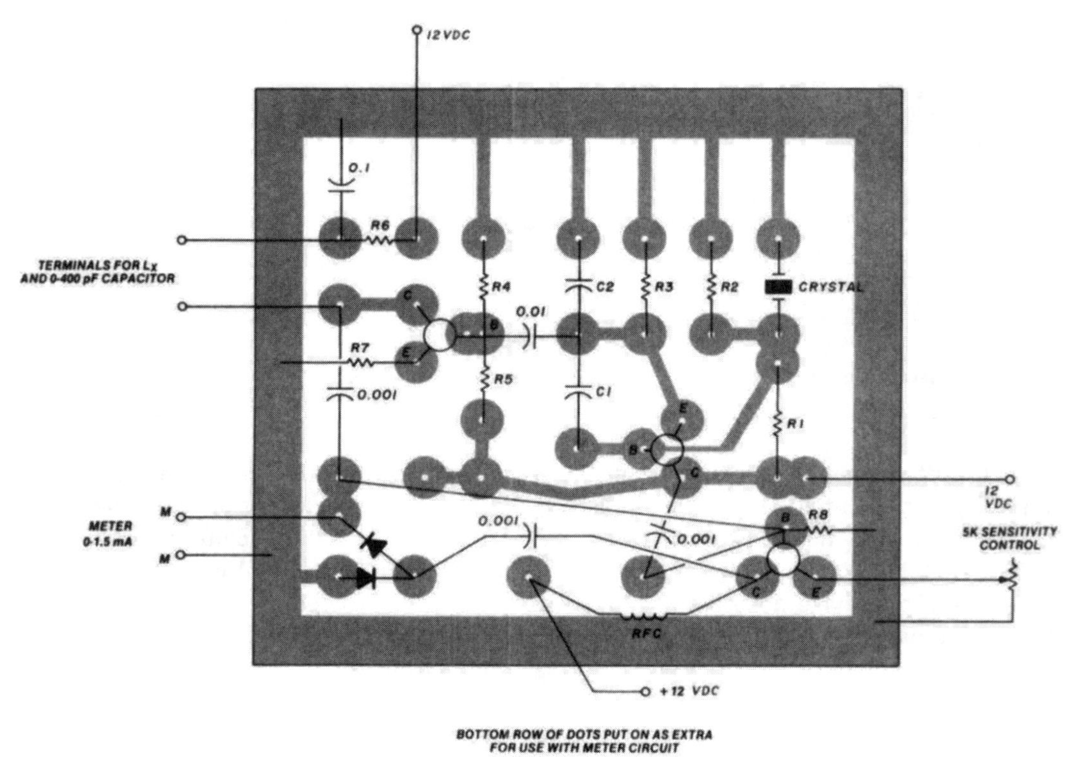

fig. 2. Printed-circuit board and parts placement for the inductance meter.

table 1. Capacitor calibration setting in μH.

pF	3500 kHz μH	7000 kHz μH	8100 kHz μH
10	200	51.6	38.5
20	165	25.7	19.2
30	70	17.2	12.8
40	52	12.9	9.6
50	42	10.3	7.7
60	34	8.5	6.4
70	30	7.3	5.5
80	26	6.4	4.8
90	23	5.7	4.2
100	21	5.17	3.8
125	17	4.5	3.0
150	14	3.4	2.5
175	12	2.9	2.2
200	16.5	2.5	1.92
225	8.6	2.3	1.6
250	8.4	2.0	1.5
300	7.6	1.7	1.2
325	6.0	1.5	1.1
350	6.5	1.4	1.1
375	5.7	1.3	1.0
400	5.5	1.2	0.9

Note: The 8100-kHz crystal was used because of its range and availability. Calibrate variable capacitor for whichever crystal you have from this formula:

f = Hertz
L = Henrys
C = Farads
$4\pi^2$ = 39.5

$$L = \frac{1}{4\pi^2 \times f^2 \times C} = \frac{1}{39.5 \times f^2 \times C} = \frac{1}{(39.5)(7 \times 10^6)^2 (10 \times 10^{-12})}$$

Example: $7000\ kHz = \frac{1}{39.5 \times 49 \times 10} = 0.0000516, \text{ or } 51.6\ \mu H$

for the ham bands. Also, if you examine the chart in **table 1**, you will see that the choice of a 10-400 pF capacitor covers a good range of inductances to be measured. Of course, it's your choice which span you want to cover. If another crystal frequency is used, calibrate the variable capacitor by using the formula in **table 1**. I was surprised to learn that this formula is not shown in most handbooks.

The accuracy of the dial reading in μH will only be as good as the accuracy with which you can measure the variable capacitor. A better way would be to check or calibrate the dial against known inductance values. Some suggestions on where to obtain inductances of known value are shown under the subheading calibration. (I found that Air-Dux coils did not correspond to the values shown in their listing, and I wouldn't use them for calibrating the meter unless they were measured first.)

construction

There is nothing difficult about building this tester once you've located the parts. Some sources for parts are listed at the end of the article. The printed-circuit layout I used is shown in **fig. 2**. I cut the board by gripping it between two pieces of angle iron and using a hack saw. Then I filed the edges, sanded the copper with steel wool, and used paper drafting tape for the masking. The dots were Avery self-adhesive color-coding labels that I bought at a stationery store. The wider tape I used for the border is Bishop precision-type 201-250-11, 1/4 inch × 20 yards (6.35 mm × 18.2 m). It's expensive, and yellow shelf-paper with stickum on the back can be used, if you prefer, by cutting it on a paper cutter.

The board was etched with ferric chloride, which takes about an hour, but the process can be speeded up by placing a lamp over the solution and juggling it around once in a while. Once the board has been etched and cleaned, by again scrubbing it with steel wool, it should be tinned with a hot soldering iron. Be careful not to apply to much heat, or the copper will come off! (There are other ways of tinning, using a solution.)

The holes can be drilled out using a No. 60 (1-mm) drill. The crystal socket was mounted using a spacer and a long 4-40 (M3) machine screw.

Before putting a unit in a box, test it out by determining if the crystal oscillator is working. Using FT-243 crystals, I found my values of C-1 and C-2 to be correct as shown. Handbooks call for other values, but they did not make the oscillator work for me. You can listen for the oscillator at 8100 kHz with a receiver or put an rf probe at the emitter output. You should have 1.5 volts of rf, enough to drive the following amplifier. The amplifier can also be checked by placing a coil in the collector, resonating it, and measuring with the probe at the collector. A coil with a value of 3-10 μH should be sufficient for this test.

I used germanium diode rectifiers to operate the meter because they provide more dc voltage than the silicon type. The 1N34A, 1N38, or 1N64 work well.

I used a Radio Shack cabinet, 5-1/4 × 3 × 5-5/8 inches (13.3 × 7.6 × 14.3 cm), to house all the parts including the power supply. Before installing the parts mount the capacitor on a piece of plastic to insulate it from the cabinet; use an insulated coupler and plastic shaft through a panel bushing for the dial knob. I used GR terminals with long shanks that went right up to the capacitor terminals. Make the leads as short as possible, because they become part of the inductance.

Once the amplifier has been finished it can be checked by advancing the sensitivity control and resonating a coil placed on the terminals. I put a 100-

table 2. **Some values of Air-Dux coils which might be used to calibrate or check the dial. Values must be checked first.**

Air Dux number	total inductance of coil (μH)
404	0.18
406	0.39
408	0.71
410	1.10
416	2.87
432	11.30
504	0.27
506	0.61
508	1.10
510	1.60
516	4.30
532	17.30
604	0.38
606	0.86
608	1.52
610	2.38
616	6.08
632	24.20
804	1.02
806	2.33
810	6.47
816	16.30
832	66.30

Other values of coils can be obtained by writing Illumitronic Engineering Corporation, 680 East Taylor Avenue, Sunnyvale, California.

ohm resistor in series with the control to limit the current reading and not ground the emitter. Most meters have the positive terminal on the left looking at the back. If your meter does not read, or reverse readings are obtained, turn the meter leads around. The meter dial plate is shown in **fig. 3**.

calibration

Several suggestions on calibrating the dial are given. I used a 0-400 pF capacitor, a BDC type out of Swan equipment, that I found at a flea market. The 0-365 pF units are more common and are available from J.W. Miller and Radio Shack.

One of the best ways of calibrating or checking your dial readings is to use the rf chokes listed below. They seem to be very accurate:

Miller rfc 420	0.22 μH
Miller rfc 220	0.82 μH
Miller rfc 144	1.72 μH
Miller rfc 50	8.20 μH
Ohmite Z-50	10.00 μH

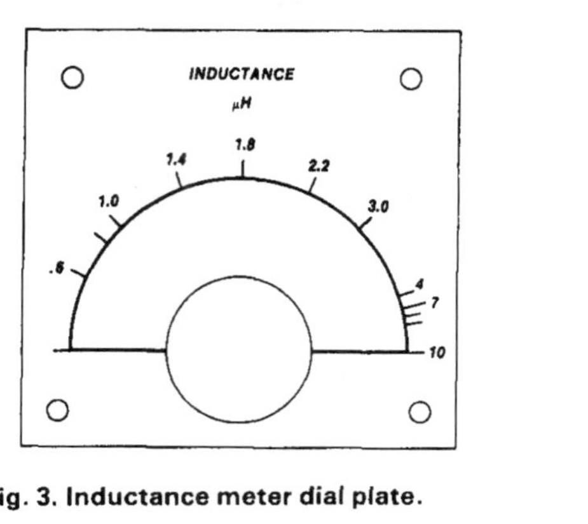

fig. 3. Inductance meter dial plate.

Other values of the type 4500 and 70F106AI chokes can be obtained from the J.W. Miller catalog (J.W. Miller, 19070 Royes Avenue, Compton, California 70224). If you know someone who can measure inductance, the Air-Dux line in the 400, 500, 600 and 800 series are also good to use. Some of the approximate values of Air-Dux are given in **table 2**.

table 3. **Reference chart for other J.W. Miller coils that could be used for calibration.**

J.W. Miller stock No.	inductance μH	Q
4580	0.10	70
4582	0.15	80
4584	0.22	95
4586	0.33	100
4588	0.47	105
4590	0.68	113
4592	0.75	115
4594	0.82	112
4602	1.00	58
4604	1.50	62
4606	2.40	63
4608	3.90	70
4609	5.50	70
4610	6.20	67
4611	8.20	67
4612	10.0	67
4622	10.0	70
4624	15.0	58
4628	39.0	93
70F106AI	2.0	
70F226AI	2.2	
70F336AI	3.3	
70F396AI	3.9	
70F476AI	4.7	
70F686AI	6.8	
70F826AI	8.2	
70F105AI	10.0	

See the catalog for a complete list of values in between. Contact J.W. Miller Co., Sales, Mr. Bill Courtney, 19070 Reyes Avenue, Compton, California 90224.

table 4. Some typical toroid values collected from magazine articles.

band	core	inductance (μH)	no. turns	enamel wire size AWG	(mm)
40 meters	T-50-2	13	50	28	(0.3)
20 meters	T-50-2	8	44	28	(0.3)
15 meters	T-50-2	4	25	28	(0.3)
10 meters	T-50-2	0.6	12	24	(0.5)

Note the red core is type 2, and the yellow core type 6.

core	inductance (μH)	no. turns	enamel wire size AWG	(mm)
T-68-1	0.5	10	22	(0.6)
T-68-6	0.5	8	20	(0.8)
T-68-6	0.7	12	22	(0.6)
T-68-6	0.8	13	26	(0.4)
T-68-6	1.0	14	20	(0.8)
T-68-2	1.2	13	20	(0.8)
T-68-2	1.8	19	22	(0.6)
T-68-2	2.0	20	22	(0.6)
T-68-2	2.1	22	22	(0.6)
T-68-2	6.0	32	24	(0.5)
T-68-2	7.0	32	24	(0.5)
T-68-2	21.0	56	24	(0.5)
T-68-2	24.0	60	28	(0.3)
T-50-2	0.57	7	26	(0.4)
T-50-6	0.60	12	24	(0.5)
T-50-2	1.70	17	26	(0.4)
T-50-2	0.9	11	20	(0.8)
T-50-6	2.4	25	26	(0.4)
T-32-2	0.28	8	22	(0.6)
T-32-2	0.37	9	22	(0.6)
T-32-2	0.8	14	22	(0.6)
T-32-2	1.0	22	22	(0.6)
T-32-2	2.6	25	26	(0.4)
T-32-2	3.0	27	26	(0.4)
T-32-2	6.0	37	28	(0.3)

Note: T-94 coil is 0.94 inch dia. (24 mm)
T-80 core is 0.79 inch dia. (20 mm)
T-68 core is 0.69 inch dia. (17.5 mm)
T-50 core is 0.50 inch dia. (12.7 mm)

Table 3 gives values for some J.W. Miller coils, and **table 4** shows some typical toroid values. All these components are available from many sources throughout the country.

Perhaps this little gadget is not the best way to measure inductance, but I find it does the job for most of my construction projects, especially for ham-band inductances.

Another reference for a final check of your dial: ARRL L/C/F Calculator Type A.

Multiple-line transformers
improve accuracy
making possible realistic
phased-array impedance measurements

a precision noise bridge

An rf impedance bridge is an important tool for anyone with more than a casual interest in measuring complex impedances. One typical application for such a bridge is in the design of a phased-array antenna.

Solid-state technology has made it possible to move the rf impedance bridge out of the laboratory and into the average ham shack, but, unfortunately, laboratory accuracy cannot usually be achieved with most of the available units. Although readings within 10 percent are acceptable for most Amateur applications, much better accuracy and discrimination are necessary when taking measurements of antenna-array mutual impedances. We must in fact be able to accurately determine resistive components of impedances which may change as little as only 3 or 4 ohms when a nearby antenna element is mutually coupled. Obviously, most noise bridges can't do the job.

Internal view of noise bridge showing coupler and PC board placement.

This article explains how it is possible to achieve the needed accuracy in the range of the hf Amateur bands. A multiple-line, distributed-impedance transformer is the critical bridge component. It is not very susceptible to fabrication variations, and it maintains a single calibration over a wide frequency range. With care in adjustment and good construction practice, it is possible to achieve an accuracy of 3 percent over most of the range — with even better results at 14 MHz and lower.

background

In an excellent article on noise bridges,[1] W6BXI and W6NKU contributed two major innovations for improving noise bridge accuracy: compensating for bridge circuit strays by adjusting inductance in one of the bridge secondary arms; and equalizing primary-to-secondary interactive effects by the addition of a dummy primary wire.

The first idea is one of those insights that seem so obvious after someone else has pointed it out: One wonders why the C_p calibration "rotation" problem at high frequency ever seemed so difficult. The second suggestion is logical, but I could confirm it only empirically. After winding a dozen different kinds of

By Forrest Gehrke, K2BT, 75 Crestview Road, Mtn. Lakes, New Jersey 07046

three-wire and four-wire toroid transformers and comparing results, I leaned toward the four-wire version. I became convinced that control of interactive effects between the primary and secondary was a key factor in achieving the accuracy I wanted.

I used the improvements and suggestions in the referenced article in three of my own bridges, with good results similar to those achieved by the authors: a spread of 2 to 3 ohms (resistance and/or reactance) for a range of accurately known impedances. But further improvement in accuracy was still needed.

W6BXI, in private correspondence, told me that he'd had no particular difficulties with the bridge transformer, either in reproducing it or with circuit adjustment. I wasn't as lucky. But he did discuss a method of matching and balancing the bridge transformer to the noise source. I tried an emitter follower that provided a low-impedance source for the noise amplifier. I also tried a balun toroid transformer, to take advantage of balanced drive to the double-ended bridge circuit from the single-ended noise source. Neither approach worked.

I have probably wound a hundred or more different configurations of the toroid transformer used in this bridge, in an effort to get a better understanding of this key component in the bridge. One possible solution lay in trying to get maximum inductive coupling with minimum capacitance between primary and secondary. So I wound a more classic type of transformer, with the primary on one side of the toroid and a single-wire, center-tapped secondary on the other. Capacitive coupling between windings was very low — 3 pF — but the transformer behaved poorly in the bridge. Calibration barely held over a single Amateur band, and there was no correspondence to the real values of the reference arm of the bridge. Next I wound a second transformer, also with the windings on opposite sides of the toroid, except that now the secondary was a bifilar winding. The bridge worked, and calibration held from 1.8 to 30 MHz. But the noise signal was too low at the high-frequency end, indicating insufficient primary-to-secondary coupling, resulting in poorly defined nulls. Different core materials did not help. This experiment showed me that the required bridge coupler was no ordinary transformer, but something akin to a distributed impedance transmission line.

Most of the trifilar and quadrafilar toroid transformers I tried had been wound with the three or four wires kept carefully parallel. I wound some with the wires twisted together at two or three turns per inch. A stranded bundle of approximately 9 inches (23 cm) was wound on the core. I decided to try connecting up a strand in the usual way, but as a large single-turn loop, without winding it on a toroid core. This worked excellently, without any of the puzzling anomalies of the toroid versions. Noise signal was as high as with a toroid, and calibration held constant from 1.8 to 30 MHz. Also, the reference potentiometer settings closely coincided with the value being measured. With this approach there was no question about the four-wire line being superior for holding calibration over a wide range, and my results were reproducible: Reproductions of these lines always produced the same results. I concluded that the toroid core had been contributing most of the anomalies — and that it was not at all essential to this bridge component.

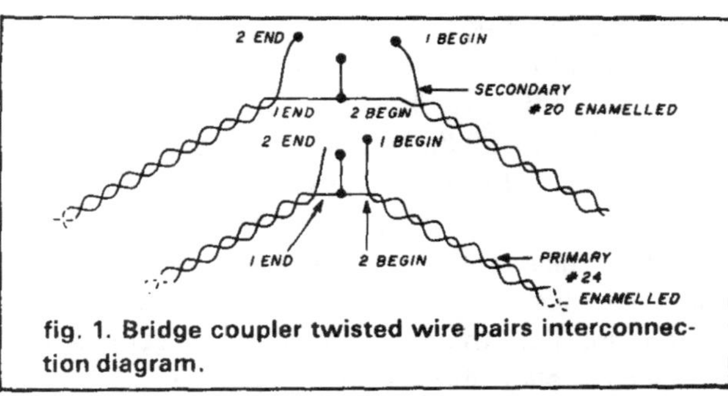

fig. 1. Bridge coupler twisted wire pairs interconnection diagram.

Having eliminated the toroid, I next worked on optimizing line length, wire size(s), and insulation thickness, and on a method of combining the individual wires into a line. The matter of whether adjacent or alternate wires of a single twisted line should be used in connecting up the primary and secondary circuits was also investigated. I tried dozens of variations: No. 24 (0.5 mm) hookup wire with PVC insulation, enameled wire sizes from No. 18 (1.0 mm) to No. 32 (0.2 mm), including mixed sizes, lengths from 4 inches (10 cm) to 16 inches (40 cm), single twisted and compound twisted, and so on. This is what I learned:

1. Close coupling between primary and secondary is absolutely required. This means using enameled wire, twisted as tight as possible.

2. Line lengths between 10 and 12 inches (20 and 30 cm) are best: If too short, noise signal is too low; if too long, a large amount of compensation is required on one secondary half for R_p adjustment, which is accompanied by a pronounced C_p calibration shift between 21 and 30 MHz.

3. Compound twist gives the most consistent results (primary and secondary wires are first twisted independently, then these are twisted into a single strand). All four wires can be twisted simultaneously, but they would then have to be interconnected with attention to whether alternate or adjacent wires are being used as the respective primary and secondary. See **fig. 1** for winding interconnection detail.

4. Enameled wire size No. 20 (0.8 mm) seems best. Smaller wire sizes are more difficult to connect, adjust, and compensate because of their fragility. Heavier wire, though effective, resulted in a line that became too stiff to work with. I found a good compromise to be No. 20 (0.8 mm) for the secondary and No. 24 (0.5 mm) for the primary. These lines can be looped into multi-turn coils, to conserve space, with no effect on calibration. In the single-turn calibration no ill effects occur unless the loop is tightly collapsed on itself, and then the most noticeable effect is a resistance calibration shift at low frequencies.

These coupling devices act like distributed-impedance transmission lines. The circuit impedance appears to be low, as it is virtually immune to stray capacitances. As with transmission-line baluns, however, the device displays sensitivity to lead dress of the ends of the lines. Unfortunately, it is not physically possible to keep these ends apart. One end of each pair has to be connected to each other, and the ends connected to the bridge arms must be as short as possible. If, indeed, this is a multiple-wire transmission line, theoretical calculations would point toward a low characteristic impedance. The effect of the re-entrant connection is difficult to assess. I would welcome letters from readers who are able to show the mathematics. The most puzzling aspect, to me, is that a line so short (in terms of wavelength) should be effective at very low frequencies. I have used a bridge *calibrated* for frequencies between 3.5 and 30 MHz to as low as 150 kHz. The calibration was unchanged and the noise signal level was as high as at 3.5 MHz.

scale extensions/expansion

The article by W6BXI and W6NKU and subsequently published letter[2] give a technique for measuring impedances beyond the basic range of the bridge by using series or parallel resistances inserted at the **Unknown** terminal. Though this approach is quite effective, the user is warned that, as with any scale multiplier, inaccuracies in the basic range calibrations are also multiplied. Improvement of the base range accuracy should make these range extenders more trustworthy — and using the minimum multiplier possible. For antenna measurements the series extender finds the greatest use; I have several, ranging from 10 to 100 ohms.

For some applications, like mine, there is a need to expand the resistance scale to be able to discriminate between small changes. Since the R_p range of interest was between 27 and 77 ohms, a 50-ohm potentiometer in series with a 27-ohm fixed resistor was used in the reference side of one of my bridge models. This affords nearly 270 degrees of scale rotation for a translated 50-ohm range. To minimize strays, the potentiometer connections should be as shown in **fig. 2**.

transformation equations

A series combination of resistance and reactance can always be found that exhibits the same equivalent impedance as any given parallel combination of resistance and reactance.

The transformations give the relationship between the elements of the series and parallel networks (shown below) when the driving-point impedances are equal. The equations for the respective impedances are:

$$Z_p = \frac{R_p (jX_p)}{R_p + jX_p}$$

$$Z_s = R_s + jX_s$$

Equating the real and imaginary parts of both expressions for network equivalency yields:

$$R_s = \frac{R_p X_p^2}{R_p^2 + X_p^2}$$

$$X_s = \frac{R_p^2 X_p}{R_p^2 + X_p^2}$$

Transforming from series to parallel:

$$R_p = R_s + X_s^2/R_s$$
$$X_p = X_s + R_s^2/X_s$$

other modifications

The other major changes from prior circuits were to go to a fixed capacitor in the reference arm of the bridge and to place the variable capacitor, C_p, across the **Unknown** terminal. Operation of the bridge remains the same, except that the meanings of the direction of rotation from "0" on the calibration scale become reversed. That is, an indicated null which *increased* the capacitance above its value at "0" means the measured impedance has an *inductive* reactance component. Conversely, a decrease in capacitance from "0" indicates a capacitive reactance. This change ensures that the reactive component being measured is in direct correspondence with that of the unknown impedance. The difference is a subtle one but, in effect, at nulled conditions the reference arm is always "seeing" the identical situation, the fixed capacitance across the potentiometer.

adjustment

Two adjustment procedures are necessary, both requiring a bit of care and patience. One is done at low frequency to enable the unknown load resistance to track with the reference potentiometer's nulled resistance. The second adjustment is done at the high frequency end of the range to enable the load resistance to continue to track with the potentiometer and to reduce the C_p calibration shift. The adjustments are performed in the order given and must be iterated since they are interdependent. They are best done with a good commercial load termination (of minimum reactance), though 50 to 100 ohm quarter-

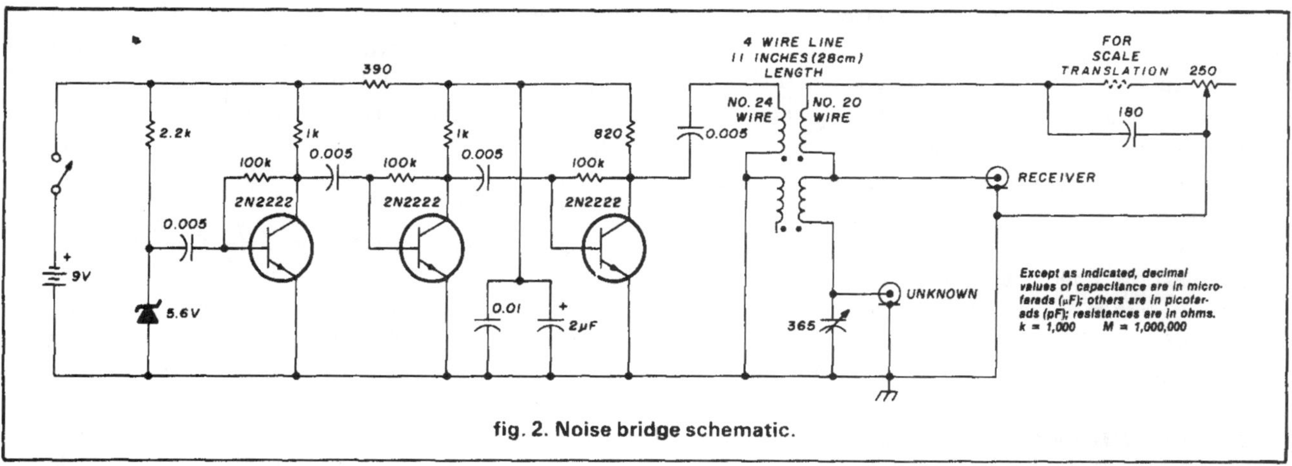

fig. 2. Noise bridge schematic.

watt carbon resistors connected by short leads are a second best alternative. These adjustments ensure that the calibration remains constant over the bandwidth and provides an accurate R_p readout. I have found multi-tester DWM's to be remarkably accurate ohmmeters and useful in measuring the calibration load resistances and the nulled potentiometer resistance.

The low-frequency adjustment is better described than explained. The adjustment cancels inductive and capacitive strays, and after building six bridge units, only two with identical layouts, I am confident of the procedure:

low-frequency adjustment

1. Set the null-detecting receiver to a low frequency (I used 3.5 MHz) and null the bridge with a known resistive load of 50 to 100 ohms. The R_p and C_p controls are interdependent; repeat for best possible null.

2. Turn off the noise source and remove the load and null detector from their terminals. Measure the potentiometer resistance without disturbing its setting (measure the resistance seen at either the **Unknown** or **Detector** terminals). Measure the load resistance.

3. If there is agreement to within 0.1 ohm go to the high-frequency adjustment procedure. More likely, the agreement will be close but not good enough. With the bridge returned to operation as in step 1, gently bend the bridge line to a new position in any direction by a few degrees. The potentiometer null will change; renull and note whether the potentiometer resistance more closely agrees with the load. If it does not, bend the line in another direction, possibly opposite to the direction of the first trial. If after a few trials it is found that no direction of movement will cause good enough correspondence of readings, go on to the high-frequency compensation procedure. The two procedures are interdependent and several iterations will be necessary.

high-frequency adjustment

This adjustment equalizes any residual imbalance of inductance that may exist in the bridge. The adjustment is made at the high-frequency end of the range because this is where the influence of an imbalance is most noticeable. It consists of adding inductance (offset) to the secondary arm found to require compensation. Which arm this is depends upon the direction the R_p null must be shifted. I have found it always to be required in the reference arm connection if the noise signal is inserted to the primary connection at the same end of the line.

1. Using the same pure resistance load as in the low-frequency adjustment, null the bridge carefully at 30 MHz. Turn off the bridge, remove the load and detector, and compare the potentiometer resistance with that of the load. If this is the same you had at 3.5 MHz you're done. More likely, compensation will be necessary.

2. Add an inch (2.5 cm) of the same size wire used for the secondary to the secondary connection at the reference bridge arm. Returning the bridge to operation, if the R_p null resistance is higher than the resistance being measured and now goes higher still, remove the wire and add it in on the other secondary arm of the coupler. If it now goes lower, reduce the amount of added wire until the R_p null resistance matches the termination.

3. In making this adjustment be careful not to disturb the position of the line found for the low-frequency adjustment. In any case, recheck the low-frequency adjustment after completing this procedure. Several repetitions of the two adjustment procedures may be necessary. If more than an inch of secondary-size is required, you may encounter an unacceptable C_p calibration shift at 30 MHz, although R_p tracks over the whole range.

These two adjustment procedures force tracking of R_p at widely separate frequencies. But this does not guarantee flat tracking at frequencies in between, so after completing these adjustments check the calibration in the middle of the range, such as 14 MHz. R_p calibration should remain well aligned and C_p shift, compared with the 3.5-MHz position (the center, or "0" of this scale), minimal (less than 1 pF). If all is well, now check with other values of resistance. If you cannot obtain commercial terminations, quarter-watt carbon resistors of the deposited type are quite good, but connect them to the unknown terminal with as short a lead length as possible. In spite of all your best efforts, you will see more C_p inductive shift at 30 MHz with these. If you used a 250-ohm potentiometer, expect to see more absolute R_p shift at 30 MHz than found at 3.5 MHz for resistance at the upper end of this range. My bridge reads 1 ohm low with a 237-ohm load at 30 MHz. At the low end of the range, using a 6-ohm resistor, the R_p reading is within 0.1 to 0.2 ohm over the whole frequency range, with one bridge reading on the high side at 30 MHz and the other on the low side. A DVM checked against a laboratory standard was used for a resistance calibration standard.

"An rf impedance bridge is an important tool . . . in the design of a phased-array antenna."

If the potentiometer is not tracking well with frequency, the circuit has layout problems. Here are three general areas to review:

1. Ground loops in the circuitry (multiple connections to ground for the same circuit path).

2. Too much inductance has been allowed to creep into the connections in known or unknown arms of the bridge. One clue is a continuous shift of C_p center as you step up in frequency bands from 3.5 to 30 MHz.

3. The noise source ground connection to the bridge has too much impedance. This is indicated by a pronounced shift in C_p center position at 30 MHz, with little or none at 21 MHz. The effect is very different from the prior case, in which there is gradual shift. Note that this same effect also arises from a requirement for a large amount of compensation at high frequency. Still, this is related; the amount of compensation is an indicator of stray reactance in the circuit.

circuit hints

Avoid creating ground loops. Have all ground returns going to a common point; this point is best located at some point midway between the circuit connections for the known and unknown bridge arms. Calibration accuracy to 30 MHz requires extreme care here. For example, in the last two units I built the detector coax terminal is isolated from the chassis. Coax (RG-174) is used to continue this line directly to the bridge connections, including ground. The bridge circuit and components are mounted on a printed-circuit board using copper foil on the circuit side. The chassis screw mountings for this board are insulated from the foil. The only connection with the chassis and the copper foil occurs via the SO-239 **Unknown** coax terminal.

Any stray reactance in this bridge creates calibration shift problems, particularly at 30 MHz, and if severe, even at 14 MHz. I found that liberal use of flat, braided shield (such as is used on RG-58/59 coax) for all bridge connections is indicated. In this regard, the miniature 365-pF variable capacitor recommended in the referenced article has too much internal inductance. The small Japanese air variable capacitors built on heavy aluminum frames have been found to be excellent and quite linear.

As a general rule disproportionate shifts in calibration with increasing frequency are indicative of stray capacitance problems, while linear shifts are due to unwanted inductance.

As for the potentiometers, Mil. Spec. linear-resistance carbon pots are recommended. I found Clarostat Type 53C1, S-taper to be best. These are available in resistances as low as 50 ohms and are very linear.

As for the noise source, obtain several samples of the zener diode for the noise amplifier. Select the one with the highest noise output over the frequency range, emphasizing 30 MHz. Some diodes are so quiet there will be difficulty obtaining enough signal. Isolate the noise amplifier and its batteries from all grounds except the connection to the common ground at the bridge circuit. In one of my units, flat braid for this ground return solved a stubborn residual C_p shift problem at 30 MHz.

Unless you have a compelling need to carry a bridge in your hip pocket, don't try to crowd this device into a snuff box. My units are housed in a standard aluminum two-piece box 5 × 4 × 3 inches (13 × 10 × 8 cm). This is small enough, gives you space to work in, and keeps down strays. With the extra space available another 9-volt transistor battery can be paralleled for extended operation.

C_p dial calibration

With R_p calibration done, you are ready for calibration of the C_p dial. Assuming you have found minimal or no C_p shift of the "0" position with a 50-ohm termination at 30 MHz, I recommend that 3.5 MHz be used for this calibration, since it minimizes the effect of lead inductance errors.

In the referenced article[1] the authors suggest the following procedures: Calibrate the variable capacitor against a standard before connecting it to the bridge circuit. Or calibrate the capacitor, with bridge operating, using known values of capacitors in parallel with a resistance at the unknown terminal (50 ohms suggested). After completing the capacitive portion of the dial, temporarily disconnect the 180-pF bridge centering capacitor and work backward toward "0" for the inductive portion of the dial.

The first method is the preferred one. But since I had no capacitor standard available, I used a variation of the second method: After calibrating the capacitive portion of the dial, connect a 50-ohm quarter-wavelength coax cable cut for 3.8 MHz to the **Unknown** terminal. At the far end connect a 39 or 68 ohm resistance with minimum lead length. Find the frequency which nulls this resistance while the C_p dial is set at "0." Note the frequency and remove the resistor. This is the exact frequency at which your cable is a quarter-wavelength. Now use the same known capacitors in parallel with 50 ohms at the end of this cable to calibrate the inductive side of the dial. The quarter-wavelength coax causes these capacitors to look "negative," that is, inductive by a nearly equal amount. Since these are parallel circuits in series with the coax, a small correction is necessary which increases with increasing capacitance. Assuming a 50-ohm coax with a 50-ohm termination at 3.8 MHz, use **table 1** correction points (interpolating for intermediate calibration points).

upper frequency limitations

Frequencies higher than about 20 MHz begin to present problems for any circuit measurement device. This is why it is not easy to achieve a constant calibration with good accuracy as we go up in frequency. Here we enter a realm in which a resistor, a capacitor, or a coil can no longer be thought of as discrete elements; indeed, each can be a combination of all three. Since this also applies to the bridge circuit, calibration correction and circuit adjustment schemes become necessary. Reactance is involved, and so these compensations are frequency dependent. This means calibration is meaningful over relatively small frequency ranges. I have checked my noise bridge as high as 100 MHz and found that, while deep nulls can still be detected, the calibration shifts considerably.

table 1. Parallel-series-parallel correction for use in calibration of inductive side of C_p dial.

capacitance (pF)	
at end of coax	at terminal
50	49.82
75	69.51
100	98.60
120	117.59
130	126.94
140	136.20
150	145.34
160	154.37
170	163.28
180	172.06

low resistance limitations

Measurement of very low R_p circuits (below 5 ohms) poses two problems for this bridge: The low R_p, coming quite close to a short circuit, reduces the C_p null sensitivity, making it more difficult to determine balance. At the same time, any accompanying reactance, since we are measuring the parallel equivalent of the circuit, results in large excursions of the C_p dial for relatively small reactances.

As R_p approaches zero, bridge circuit strays become more significant. Do not depend upon any measurements made with this type of bridge with the R_p dial setting near zero, since many potentiometers do not actually reach zero resistance at their mechanical stop. I have heard of attempting to determine a quarter-wavelength of coax this way. This bridge can do that and more, but not in this way. If measurement of very low impedances is an objective, use a series extender, or consider a series-type bridge.

detector considerations

One of the reasons the noise bridge is a relatively simple circuit is that no null detector is included. For this an ordinary receiver is used. This can be the station receiver, and, in these days of accurate receiver frequency readout, it can be a frequency standard as well. For purposes of bridge calibration an Amateur-band-only receiver is adequate, but for most measurement needs a general-coverage receiver is better. The presence of an S-meter is helpful but not necessary.

When making measurements on antennas, a battery-operated receiver is convenient. I use an inexpensive all-band portable and a transistor crystal oscillator for marker frequencies. Since the noise signal is considerable, (S9 +20dB, off null), receiver sensitivity is no great consideration; too much can lead to difficulty in finding a null because of receiver

AVC. The sharpness and depth of nulls provided by this bridge requires getting used to!

I have used as much as 30 feet of coax between the bridge detector terminal and the receiver with no effect upon calibration.

An fm receiver will not work well with this device, as the noise is almost entirely a-m. If the receiver has AFC it won't work at all, as AFC always shifts the receiver local oscillator off the null frequency.

applications

Before discussing applications, I want to mention the difference between parallel and series circuits. Each type can be transformed into the other. This bridge measures parallel-circuit equivalent values for both real parallel and series circuits. If a real parallel circuit consisting of a 50-ohm resistor and a 180-pF capacitance is measured, then that is what is read out, regardless of the frequency. But a series circuit of 47.8 ohms and 4078 pF at 3.8 MHz also will be read by the bridge as 50 ohms R_p and 180 pF C_p. This point must always be considered when using this bridge.

An excellent tutorial on bridge applications and calculations using complex algebra exists.[3]

coax cable measurements

If you used my method to calibrate the C_p dial, you have learned how to use this bridge to find very accurately the frequency for a quarter-wavelength of coax. If you have R_p, then taking the square root of the product of R_p and the resistive termination yields the characteristic impedance of this line. The ratio of the physical length of the line to the free-space quarter wavelength for the frequency measured is the velocity factor. If you were working with foam cable, you might be in for some surprises at this point. Neither of your calculations may agree very closely with the nominal values usually quoted. Characteristic impedance can vary as much as 10 percent, and velocity factor is seldom as high as the 0.82 usually given; 0.70 is more likely.

When using this method, choose a termination resistance different from the characteristic impedance, but one which will yield a transformed resistance within the range of the bridge. As a check on your work, try a resistance termination on the line equal to your calculation for characteristic impedance. You should find that the bridge, after being nulled for this resistance, maintains that null over a wide range of frequencies.

I have found that quarter-wavelength cables determined via grid-dip methods are 1 to 3 percent too long, when compared with my results. The grid-dip method introduces error because of the shorting link and the pickup coil.

antenna measurements

An obvious application for this bridge is the measurement of self and mutual impedances of antenna elements and matching adjustments. Measurement of mutual impedances must be done indirectly and involves fairly complex calculations. (This will be discussed at greater length in forthcoming articles.)

Self impedance measurement is straight-forward: Simply connect the antenna leads to the bridge and adjust for null. Remember to keep lead length the same as in the actual installation, and remember too that the values being read are the parallel-circuit equivalents.

A very nice application for an accurate bridge is measuring the impedance of antennas right in the shack, having previously obtained an accurate measurement of the feedline length and its characteristic impedance. Using a Smith chart or programmable calculator to rotate the measurement back to the antenna saves a lot of legwork.

One point particularly applicable to antenna measurements is, make doubly sure of all connections and joints. This bridge operates with noise power measured in microwatts. Poor connections, which do not show up in normal operations, even when driven QRP, will become evident. A few watts may temporarily "weld" poor connections; the bridge hasn't enough power to do that.

impedance transformers and networks

Another useful application for this bridge is measurement of the input and output impedances of transformers and networks. Remember that the readings are in parallel circuit form, and that the terminations may be in either form. It's sometimes easier to arrive at one than the other with available components.

references

1. Robert Hubbs, W6BXI, and Frank Doting, W6NKU, "Improvements to the RX Noise Bridge," *ham radio*, February, 1977, page 10.

2. Robert Hubbs, W6BXI, and Frank Doting, W6NKU, "Antenna Noise Bridge" (letter), *ham radio*, September, 1977, page 100.

3. Leonard Anderson, "Antenna Bridge Calculations," *ham radio*, May, 1978, page 34.

bibliography

Gehrke, Forrest, K2BT, "Impedance Bridges" (letter), *ham radio*, March, 1975, page 60.

Gehrke, Forrest, K2BT, "Antenna Noise Bridge" (letter), *ham radio*, September, 1977, page 100.

Howard W. Sams Co., Inc., "Reference Data for Radio Engineers," 5th Edition, page 11-5.

Pappot, G., YA1GJM, "Noise Bridge for Impedance Measurement," *ham radio*, January, 1973, page 62.

Measure over
a 30 dB range
with confidence

wide-range RF power meter

Some time ago I decided to build a small antenna range. One of the key items I knew I'd need was an RF power meter with good stability and wide range. Most commercial units I found were beyond repair or the limits of my budget, and the homebrewed units were either limited in range or used modulation to avoid a drift problem.

I had used an LM11 operational amplifier in designing an earlier project and a friend later introduced me to an even better one. Some of the new chips coming on the market offer unbelievable performance and are slowly making system designers out of us circuit designers. A chip here, a chip there, follow the spec sheet as to optimum feeding — and we have a piece of test equipment that rivals commercial units.

I combined some of these into an RF power meter that features a 30 dB (useful to 35 dB) range from −15 dBm to −45 dBm, remote control, and good temperature stability. Although the antenna range is still in the future, the power meter has been used on the bench for evaluating hybrid couplers, helical filters, cavity filters, IF amplifiers, and such. I plan to use the power meter on the 70 cm band. But it can also be used from the HF band up into the GHz range.

theory of operation

The heart of the unit is the Hewlett-Packard HSCH-3486 zero-bias Schottky diode used as the detector. This device offers high voltage sensitivity and doesn't need the biasing featured in other detection schemes. The response curve is logarithmic from −50 dBm to −20 dBm; above −20 dBm the diode becomes increasingly nonlinear in detection response. The lower end is limited by the amplifier used.

To avoid using a modulation method of detection, a chopper stabilized operational amplifier was used. (The schematic is shown in **fig. 1**). The Intersil ICL7650 features an extremely low input offset voltage of 1 μvolt over a wide temperature range. The chopper op-amp basically converts the input DC voltage to AC, amplifies it, and converts it back to DC. Amplifying the DC output from the detector 150 times with a chopper op-amp puts the signal at a level that simpler op-amps such as the LM11 can handle. The National Semiconductor LM11 is a precision DC amplifier that combines the best features of existing bipolar and FET op-amps. Offset voltage is 100 μvolts and drift is 1 μv/ °C. Six ranges in 5 dB steps are accomplished by this circuit by changing the gain of the amplifier. Each range is controlled remotely by reed relays. Offset voltages in the amplifier are nulled with two pots, one for the high range and one for the lower three ranges. These three devices — a diode which converts RF power into a logarithmic output equal to a dB scale and a pair of operational amplifiers — amplify AC microvolt level signals to volt levels, while introducing little drift.

construction

Originally the unit was to be mounted directly at the antenna and was therefore constructed in a diecast box for good shielding. Power is supplied remotely from a separate box, which also contains the meter and scale change (**fig. 2**). A schematic of the power supply is included in (**fig. 3**). When purchasing a dB scale meter make sure that the −3dB point falls exactly at half scale. Some meters have been "fudged" to accomodate circuit nonlinearities.

The inside of the box is shown in **fig. 4**. Its detection circuitry, visible on the left side, is shown in an enlarged view in **fig. 5**. The parts are mounted on a small piece of 0.015 inch brass shim stock and held in place by the TNC connector. Note the chip capacitor on the left, supporting the 50-ohm resistor. A value of 100 pF is adequate down to 10 MHz; below 10 MHz this value should be increased. For work above 70 cm up to 4.2 GHz, a coaxial-mounted detector is recom-

By Rudolf E. Six, KA8OBL, 30725 Tennessee, Roseville, Michigan 48066

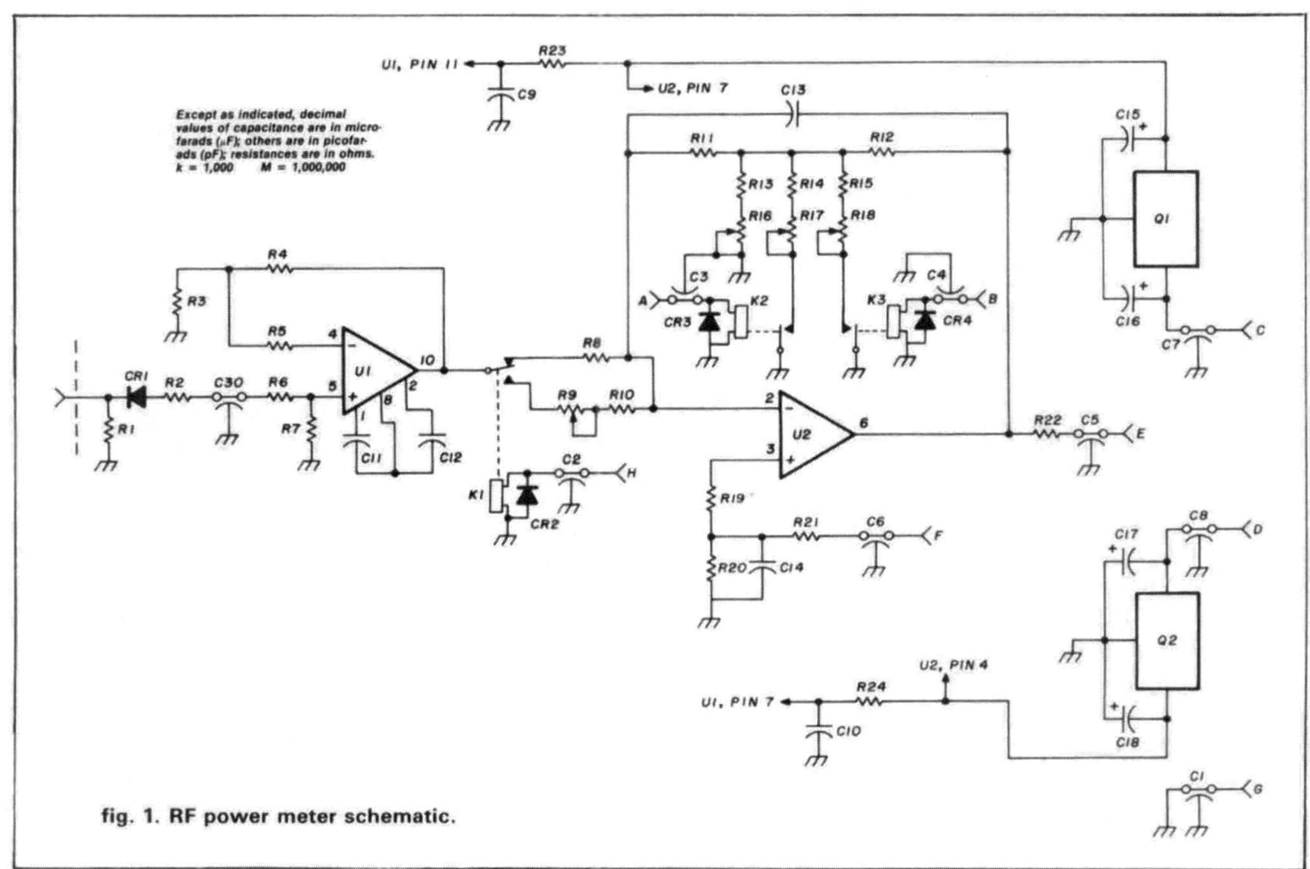

fig. 1. RF power meter schematic.

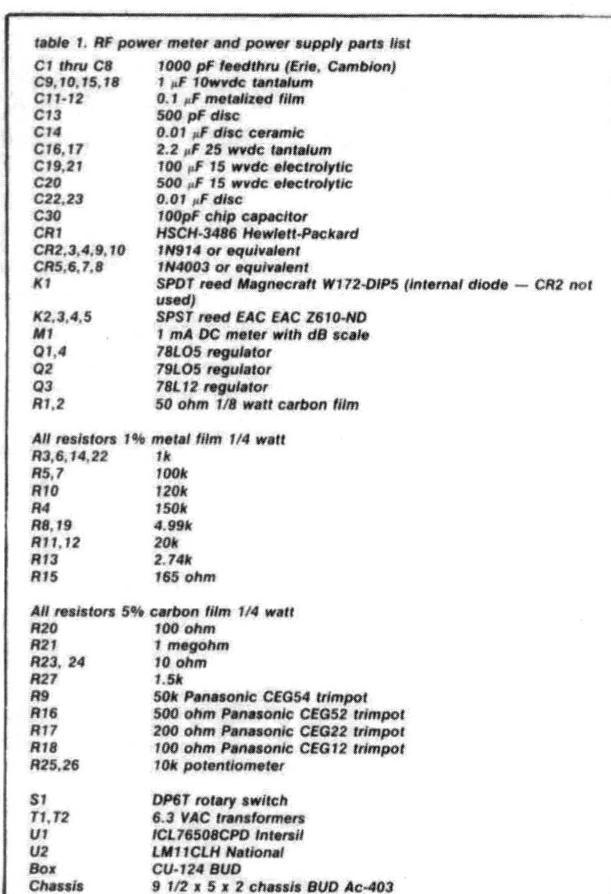

table 1. RF power meter and power supply parts list

C1 thru C8	1000 pF feedthru (Erie, Cambion)
C9,10,15,18	1 μF 10wvdc tantalum
C11-12	0.1 μF metalized film
C13	500 pF disc
C14	0.01 μF disc ceramic
C16,17	2.2 μF 25 wvdc tantalum
C19,21	100 μF 15 wvdc electrolytic
C20	500 μF 15 wvdc electrolytic
C22,23	0.01 μF disc
C30	100pF chip capacitor
CR1	HSCH-3486 Hewlett-Packard
CR2,3,4,9,10	1N914 or equivalent
CR5,6,7,8	1N4003 or equivalent
K1	SPDT reed Magnecraft W172-DIP5 (internal diode — CR2 not used)
K2,3,4,5	SPST reed EAC EAC Z610-ND
M1	1 mA DC meter with dB scale
Q1,4	78LO5 regulator
Q2	79LO5 regulator
Q3	78L12 regulator
R1,2	50 ohm 1/8 watt carbon film
All resistors 1% metal film 1/4 watt	
R3,6,14,22	1k
R5,7	100k
R10	120k
R4	150k
R8,19	4.99k
R11,12	20k
R13	2.74k
R15	165 ohm
All resistors 5% carbon film 1/4 watt	
R20	100 ohm
R21	1 megohm
R23, 24	10 ohm
R27	1.5k
R9	50k Panasonic CEG54 trimpot
R16	500 ohm Panasonic CEG52 trimpot
R17	200 ohm Panasonic CEG22 trimpot
R18	100 ohm Panasonic CEG12 trimpot
R25,26	10k potentiometer
S1	DP6T rotary switch
T1,T2	6.3 VAC transformers
U1	ICL76508CPD Intersil
U2	LM11CLH National
Box	CU-124 BUD
Chassis	9 1/2 x 5 x 2 chassis BUD Ac-403

fig. 2. RF detector and amplifier mounted in a shielded enclosure. Range selection, meter and power supply are in a separate unit.

mended. A suitable unit, Model CD-51, is available from Elcom Systems Inc., 4032 Clint Moore Road, Boca Raton, Florida 33431-2895. The printed circuit board is suspended in the box (**fig. 6**). Two hangers made from 0.015-inch (0.038 cm) brass shim stock are soldered to the ground foil of the printed circuit board and are held by the feedthrough capacitors. Metal and

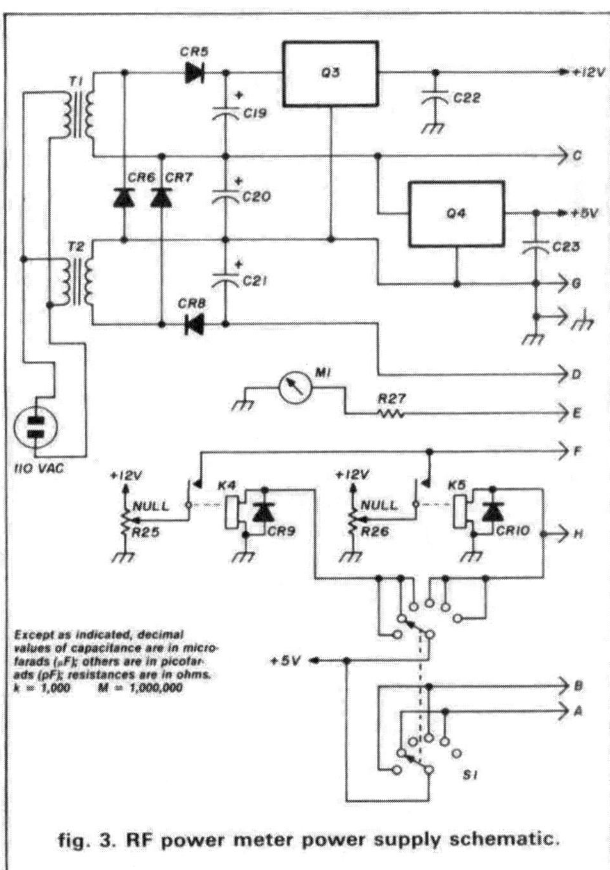

fig. 3. RF power meter power supply schematic.

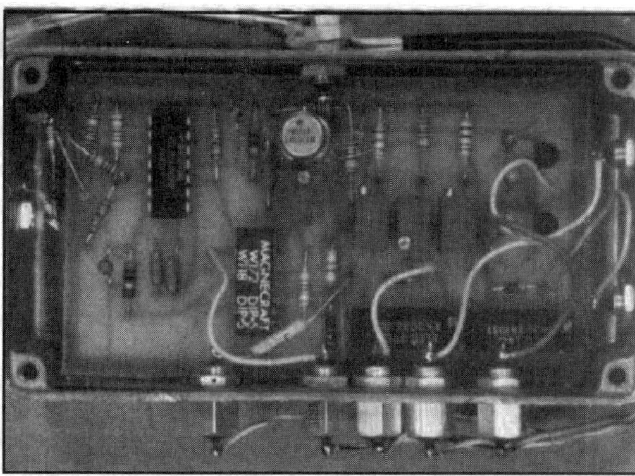

fig. 4. Parts layout in the detector-amplifier. R21, 22, and 6 are soldered between the box and the pc board. C13 is soldered on the back of the pc board under U2.

carbon film resistors are used for accuracy and low noise. The PC board artwork and components layout are shown in **figs. 7** and **8**, respectively.

calibration

Calibration depends on the accuracy of the standard used. If you have no fixed attenuator, purchase the Model AT-51 5 dB TNC from Elcom ($14). Set the meter to the −15 dB range, check and adjust for zero with no signal applied. The meter zero pot has little control on this scale and if the meter doesn't read zero, there's something wrong with the circuit. Adjust a signal generator for a +30 dBm output level and turn R16 for full scale or 0 dB on the meter. The frequency of the generator is not important — in this case, 150 MHz was simply convenient. If the signal generator has no dBm scale, turn R16 to midpoint and adjust the signal generator for 0 dB. Insert a 5 dB attenuator. The meter should read −5 dB. Turn to the −20 dB scale while momentarily disconnecting the signal generator, then check and adjust for zero. The meter zero pot should show more control. Reconnect the signal generator and adjust for 0 dB with R17. Insert 5 dB of attenuation and the meter should again read −5 dB. Turn to the −25 dB scale and repeat the above procedure. The meter zero pot will have quite a lot of control. Note that on the −25 dB scale the needle shows some jitter or drift. This is circuit noise. This drift should be less than ±1/10 dB at full scale. Return to the −15 dB scale, insert 5 dB of attenuation and increase output for a 0 dB reading. Turn to the −10 dB scale, remove the signal generator and adjust for zero with the right side meter zero pot. Remove the attenuator and reconnect the signal generator. Adjust R9 for 0dB. Insert attenuator; adjust signal generator

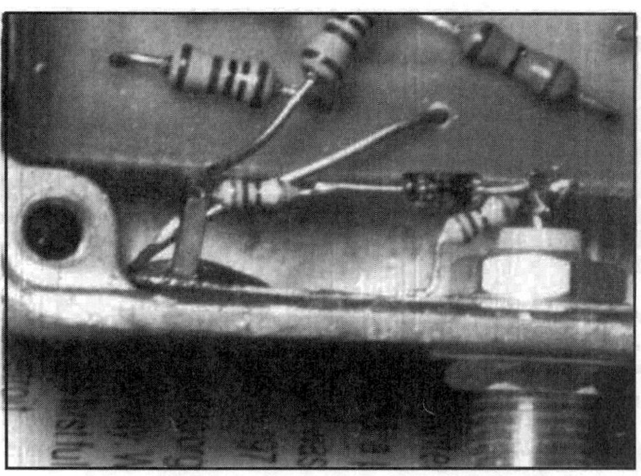

fig. 5. Close-up of the detector circuitry.

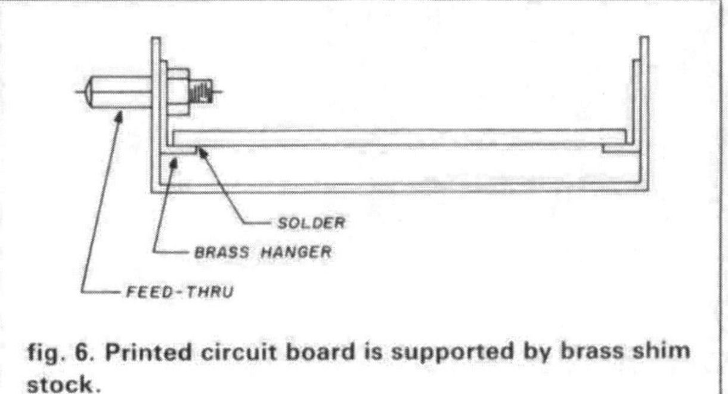

fig. 6. Printed circuit board is supported by brass shim stock.

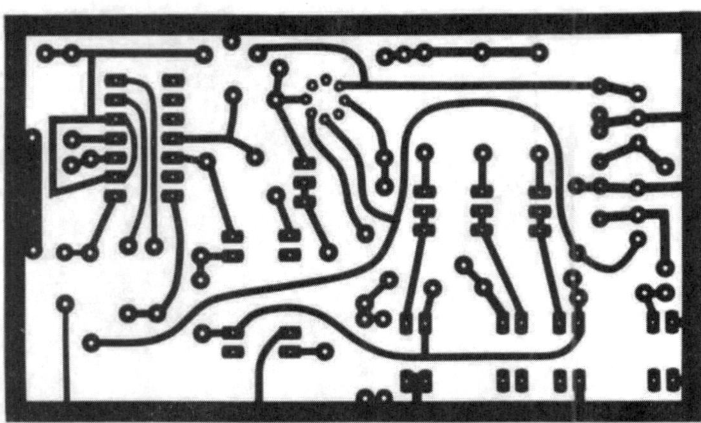

fig. 7. RF power meter artwork.

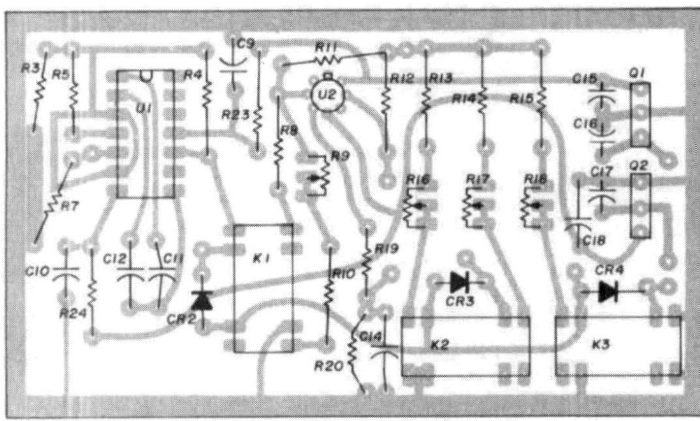

fig. 8. Component layout.

fig. 9. RF power meter is used, in a typical application, to determine hybrid coupler isolation value.

table 2. Parts suppliers.

C1 thru C8, C30, R1, 2, TNC connectors
Microwave Components
11216 Cape Cod
Taylor, Michigan 48180

U1, CR1
Hamilton-Avnet
Local outlets in most states. Distributor will ship COD to individuals. Cost of part approximately $25.

U2
Jameco Electronics
1355 Shoreway Road
Belmont, California 94002

K1 and Bud chassis and box
Newark Electronics
Distributors in most states.

All other parts
Digi-Key
701 Brooks Avenue, South
Thief River, Minnesota 56701

Printed circuit is available from author for $11.00. Please add $1 for shipping.

for 0dB. Turn to the −5 dB scale and remove the attenuator. The meter should read 0 dB. Insert the attenuator again and adjust the signal generator for 0 dB. Turn to the 0 dB scale and remove the attenuator. Note that the meter doesn't read 0 dB, but it should be within 1/4 dB of full scale. We are now starting to run into the nonlinear portion of the detector diode.

using the power meter

Figure 9 shows a typical set-up in which the power meter is used. A 70 cm hybrid coupler is checked for isolation between port 1 and port 2. The ICOM-471A provides the signal with its output reduced by a 10dB-10 watt attenuator to less than 1 watt. Further attenuation is introduced by a step attenuator.

an rf voltmeter

Don't let its apparent simplicity fool you — this instrument has many uses

Many of us who experiment with circuits need to measure the level of signal sources such as oscillators, amplifiers and multipliers in transmitters, and local oscillator systems in receivers. The voltmeter design described in this article came about when I wanted to measure the voltage reflection coefficient of antenna systems using a return loss bridge at low levels so as not to interfere with other band users.

The common method of measuring signal levels is through the use of a simple diode detector. In its basic form, however, it has a number of shortcomings, some of which can be easily overcome.

voltmeter requirements

This voltmeter covers a range from less than 70 millivolts to greater than 3 volts rms (equivalent to −10 to +23 dBm in a 50-ohm system), covers a frequency range from 10 kHz to 150 MHz, and provides readings accurate to within ±2 dB without calibration — i.e., as built and tested. Its input impedance is set by the input resistor; a value of 50 ohms was used in the models shown. Its output is linear; if an analog meter is used, no special marking of the meter scale is necessary. An external general-purpose meter can also be used. The meter draws less than 15 mA from a pair of 9-volt transistor batteries.

diode detectors

The characteristics of an ideal linear voltage detector are illustrated in **fig. 1**. This mythical device conducts current in one direction only, with a low and constant resistance when forward biased and an infinite resistance in the reverse direction. The constant forward resistance is maintained right down to 0 volts,

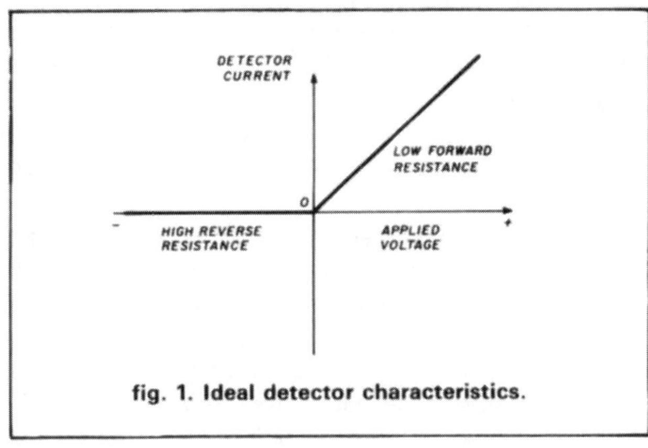

fig. 1. Ideal detector characteristics.

By Ian Braithwaite, G4COL, 28 Oxford Avenue, St. Albans, Herts, AL1 5NS, England

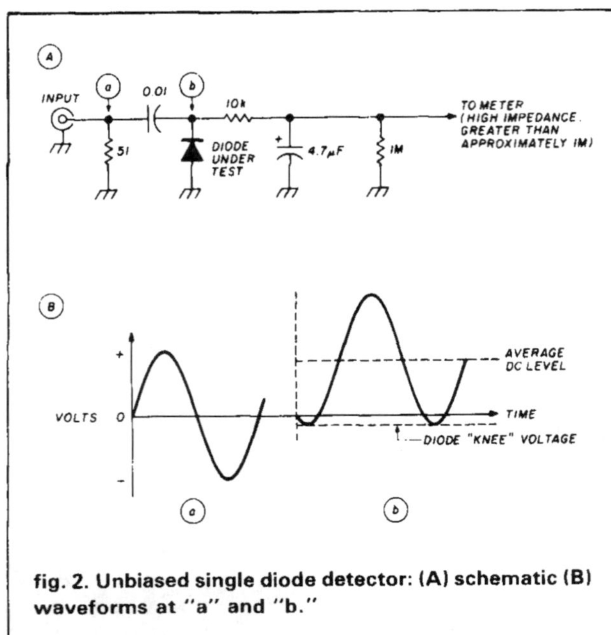

fig. 2. Unbiased single diode detector: (A) schematic (B) waveforms at "a" and "b."

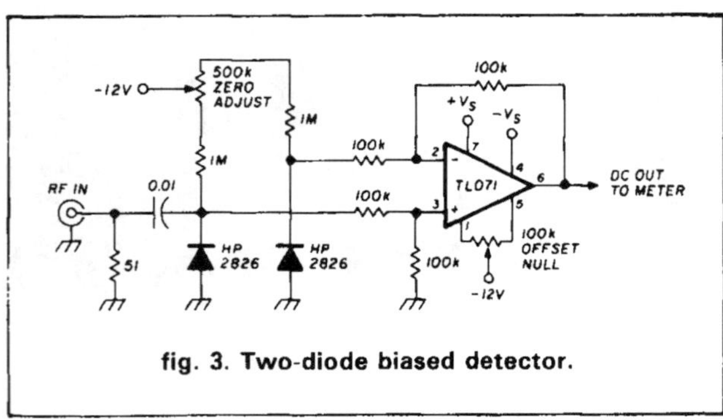

fig. 3. Two-diode biased detector.

with an abrupt transition to the reverse region. Such a device used as a rectifier would deliver a dc output proportional to the applied ac.

Real diodes, however, don't behave this way. Most do not conduct appreciably in the forward direction until the input voltage across them exceeds a threshold or "knee" voltage, which for an ordinary silicon junction diode is around 0.7 volts. The threshold voltage for germanium and Schottky diodes is lower — 0.2 to 0.4 volts. Real diodes also conduct slightly in the reverse direction (the so-called reverse leakage current).

The transition between the conducting and nonconducting states is not sharp, but occurs over a region where the diode is said to have "square law" behavior and the dc output is proportional to the applied *power* (voltage squared), rather than the signal voltage. This is used to advantage in low-level diode power meters.

To see how real detectors behave, I made some measurements on a few types using a crystal oscillator signal source at 10 MHz, and a power meter and attenuator to give a range of calibrated levels. **Figure 2** shows the first test circuit, a simple peak detector using an HP2826 Schottky diode. The right-hand plate of the input capacitor is clamped at the diode knee voltage below ground on negative input swings. If this knee voltage were actually zero, the average voltage on the diode would equal the peak of the input voltage, but the real diode produces less. The resistor and capacitor filter the rf present on the diode, leaving the dc component. This can be measured by a high impedance meter which reads the peak of the rf voltage minus the diode knee voltage.

A dc forward bias current can be used to improve the sensitivity of the diode detector. If the diode is fed from a high resistance with a current of a few μA, its forward junction voltage will sit around the knee voltage. This potential no longer has to be supplied by the rf, which sweeps the diode's nonlinear characteristic and is detected. Direct current bias is used in the more sophisticated circuit shown in **fig. 3**. Two diodes are used. Both are biased, but rf is fed to only one of them. An op-amp subtracts the diode voltages so that the output of the circuit can be set to zero in the absence of an rf signal. With the diodes connected together, with no rf, the op-amp offset is nulled. The 500-k pot is then adjusted to give zero output with the circuit exactly as drawn. The circuit works best with matched pairs of diodes, since these track well with temperature.

The performance of these detectors is described graphically in **fig. 4**. This shows the improvement in sensitivity achieved with bias. Also shown is the curve for the voltmeter design, which indicates further improvement in sensitivity and linearity, gained by using just one additional technique.

The complete rf voltmeter is shown in the block diagram in **fig. 5**. Two detectors are used. One receives the incoming rf signal, while the other, a "mimic" detector, is fed with a low frequency signal. This signal is an internally generated sinusoid, derived by chopping the dc output of an integrator to give a square wave which is then filtered, leaving the fundamental frequency component.

The integrator input is the difference of the two detector voltages, and its output will change (or "slew") when this difference (the feedback loop's error signal) is other than zero. The action of the negative feedback loop around the mimic detector in fact causes the integrator to try to achieve this zero-error condition, at which point, if the detectors are well matched, the low frequency signal will have the same amplitude as the rf signal. Because the low frequency signal is produced by chopping and filtering the integrator's dc output, this latter voltage is proportional to the rf input voltage, and can be scaled and metered to provide readings in "rf" volts.

Through the use of well-matched and closely spaced

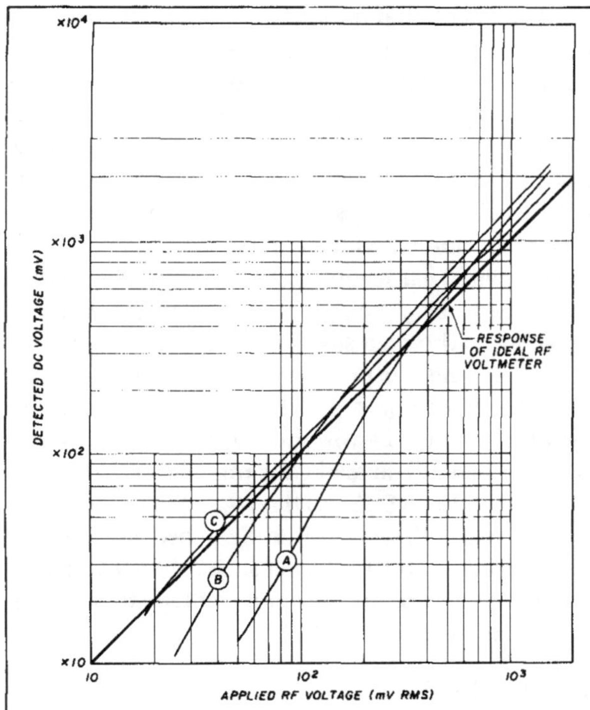

fig. 4. Detected dc output versus applied rf input: (A) single unbiased diode detector; (B) biased diode detector; (C) rf voltmeter; (D) ideal rf voltmeter.

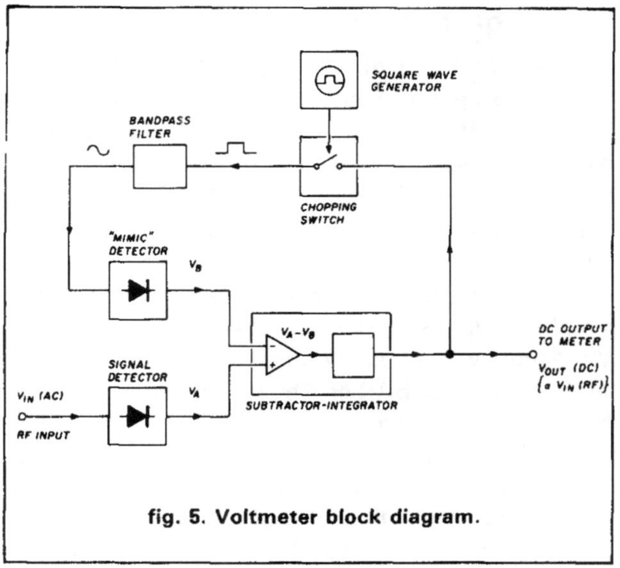

fig. 5. Voltmeter block diagram.

diodes, their temperature and I-V curve variations are minimized. Of course, since the mimic detector measures only a fixed low frequency signal, no frequency response compensation for the input detector is provided. Therefore, it's best to choose the diode that has the flattest possible response.

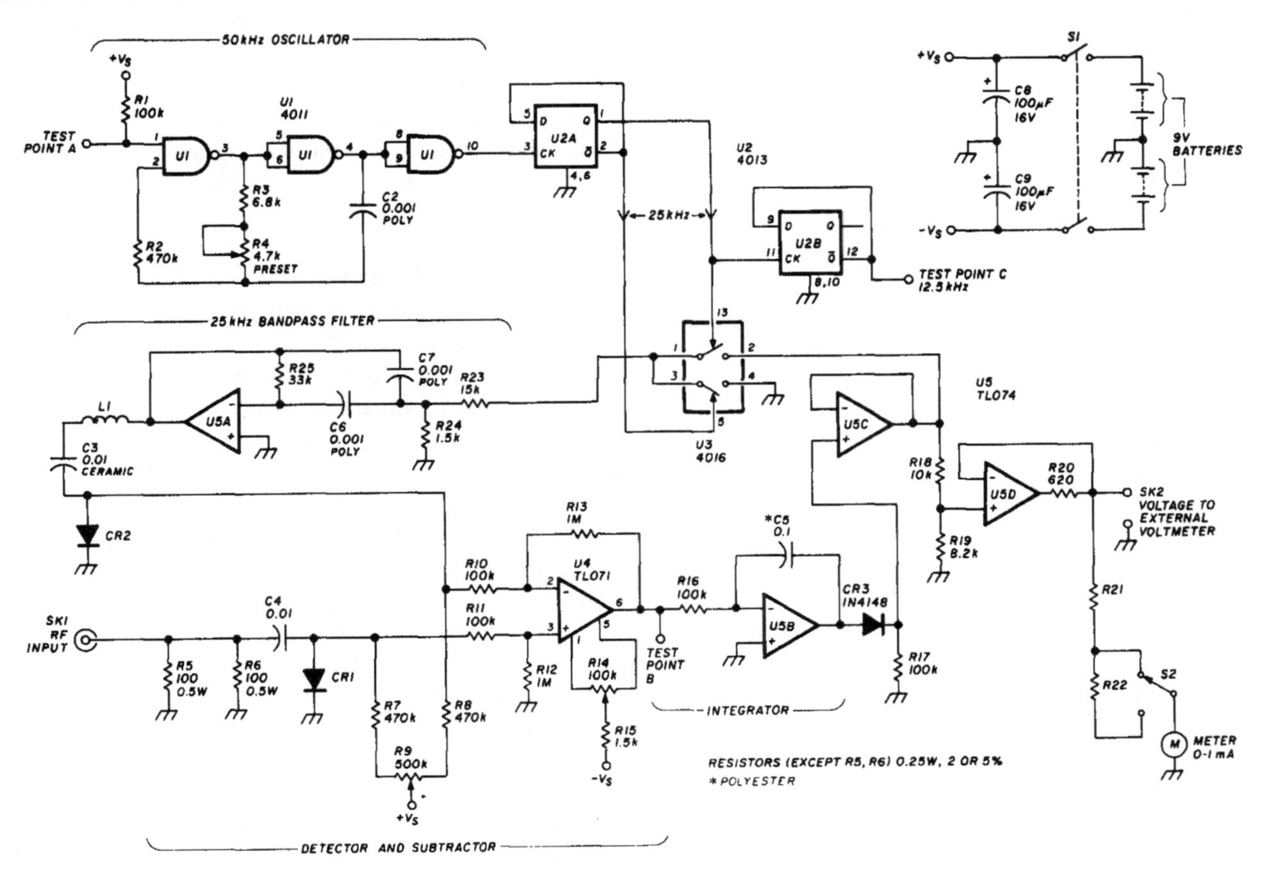

RESISTORS (EXCEPT R5, R6) 0.25W, 2 OR 5%
* POLYESTER

fig. 6A. Voltmeter schematic diagram.

circuit description

RF enters the instrument via socket SK1. R5 and R6 provide a good impedance match to 50-ohm cable. Two 0.5-watt metal film resistors (or any other combination providing 50 ohms and a 1-watt rating) should be used. Diodes CR1 and CR2 form the detectors: CR1 is supplied with rf, and CR2 with the internally generated 25-kHz sine wave. They should be Schottky diodes and, if possible, should be reasonably well matched in terms of forward voltage at around 10μA. Many types will do, among them the HP2800 and 2826 and the Thomson BAR28. The forward voltage will be in the region of 250 mV, and a pair matched to within a few millivolts can often be found from a small batch. Circuitry around U4 performs subtraction of the two detector outputs with a gain of 10. R14 allows U4's offset voltage to be nulled. Because this is a relatively high gain stage, the remaining stages do not need to be nulled, and can be grouped into a quad package. U5B is wired as an integrator, and CR3 and buffer U5C prevent the CMOS switch U3 from being driven negative. The buffer's dc output is chopped by switch U3, which is operated by 25-kHz square waves from U2, a divide-by-2 flip-flop fed from U1, a 50-kHz oscillator. U2 could be omitted and U1 run at 25 kHz, but U2 does achieve a perfect square wave (1:1 mark-to-space ratio) at small cost. U2B provides a 12.5-kHz output signal so that a rough check can be made on the switching signal with a crystal earpiece at test point C. The chopped dc from U3, now a 25-kHz square wave with a peak-to-peak amplitude equal to the U5C dc output, is filtered by active filter U5A (second order bandpass) so that CR2 receives a fairly sinusoidal signal. L1 was included to stop hf oscillations in the output stage of U5A when the connection to the detector was completed via several inches of ribbon cable (the adjacent wire being grounded). It consists of four turns of enameled wire on a single-hole ferrite bead (I used an FX1115), and has no measurable effect at 25 kHz. R18 and R19, buffered by U5D, attenuate the dc by a factor of 0.45 (see **appendix** for derivation), which provides scaling to units of volts rms. An external voltmeter plugged into SK2 will then read the rf input voltage. R21 and R22 with switch S2 allow the use of a meter to read 1 and 10 volts full-scale. With 9-volt supplies from batteries, the maximum voltage that can be read will be around 3 volts.

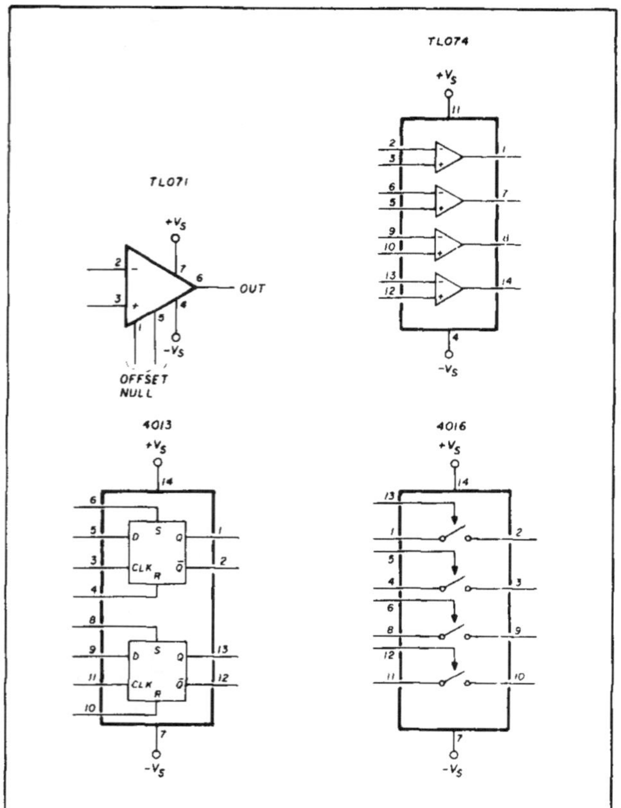

fig. 6B. Voltmeter integrated circuit pinouts.

construction

I have built three instruments according to the design described in this article (see **fig. 7**). The outer two are battery powered (internally); the "economy model" in the center uses an external power supply and meter. Details of the unit on the left are shown in **figs. 8A, 8B,** and **8C**. Construction is straightforward and can be done with ordinary hand tools. The only critical area is the detector, which carries rf. The other areas involve only low frequency circuitry. As shown in **fig. 8C**, I built the input circuitry, consisting of the two detector diodes and U4, on a small piece of double-sided, copper-clad glass-fiber board, using a counterbore tool to provide pads for the components. Frankly, this method of construction — with the components mounted rats-nest style above a copper ground plane, will work at least as well and probably better than a pc board, and is certainly much faster. Those willing to make a pc board for the voltmeter are welcome to do so — I'm afraid I'm too lazy!

The rest of the circuit is wired on a perforated breadboard with copper strips on the underside (known in the UK as "Veroboard"). As the photo of the detector board shows, the signal connection from the front panel socket was made using RG-178 coaxial cable. The only place it's important to keep leads as short as possible is in the detector area. Try to mount the two diodes close to each other for good thermal tracking.

alignment and testing

Check the 50-kHz oscillator and divider by placing a crystal earpiece or high impedance audio amplifier between test point C and ground. The frequency can now be adjusted to coincide with the bandpass filter.

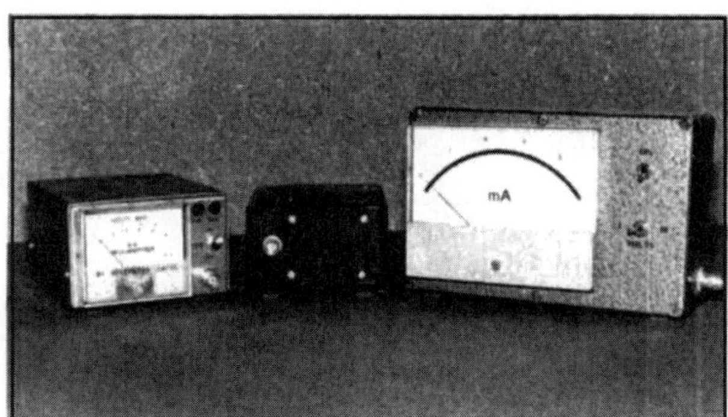

fig. 7. Three different packaging approaches for the same voltmeter.

Turn R9 or R14 so that U4 output is slightly negative, which will cause the integrator U5B to slew to the positive limit. This should result in a healthy square wave output from U3 (pins 1 and 2). A high impedance meter should read a voltage (dc) here that is half that at the U5C output. If the meter is transferred to the anode of CR2, it should be possible to peak this voltage by adjusting the oscillator frequency control R4.

The detector circuit can now be set up with no input. Ground test point A to stop the 50 kHz oscillator. Connect CR1 and CR2 so that the subtractor sees the same voltage at both its inputs. Set test point B to zero volts with R14. Remove the connection between the diodes, and again zero-test point B, this time with R9.

If rf is now applied, U5C should go positive, and the voltage at the output socket SK2 should be 0.45 of this. The meter is now ready to use.

performance

The absolute accuracy and linearity of this meter is illustrated in **fig. 4**, which was constructed from measurements made at 10 MHz. The flatness with frequency was measured at 1 mW, 224 mV rms (0 dBm), and the results are shown in **fig. 9**, which represents a respectable performance of within ±3 percent up to 150 MHz. This could no doubt be improved to extend the useful range to 70 cm and beyond.

To verify the repeatability of these measurements, I tested the three units against each other using the same source, a 10-MHz crystal oscillator. Referring to the units by position in the photo, the results were:

Unit	Reading on external voltmeter
Left	268 mV
Center	244 mV
Right	258 mV

Obviously, three is only a small sample, but considering that the voltmeters had received only the simple dc setup procedure described earlier, I was quite pleased with the outcome, and I hope that this sort of performance will be adequate for your applications.

further suggestions

I hope that readers who build this voltmeter will find it a handy instrument to have around the shack. Those who like to experiment and develop their own hardware might enjoy exploring the following options:

A

B

C

fig. 8. Internal views of the rf voltmeter: (A) internal power provided by two 9-volt batteries; (B) "clean" construction enhanced by use of Veroboard; (C) "rf" section of voltmeter.

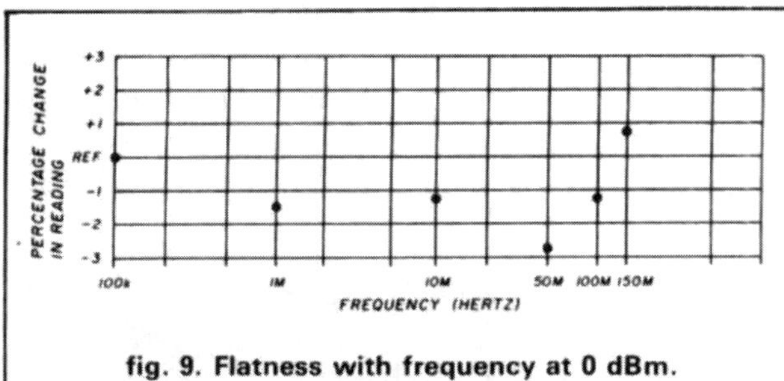

fig. 9. Flatness with frequency at 0 dBm.

• The design can be simplified by omitting the divider U2. The oscillator could then be run at 25 kHz, or the filter redesigned for 50 kHz. (The choice of frequency was somewhat arbitrary, being high enough to use small coupling capacitor C3 and low enough for active filtering. U2 does guarantee an excellent square wave, but the oscillator alone may well be adequate.)

• The detectors can be built into a high-impedance probe for circuit tracing, rather than a 50-ohm instrument. Keep CR1 and CR2 physically and electrically close together, though.

• By paying attention to the detector matching and circuit offsets, particularly around U4, the useful range could be extended downwards. With attenuators, the range could be extended upwards.

• Careful selection of devices and construction could greatly extend the frequency range.

• The filtering of the square waves could be improved. The units I have built tend to read slightly high, and this could be because the active filter output is not a pure sinusoid, giving a slightly wrong scaling factor. Why didn't I just feed CR2 with a raw square wave? Well, when I tested a diode detector using an accurate function generator, the peak readings were different between sine and square waves — i.e. the diode appeared to clamp at slightly different voltages, depending on the waveform. I wish I knew why; in any case, the results might be worth repeating. If CR2 gave the same response to square waves, the active filter could be omitted. U5C output would then be the peak input voltage, and scaling by 0.707 would give readings in volts rms.

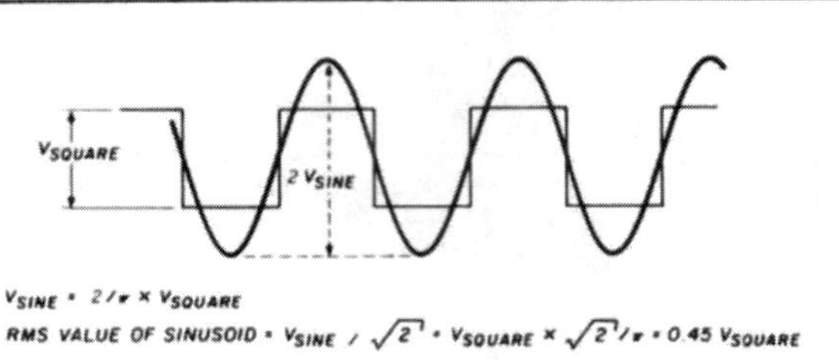

fig. A1. Relationship between a square wave and its fundamental sinusoidal component.

appendix

how readings are scaled to volts rms

The voltmeter works by making an internally generated sine wave derived from filtering a square wave equal to the rf input. The square wave, then, is generated by chopping a dc voltage. As illustrated in **fig. A1**, Fourier theory tells us that the fundamental (sinusoidal) component of a square wave has a larger peak amplitude than the square wave itself (don't worry, there is less power in this sine wave). If we call the peak amplitude of the sine wave V_{sine} and the peak-to-peak amplitude of the square wave V, then;

$$V_{sine} = \frac{2V}{\pi}$$

But in the voltmeter circuit, the peak-to-peak square wave amplitude is equal to the integrator's dc output voltage V_{dc}, that is:

$$V = V_{dc}$$

We want to make the voltmeter read rms volts. If the applied rf has an rms voltage V_{in}, then the feedback loop makes:

$$V_{sine} = \sqrt{2}\ V_{in}$$

So, the quantity we want to measure, V_{in} is given by:

$$V_{in} = \frac{V_{sine}}{\sqrt{2}} = \frac{\sqrt{2}\ V_{dc}}{\pi} = 0.45\ V_{dc}$$

This is why the dc produced by the integrator is scaled by 0.45.

THE WEEKENDER

Build this simple L-C checker

This simple capacitance-inductance checker measures capacitance to approximately 1000 pF and inductance to 50 µH. I first saw it described in *QST* some 36 years ago.[1] That version used a filament-type tube (3A5) in a self-rectifying oscillator powered from 115 volts AC. I built one in a cigar box lined with metal foil. It's seen constant use in my shack with the original tube!

The initial L-C checker measured capacitance only. I modified it to add inductance measuring capability when I transistorized the unit.

The capacitance circuit measures to 5000 pF or more; however, accuracy and resolution are reduced above 1000 pF because of circuit limitations. But, when you pick a mica capacitor marked with two red dots from your junkbox, this unit will quickly tell you if it's 2.2, 22, or 220 pF. The device has the advantage of applying no voltage or current (other than a few millivolts of RF) to the capacitance or inductance under test.

How it works

The circuit is based on the "grid-dip" or absorption effect, which occurs when a parallel resonant circuit is coupled to an oscillator of the same frequency. If you look at **fig. 1**, you'll see that Q1 operates in a conventional Colpitts oscillator circuit at a fixed frequency of approximately 4 MHz. The exact frequency isn't critical. A meter connected in series with the transistor's base-bias resistor serves as the "dip" or absorption indicator.

By Jack Najork, W5FG, 723 Flamingo Way, Duncanville, Texas 75116

FIGURE 1

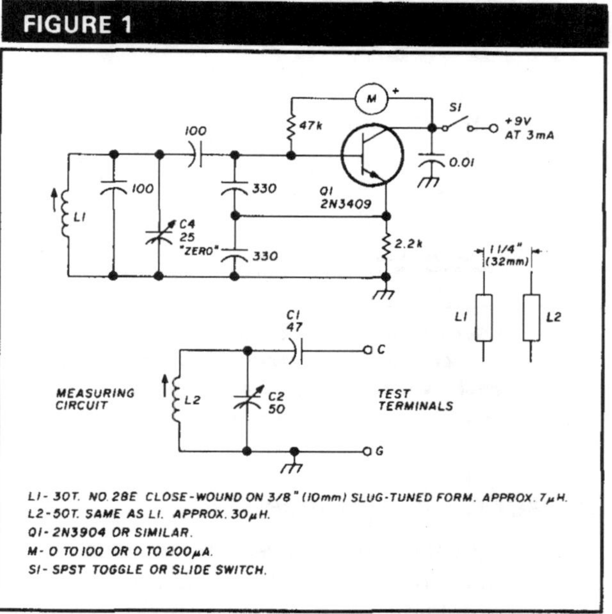

Schematic diagram of L-C checker with measuring circuit for measurement of capacitance only. L1 and L2 should be spaced as shown for optimum electrical coupling.

The variable measuring circuit consists of C1, C2, and L2 and is connected to panel terminals as shown. L2 is loosely coupled to L1 in the oscillator circuit. This measuring circuit is tuned to the oscillator frequency with variable capacitor C2 set at full or (maximum) capacitance. When power is applied to the oscillator the meter shows a dip caused by power absorption by the measuring circuit.

Connecting an unknown capacitor across the test terminals lowers the resonant frequency of the measuring circuit. To restore resonance, tune capacitor C2 lower in capacitance. The meter will dip again when you reach this point. Determine the capacitance across the test terminals by calibrating the dial settings of C2. More on this later.

Capacitor C4, a small variable trimmer in the oscillator circuit, compensates for drift or other variations and is normally set at half capacitance. It's a panel control, labeled "ZERO", and is used to set the oscillator exactly at the dip point when C2 is set at maximum capacitance. This corresponds to zero on the calibration scale.

You can also use C4 to compensate for the capacitance inherent in long leads running from the test terminals to the unknown capacitor. The leads are connected to the test terminals and dressed close to the capacitance to be measured, but not connected to it. Adjust C4 for a dip with C2 at zero. Then connect the leads and make your measurement.

FIGURE 2

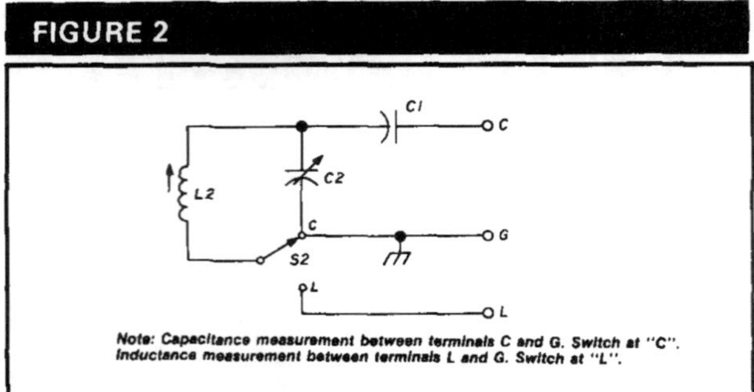

Addition of S2 to the measuring circuit enables measurement of both capacitance and inductance. Note: For capacitance measurement between terminals C and G, switch at "C". For inductance measurement between terminals L and G, switch at "L".

Measuring inductance

After I got the transistorized version working, it occurred to me that I should be able to use the circuit to measure inductance as well. **Figure 2** shows how. The oscillator circuit remains unchanged. Add an SPDT switch (S2) to the measuring circuit. With S2 in the "C", or capacitance measuring position, the circuit is as before. In the "L", or inductance measuring position, the switch disconnects the bottom of L2 from ground and connects it to a third panel terminal marked "L".

Connecting an unknown inductance across the "L" and "G" terminals again lowers the resonant frequency of the measuring circuit because the unknown inductance is now in series with L2. You must tune C2 lower in capacitance again to restore resonance. As with capacitance measurements, the best resolution occurs at the lower inductance values.

To measure inductance, set S2 to "C" and adjust the oscillator (ZERO) control for a dip with C2 at zero. Set the switch to "L" and connect the unknown inductance across the "L" and "G" terminals.

Construction notes

You'll need solid construction for consistent calibration accuracy. Use the VFO construction techniques from any handbook. Choose a good-sized, sturdy metal cabinet so that you can use a large dial or pointer for the measuring capacitor, C2. I glued a 2-inch pointer to my dial and spread the calibration marks across the largest dimension of the cigar box.

The spacing (electrical coupling) between L1 and L2 isn't critical, though it can be adjusted if necessary. Wider spacing produces a shallower dip, but improves the measuring resolution. Conversely, closer spacing (coupling) produces a more pronounced dip, which lowers resolution.

Make the wiring from the bottom of L2 to switch S2, and from S2 to the panel terminal "L", short and low inductance. Use heavy wire or copper ribbon, because these connections form a portion of the unknown inductance placed across terminals "L" and "G". Too much inductance in these connections produces erroneous inductance readings when measuring very small inductances.

Calibration

Calibrate the capacitance range by placing known values of capacitance across the "C" and "G" terminals and marking these values on the C2 dial or pointer. Most of us have enough well-marked mica and ceramic capacitors in our junkbox to make this easy. Don't forget, you can also use these capacitors in parallel or series, as needed.

Inductance scale calibration is a bit tougher. You might want to beg, borrow, or otherwise obtain a collection of small, molded RF chokes. I had a half dozen 10-μH chokes and used them in series and parallel to calibrate. This method may not result in laboratory-type accuracy, but it'll bring you right into the ballpark. **Figure 3** shows my unit's calibration.

FIGURE 3

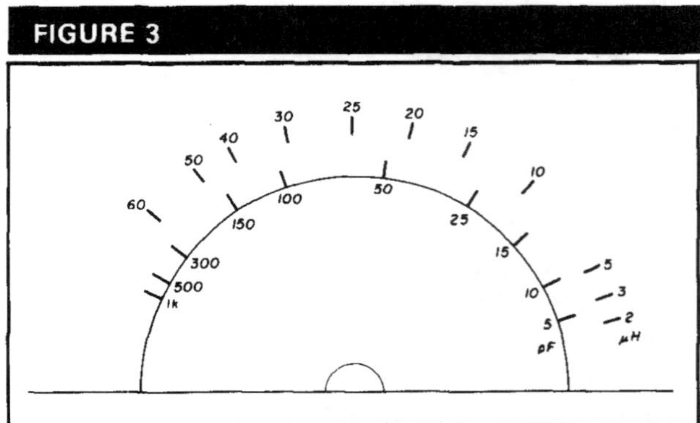

Typical range scale calibration made with C2 and 50 pF, Hammarlund HF 50 with semi-circular plates. Capacitors with a different plate shape will yield a different calibration.

Using other parts

You engineering types will by now have observed that juggling the values of C1 and C2 alters the range and linearity of the measuring range. For example, increasing C1 expands the calibration at the low picofarad end but limits the upper capacitance measurement. If L2 is an adjustable coil, as shown, select C1 and C2 for the desired results, retuning L2 as needed to restore resonance.

If your junkbox dictates, use a larger value capacitor for C2. This requires increasing the value of C1. One possible combination is 150 pF for C2 and 130 pF for C1, with L2 reduced to approximately 10 μH. To limit the inductance measuring range with this combination it's necessary to add a small fixed capacitor in parallel with C2. Try 10 to 15 pF.

If the required microammeter isn't available, take heart. Since the unit draws around 3 mA at 9 volts, you can use a zero to 5 mA meter in series with the + lead to the power source. Resonance will now be indicated by a *rise* in the meter reading instead of a dip. The variation between dip and non-dip won't be as pronounced as with a base current meter, but it will be usable.

If 3/8-inch diameter coils aren't available you can use 1/4-inch forms. Scrounge them from the i-f section of a defunct TV set. You'll need more turns, and the "Q" will be a bit lower, but they'll work. Conversely, for those who want improved resolution, substituting larger diameter coils wound with heavier wire will increase "Q". This enables looser coupling which in turn produces a sharper dip.

References

1. *QST*, March 1952.

Article B

WIDEBAND FREQUENCY ANALYZER

Analyzer checks for spurious oscillations and amplitudes, and assists in tuning

By Adelbert Kelley, AA4FB, 2307 S. Clark Avenue, Tampa, Florida 33629

As a result of interest generated by articles I'd read in *QST*[1] and *Ham Radio*,[2] I decided to build the wideband frequency analyzer shown in **Figure 1**. I needed a test instrument that would show spurs and relative signal amplitudes, and would help me with alignment when I experimented with VHF and UHF circuits. The idea of using the shop oscilloscope for a panoramic display of the signals from 20 to 400 MHz looked attractive. K2BLA's analyzer in *QST* had an ingenious circuit which gave the frequency on a digital display. He used the scaler output of a cable TV converter in a frequency counter that subtracted the oscillator offset, then counted and presented a three-digit readout of the frequency of any signal present at the center graticule in the oscilloscope display. He measured unknowns by tuning the analyzer until the signal crossed the center line.

You can use my analyzer to check circuits for spurious oscillations and amplitude, and assist in tuning. It needn't be tuned precisely to display a signal, doesn't require close coupling, and has a wider frequency range than a grid dip meter.

My unit has the following specifications:

- Frequency coverage of 54 to 900 MHz in two bands with no gaps in coverage.
- A digital readout which displays the frequency of the center graticule of the oscilloscope to + or − 1 digit. It continues to give a useful reading even when the frequency sweep is reduced to zero and only the manual tuning knob is used.
- Two steps of selectivity, good sensitivity, and uniform frequency response. It will display all the FM, TV, Amateur VHF/UHF, and communications services in a metropolitan area when connected to a TV antenna or even a random wire.
- FM audio monitoring to identify unknown stations, and a simple design with parts count at a minimum consistant with good results.

Of course this isn't a spectrum analyzer. As a homebrewer, I don't have the resources to produce a true spectrum analyzer. The unit only approximates a logarithmic amplitude response, which is good enough for me. I use a decade attenuator[3,4]* to read relative dB amplitudes when I need more accuracy. The voltage versus frequency response of varactor diodes in cable and TV tuners is inherently nonlinear. This matters only if you want to read the frequency off the graticule of the scope. A better method is to rotate the fine tuning until the signal in question is at the center of the graticule, and read out the frequency on the digital display.

TV tuners have been evolving away from switched inductor and capacitor to solid-state varactor tuning. The varactor tuner uses the capacitance change across a silicon diode that takes place when a reverse DC voltage is varied. The capacitance increases as this control voltage is lowered. If a sawtooth waveform of sufficient amplitude and correct average DC is substituted for this DC control voltage, the tuner tunes its frequency range rapidly, and very little circuitry is needed to obtain a panoramic display. The combination of solid state and no moving parts means greater reliability, smaller assemblies, and easy interfacing. This, in turn, results in low cost and high performance.

Tuner construction details

The NEC cable tuner requires no modifications. It's well shielded and appears to have a good frequency response. The 53-channel display was almost flat when I checked our local cable system on the completed analyzer.

I wanted higher frequency coverage than 400 or 450 MHz and had a UHF tuner left over from a previous project,[5,6] so I decided to try it. However, UHF tuners don't have built-in scalers. I solved this problem by adding an RCA CA3179E chip to the tuner in a circuit that should also be easy to add to any make of solid-state tuner.

Most UHF TV tuners cover channels 14 through 88 only, but the Mitsumi covers more — 500 MHz, actually. It's worthwhile to try to find one. Look for a UES-A55F or UES-A56F. Radio Shack sold one (stock no. 277-220) in 1983, but I

---

* Available as a kit from Circuit Board Specialists, (303)542-4525.

FIGURE 1

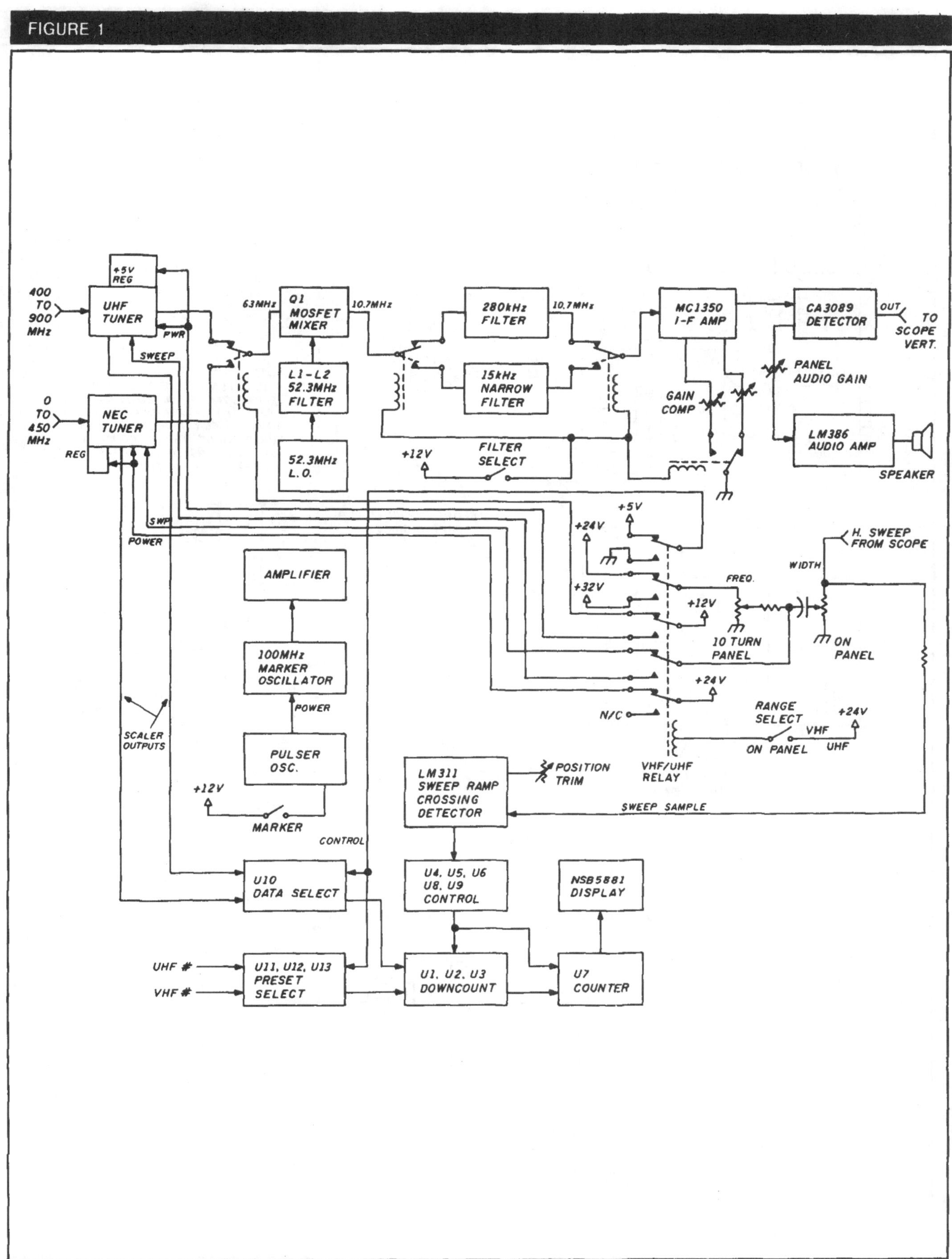

One line diagram frequency analyzer.

FIGURE 2

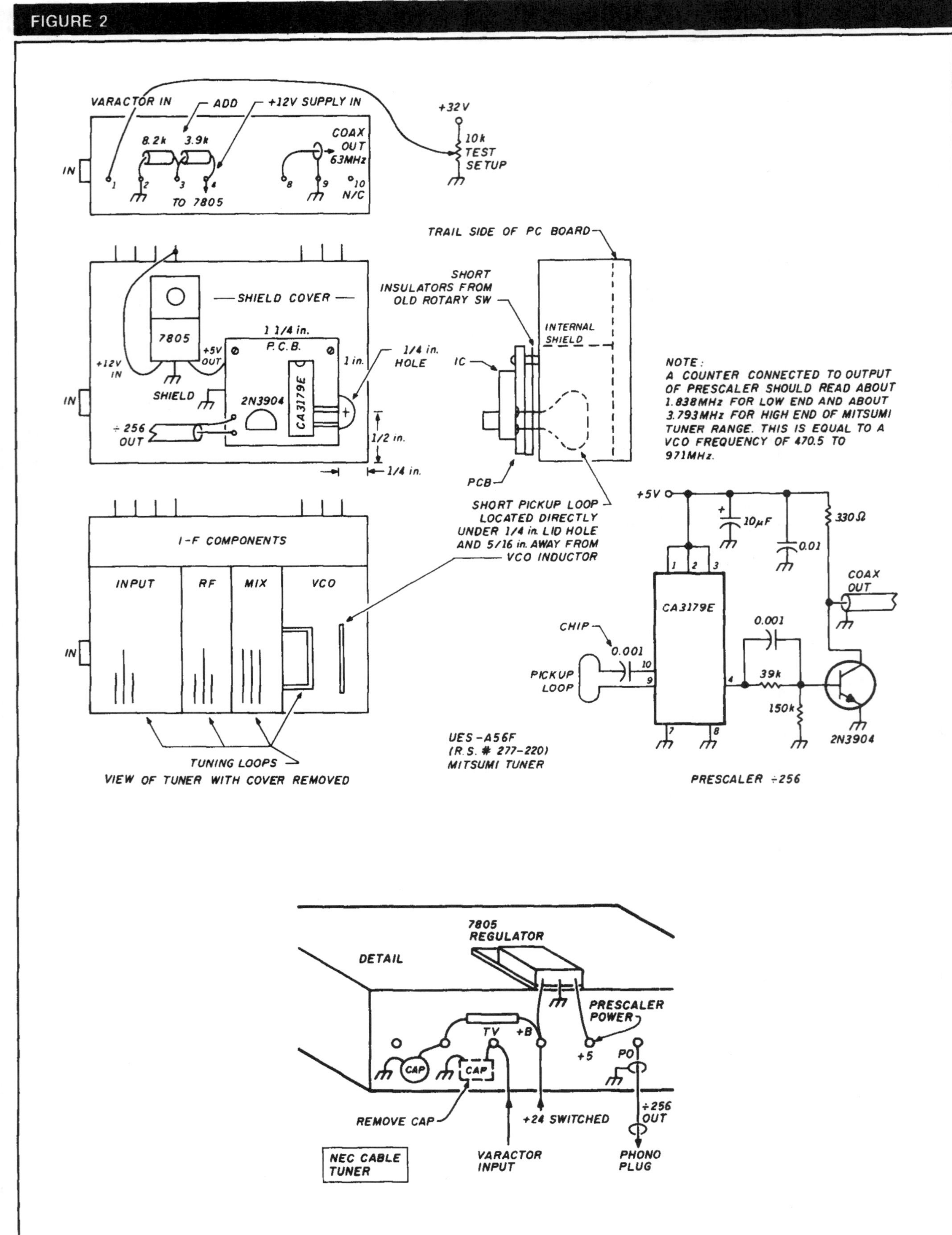

Mitsumi tuner prescaler installation detail.

FIGURE 3

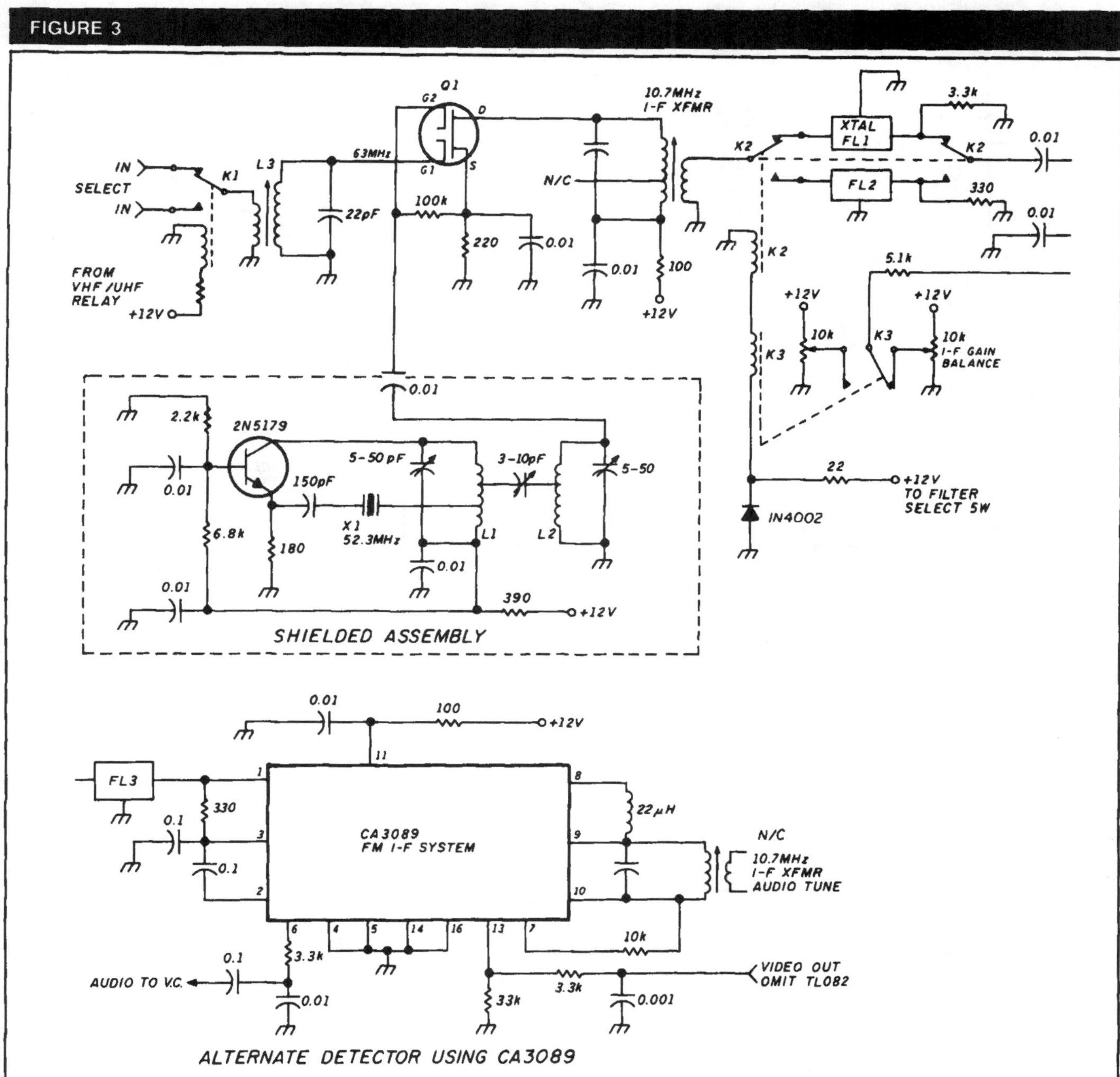

Spectrum analyzer 10.7-MHz IF assembly.

haven't seen any recently. If you can't locate a Mitsumi, use any available tuner and settle for its top frequency. Modifications covered in **Reference 5** outline how to wind a simple transformer on a ferrite bead so you can use the tuner at 63-MHz output. There's also information on how to modify the input so it's unbalanced and shielded. This is easy to do. Many tuners are already fitted for unbalanced input. Use phono plugs and jacks and miniature coax for connections and signal routing. This Mitsumi tuner was designed for an IF output of 44 MHz, and operates well at 63 MHz — the same as the cable tuner.

Figure 2 gives details of the prescaler installation. I mounted it on a small pc board and coupled a short, single-turn pickup loop to the local oscillator in the tuner. Oscillator operation wasn't affected because very little energy is required to drive the CA3179. The CA3179E is readily available and inexpensive.* If you're adding the scaler to another type of tuner, use the minimum coupling necessary. Make it mechanically secure, and keep the lead lengths short.

Tuner checkout

Attach a 7805 three-terminal regulator to the UHF tuner so it's powered when the tuner is on. Check the operation of the scaler after the scaler installation is complete and the resistors are mounted on the tuner as shown. Connect the tuner to a temporary test source of +12 volts. Then connect the output of the scaler to a frequency counter, and monitor it while you adjust the varactor tuning voltage over the range of 0 to +32 volts. Look for any dropout in count. If any loss occurs, make a minor adjustment to the

* Jameco Electronics, (415)592-8097. Stocks all the ICs and most of the small parts used in this project. $20 minimum order.

FIGURE 3

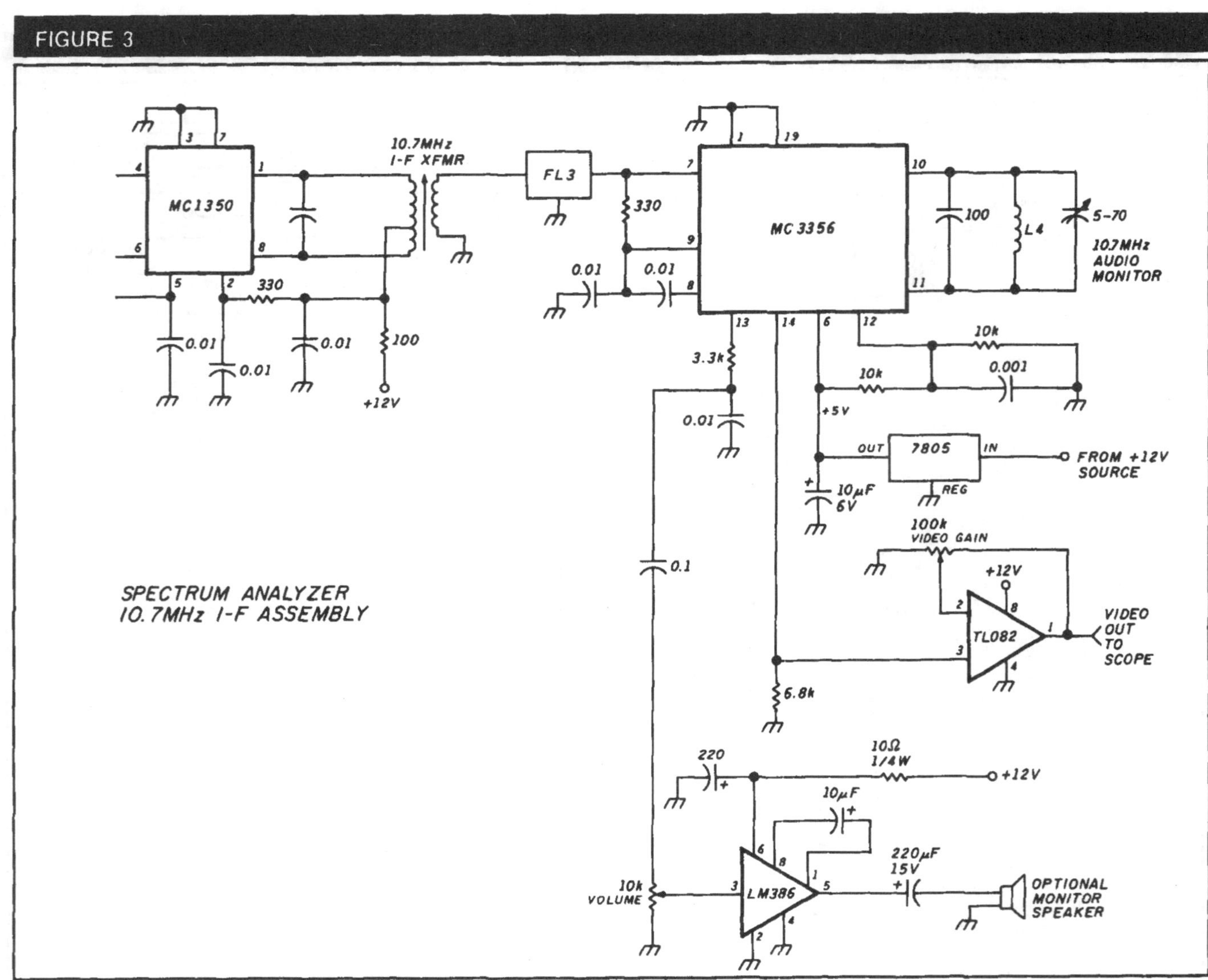

Spectrum analyzer 10.7-MHz IF assembly continued from page 12.

PARTS LIST

FL1	15-kHz crystal filter, optional
FL2,FL3	Murata ceramic filters, 280 kHz
K1	DPDT miniature relay, contacts parallel, for pc mounting
K2	DPDT miniature relay, for pc mounting
K3	SPDT relay
L1	14T no. 22 tapped at 3 and 10 turns, 52 MHz
L2	9T no. 22 tap at 6T from bottom, 52 MHz
L3	Amidon L33-10 1T primary 10T secondary, 63 MHz
L4	12T no. 22, 10.7 MHz or 10.7 MHz IFT
Q1	40673 MOSFET
X1	52.3-MHz overtone crystal HC-18 case
IFT	10.7-MHz miniature IF transformers, interstage

VCO/scaler coupling and recheck until the scaler operation is reliable over the entire varactor voltage range. The VCO frequency is the counter reading multiplied by 256; the signal frequency is the VCO frequency minus the first IF frequency, 63. You now know the exact tuning range of your tuner. I obtained counter readings of from 1.838 to 3.793 MHz, equal to a VCO frequency of 470.5 to 971 MHz. The high end signal frequency was 971 – 63, or 908 MHz; the low end limit was 470.5 – 63, or 407.5 for 63-MHz IF frequency.

The IF circuit

Figure 3 shows the schematic of the dual conversion IF strip. This is a crystal-controlled, double conversion receiver, which uses either a Motorola MC3356 or an RCA CA3089 FM detector. The AM output of the detector is the analyzer output to the oscilloscope; the FM output gives a speaker monitoring capability. I have built IF strips using both detectors and they work equally well. I would recommend the CA3089 because it's easier to find and doesn't require the TL082 op amp. I installed the narrow 15-kHz filter because I happened to have one in stock. It isn't used often.

The sweep circuit

Most of the sweep circuit already exists in any oscilloscope. The circuit is WA2PZO's idea.* Find a point in the oscilloscope wiring where the horizontal deflection waveform has zero DC component. This would be the output of the complementary transistor pair that drives the H deflection plates. Usually there's an unused Z axis jack that you can use to bring this signal out of the scope. Proceed with care when making this modification. High voltages are present in any oscilloscope. Measure the p–p voltage at this point; it will probably be well over 100 volts.

Figure 4 shows the analyzer's sweep circuit. There should be about 35 volts p–p across the sweep width potentiometer. Change the value of R2 to obtain the correct voltage. A multicontact relay lets you switch the various voltages and change tuners with a single panel switch. The diode clamps in the sweep potentiometer circuit prevent the sweep voltage from exceeding the supply voltage on either band.

The sawtooth ramp from the oscilloscope is really a

* Science Workshop, Box 310, Bethpage, New York 11714. Carries the NEC cable tuner and some other parts

FIGURE 4

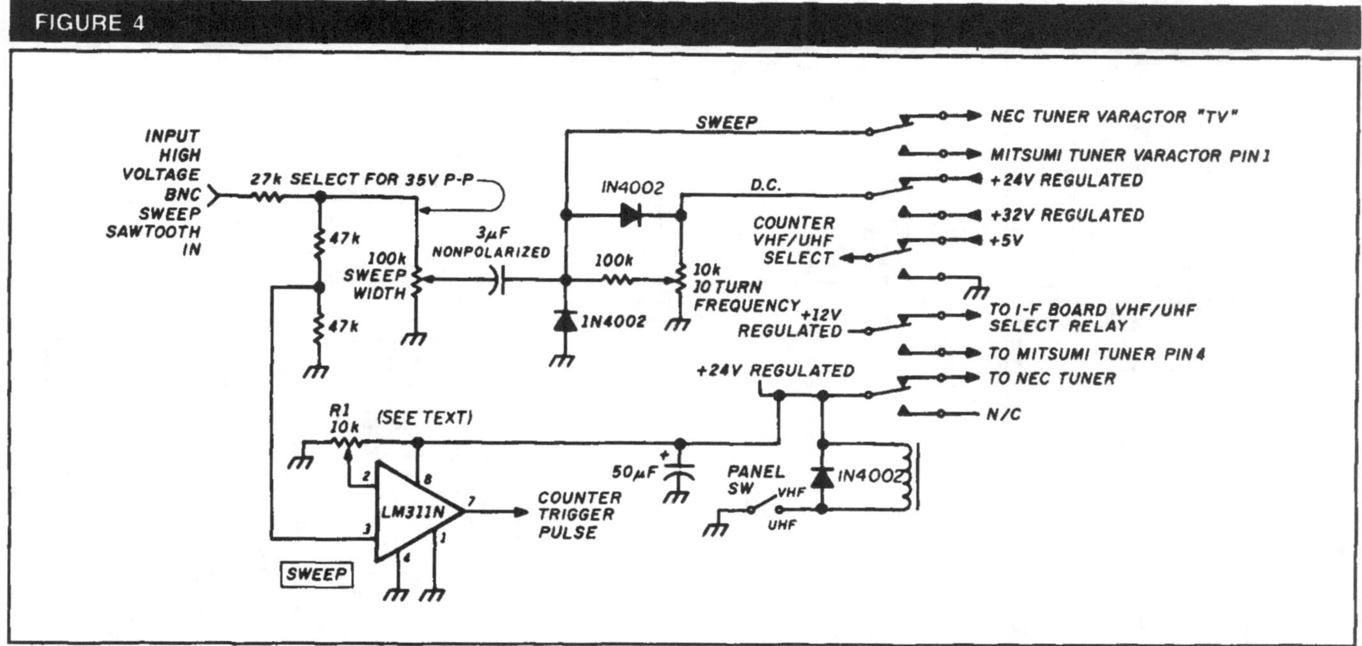

Sweep circuit of the analyzer.

FIGURE 5

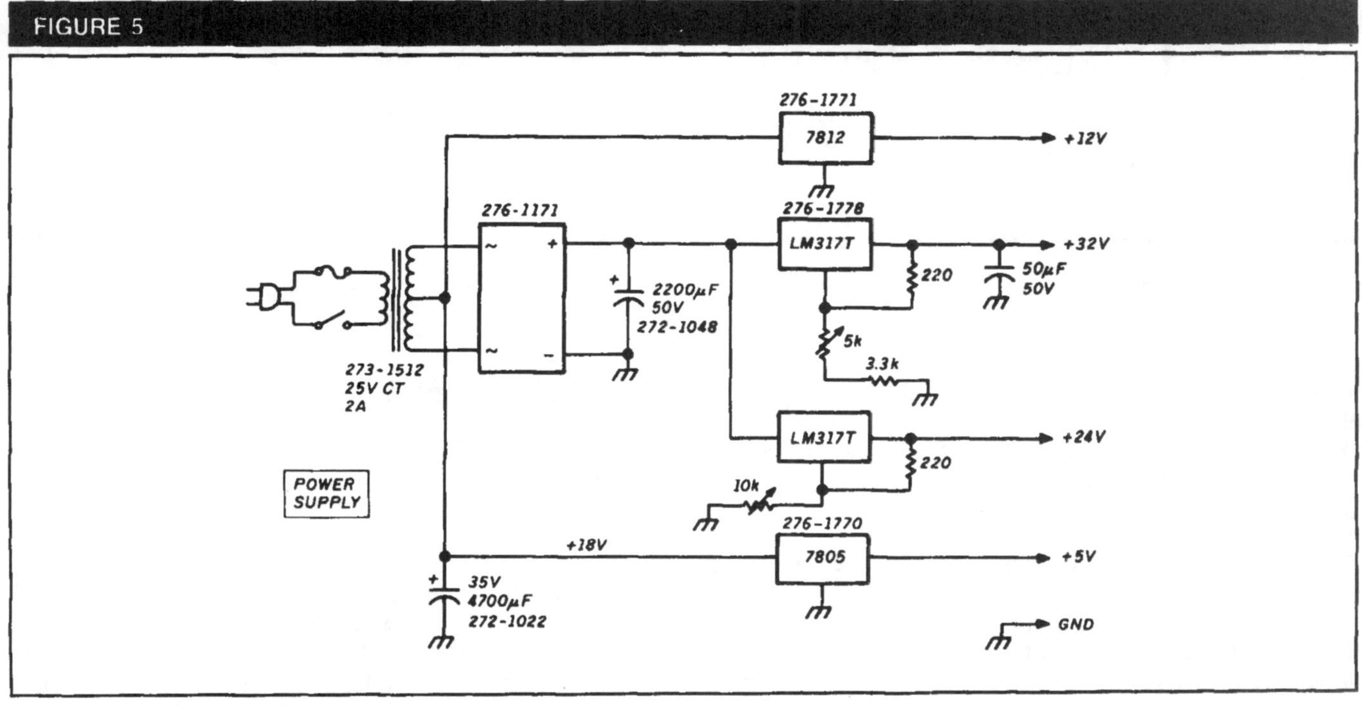

Multiple voltage power supply for the analyzer.

FIGURE 6

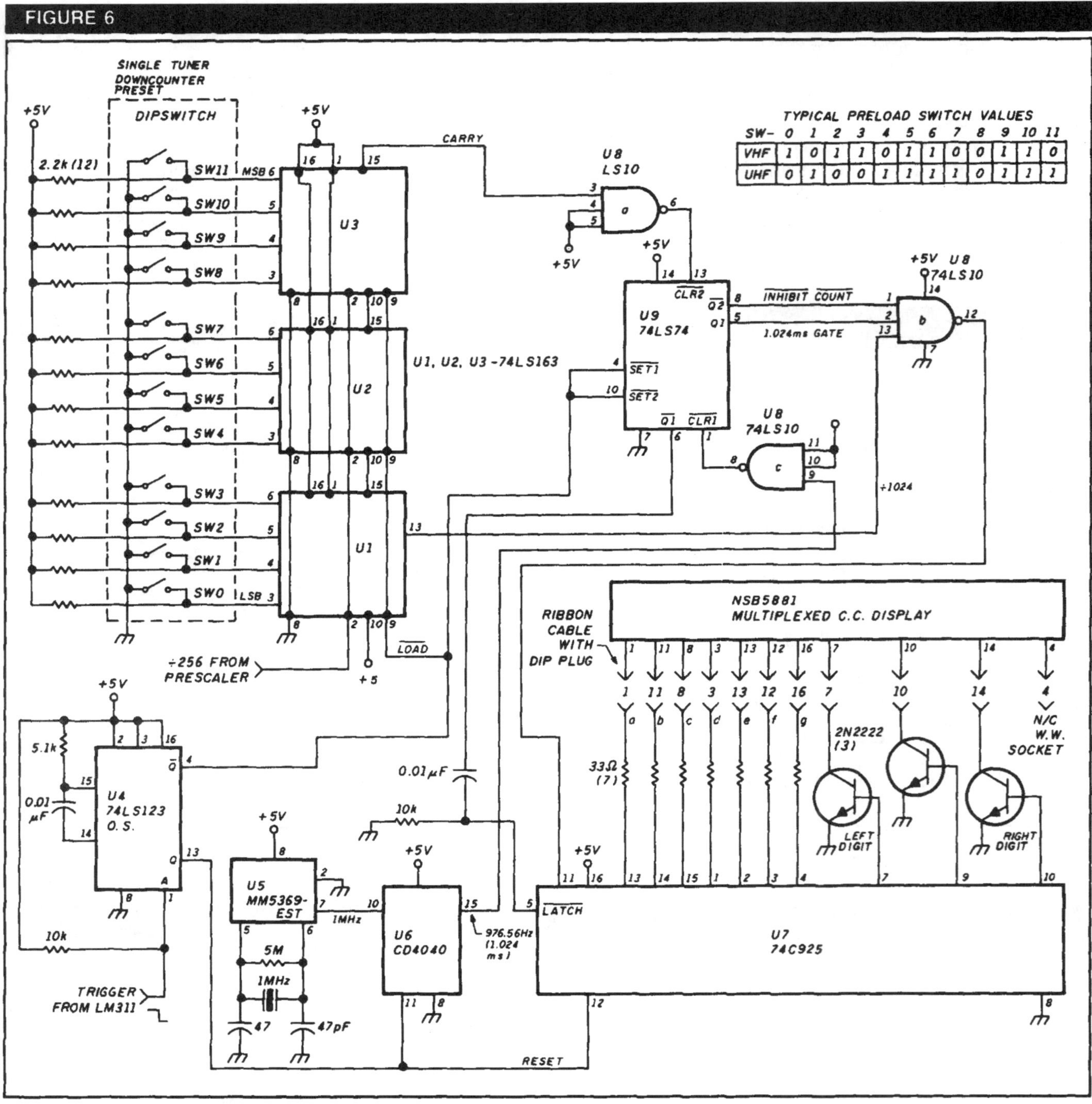

Basic counter drawing.

rapidly changing DC voltage. The LM311 switches its state at an adjustable point (set by R1) on this ramp, which sets the exact time during the sweep that the frequency display samples the scaler outputs.

The power supply

The analyzer required four voltages: +5, +12, +24, and +32. The easiest way to get these voltages is to use the circuit in **Figure 5**. I mounted all the power supply components, with the exception of the transformer, on a 2-3/4 × 5" single-sided pc board. Radio Shack stocks most of the parts.

The counter circuit

The timebase of 1.024 milliseconds in the counter is a little unusual. It's in binary instead of the usual BCD because the scalers divide by 256 instead of powers of 10. Consequently, all the rest of the timing values in the counter are binary.

Figure 6 shows the counter circuit. This is the basic circuit, with typical preload switch values for a 12-bit dip switch. I used this circuit while working on the 0 to 400-MHz tuner frequency display. The switches made it easy to adjust the counter to compensate for the unknown frequency offset

FIGURE 7

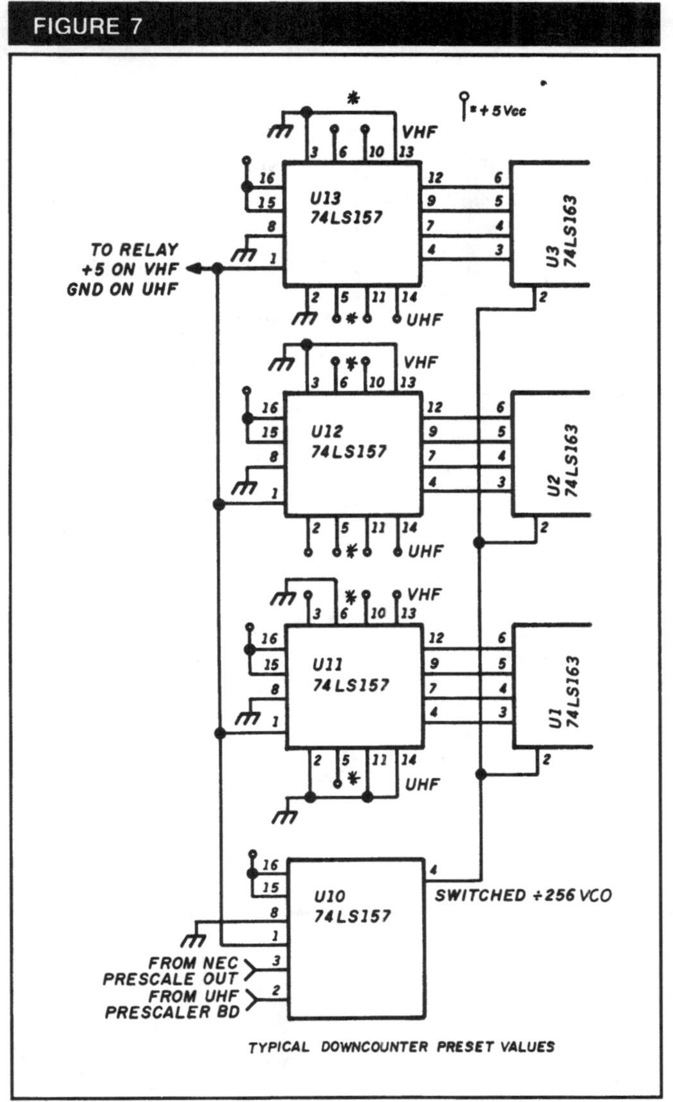

Adding automatic preset switching for VHF and UHF to basic counter drawing.

of the local oscillator in the NEC. (I later found it to be 610 MHz.) I determined the correct switch settings for the UHF tuner, and then had to find a way to enter these binary values into the system when switching between tuners.

Data selectors U11, U12, and U13 do this by acting as a 12-pole two-position switch. The circuit in **Figure 7** shows the data selectors hard wired so the correct binary switch presets are present at the IC inputs. One section of U10 selects the correct scaler. This is a flexible arrangement which can be used with another brand of cable converter and/or another UHF tuner. Just find the preload values with dip switches, then replace the switches with data selectors.

The counter circuit works as follows. A value equivalent to the oscillator's offset in the tuner is preloaded into U1, U2, and U3 when one-shot U4 is toggled as the LM311 detects center line crossover of the sweep. U4 resets U6 and U7, and sets both sections of U9 — a dual RS flip-flop. At this point, U8 pin 2 is enabled and the count signal is at pin 13, but the gate doesn't pass the signal because pin 1 is low and the NAND gate is inhibited. U6 starts counting clock pulses to establish a 1.024-ms timing interval. U1, U2, and U3 count down from the preset value until they

FIGURE 8

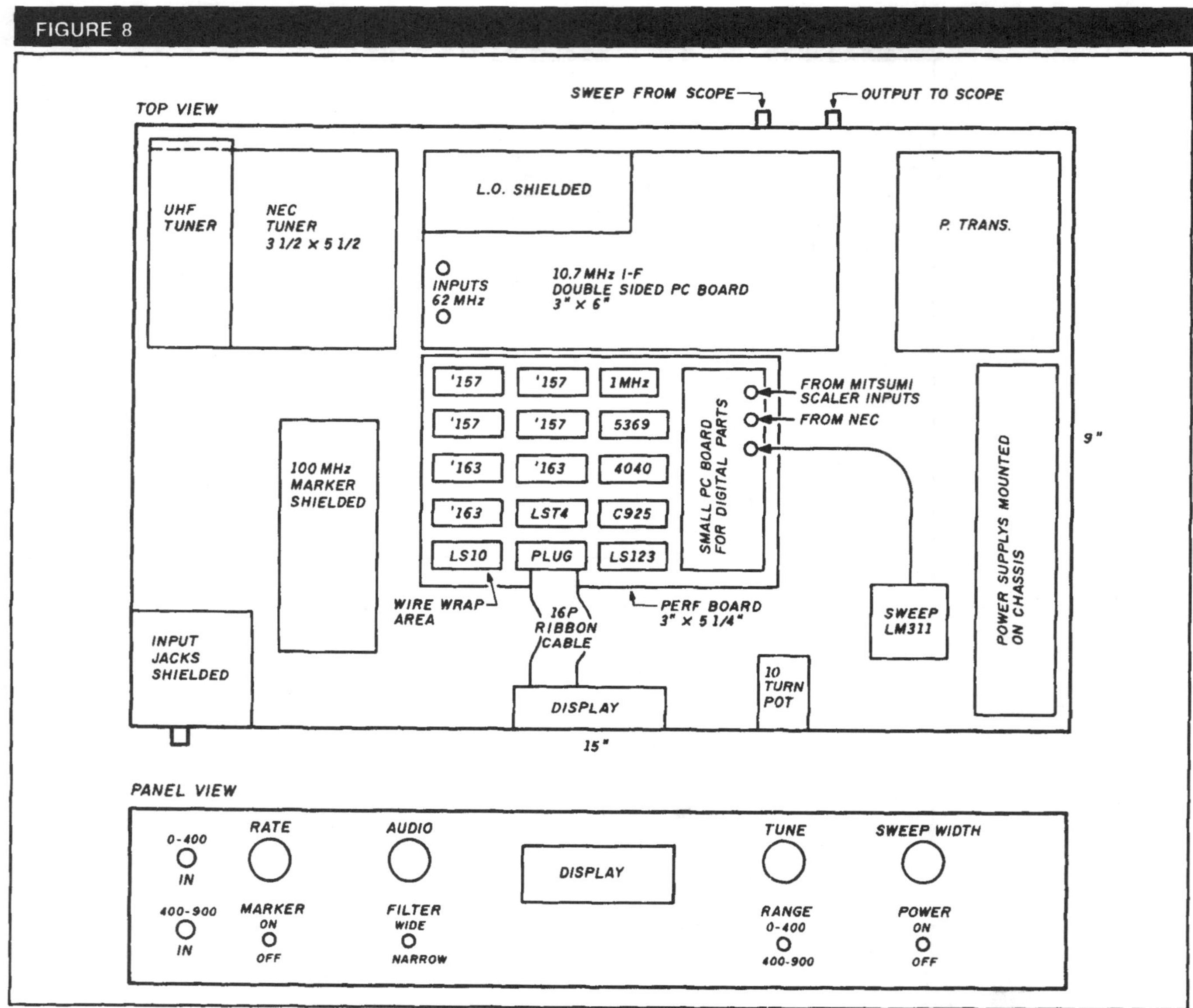

Chassis layout. (A) Top view. (B) Panel view.

reach zero, and output a borrow pulse at U3-15. The borrow is inverted and clears one section of U9, enabling U8-1.

U8 starts to pass prescaler pulses, which are divided by 4 in U1. This continues until the 1.024-ms time interval has elapsed. At this point, U6-15 goes high, inverted in U8-8. The section of U9 that has been keeping U8-2 enabled suddenly goes low, and stops the counting in U7. When U9-1 is cleared, U9-6 generates a pulse transferring the new count to U7's latch and display circuitry. U7 displays this value until the next count is completed and transferred into the latches. U7 supplies current to the multiplexed seven-segment display in a standard circuit using transistors to switch digits.

Mechanical construction

The unit is housed in an inverted 15″ × 9″ × 3″ aluminum chassis. The circuit boards are mounted on standoffs similar to those in the TVRO in the Gibson articles.[5,6]

The counter circuit was wire wrapped and the layout is shown in **Figure 8**. It's advisable to shield the counter assembly completely because of the digital noise. I built the IF/detector circuit on a double-sided pc board because it provided an excellent way to eliminate feedback and reduce shielding problems. I now use wide masking tape as resist, and trim it with an X-acto® knife to make the pc trails.[7] It's inexpensive and easy to work with.

There are two BNC jacks on the panel for the two signal inputs. I shielded these jacks with scrap pc board material. I used two more BNC jacks at the back of the chassis for sweep in and signal out.

The marker generator

You'll need a marker generator (shown in **Figure 9**) to verify that the counter is working correctly and to help in initial setup. Even though everything now works properly, I still use the marker generator when I start out to verify that

FIGURE 9

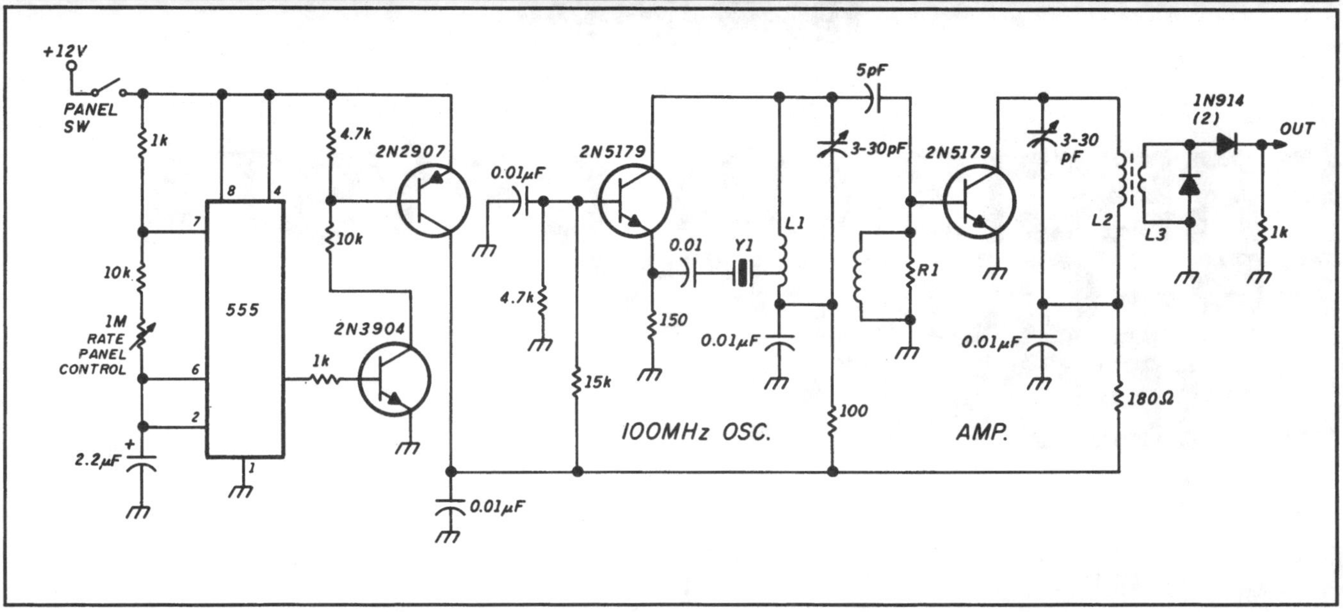

100-MHz marker.

PARTS LIST

L1	***Air wound 7T no. 16 tapped at 1T 1/4" diameter, 1/2" length***
L2	***T50 toroid 7T no. 22***
L3	***4T no. 22 over L2***
R1	***47 ohm 1/2-watt with no. 30 wire wound full***
Y1	***100-MHz crystal (Texas Crystals)***

All other resistors 1/4 watt
0.01 μF-capacitors are miniature disc

all is well. I used a circuit which pulses the marker on and off to make it stand out.

The marker generates harmonics of 100 MHz up through 900 MHz. You may have to experiment a little to get the oscillator circuit that works best with your crystal. The one shown is the best of the several I tried. Build the marker last, and use your working analyzer. You'll wonder how you got along without it.

REFERENCES

1. Al Helfrick, K2BLA, "An Inexpensive Spectrum Analyzer for the Radio Amateur," *QST*, November 1985, page 23.
2. Robert Richardson, W4UCH, "Low Cost Spectrum Analyzer with Kilobuck Features," *Ham Radio*, September 1986 page 82.
3. Shriner and Pagel, "A Step Attenuator You Can Build," *QST*, September 1982, page 11.
4. *The ARRL Handbook*, ARRL, 1987, pages 25-43.
5. Stephen Gibson, "The $100 TVRO Receiver — Satellite Central, part VIII," *73*, August 1982, page 60.
6. Stephen Gibson, "The $100 TVRO Receiver — Satellite Central, part IX," *73*, September 1982, page 60.
7. Adelbert Kelley, AA4FB, "How To Make Your Own Printed Circuit Boards," *Ham Radio*, April 1973, page 58.

PRECISION CRYSTAL FREQUENCY CHECKER

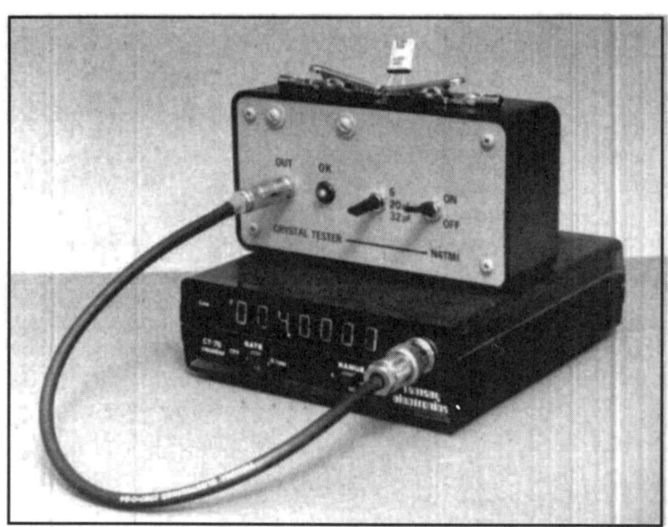

By Michael A. Covington, N4TMI, 285 Saint George Drive, Athens, Georgia 30606

Most crystal checkers perform a simple "yes/no" quality test or may give a relative indication of activity. I've designed one that teams up with a frequency counter to give precise frequency readings under three different load capacitances: series, 20 pF, and 32 pF. It can also measure inductance, though with a bit less accuracy. And even without the frequency counter, this checker will tell you whether or not a crystal oscillates.

This project really is a "weekender." You can find all the parts at Radio Shack, if they're not already in your junkbox.

The circuit is a Colpitts oscillator (see **Figure 1**); capacitors

FIGURE 1

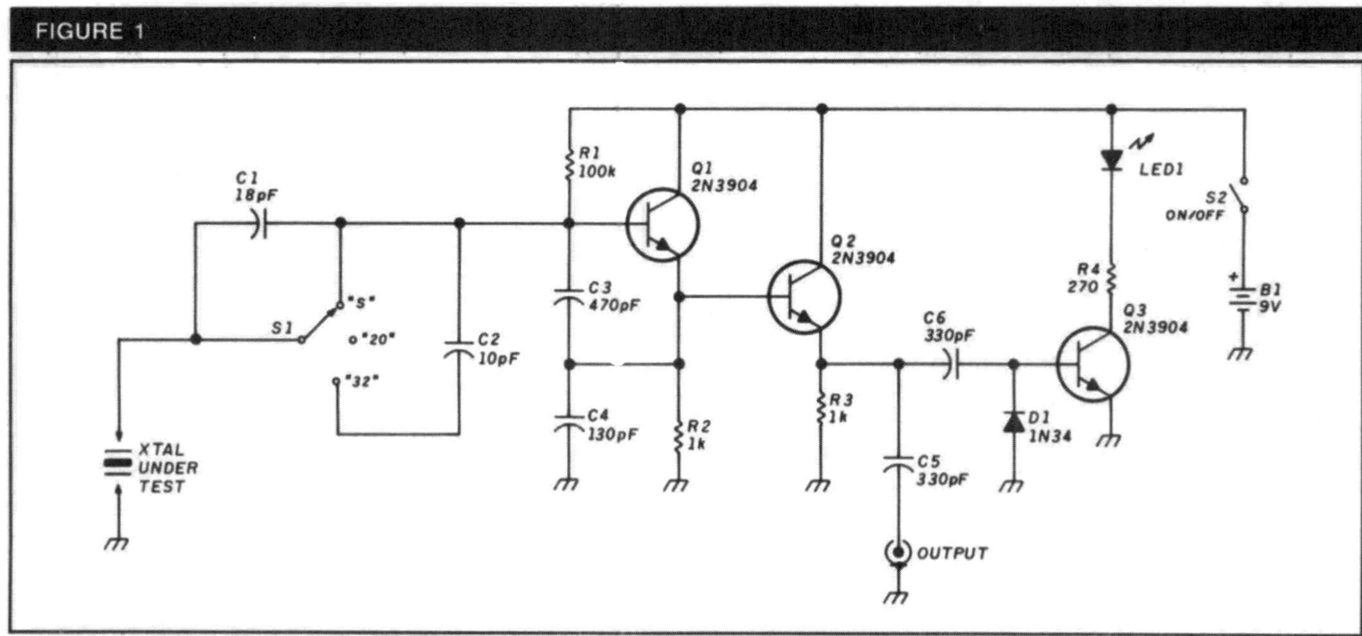

Circuit is a Colpitts oscillator with buffer amplifier.

FIGURE 2

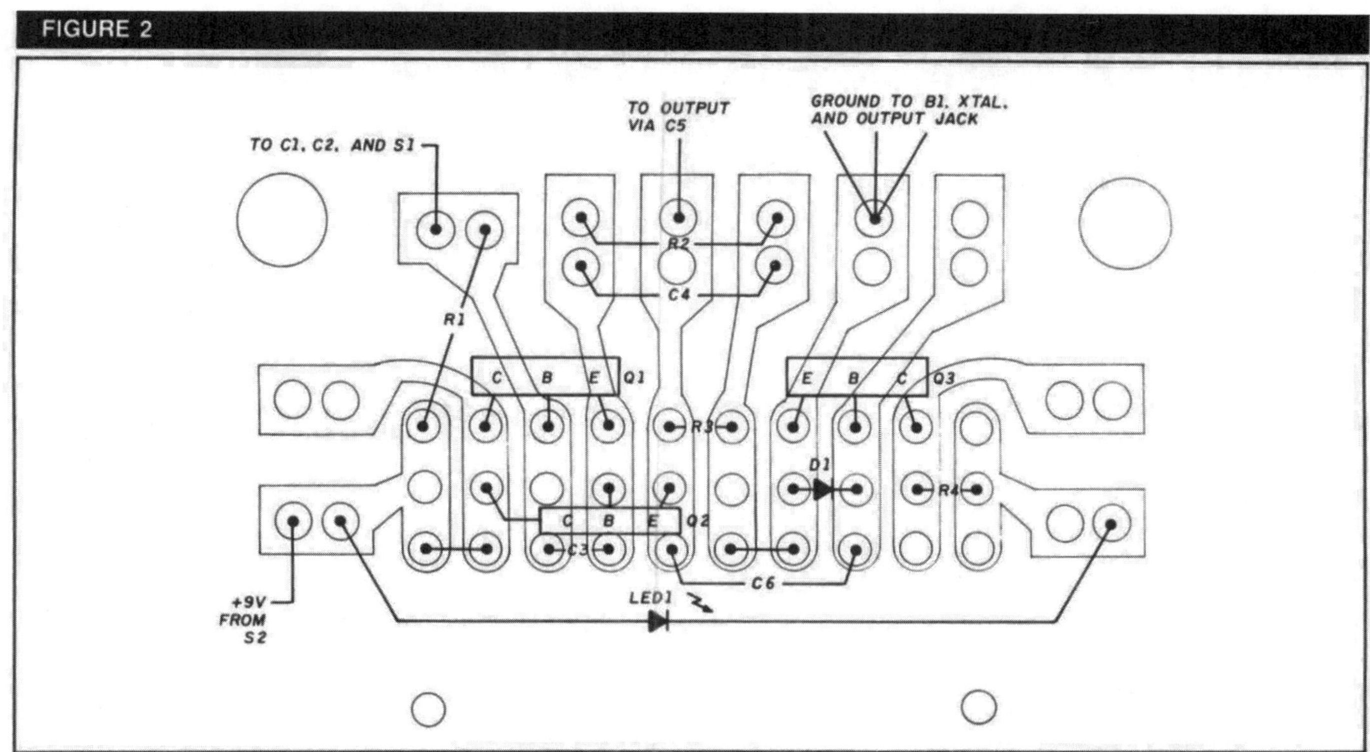

Parts placement diagram. View is from top (component side).

PHOTO A

Circuit board is half of a ready-made board from Radio Shack (RS no. 276-159A).

PHOTO B

To test a coil, measure resonant frequency with 32 pF.

were chosen to work with most crystals from 2 to 20 MHz. Q1 oscillates and Q2 buffers its output. For the yes/no test, D1 and Q3 rectify the signal from the oscillator and use it to light an LED. It's normal for this LED to dim or go out when there's a load (like a frequency counter) connected to the output jack.

When setting the load capacitance, the SPDT center-off switch S1 lets you connect the crystal directly to the oscillator, through an 18-pF capacitor, or through parallel 18 and 10-pF capacitors. Allowing for 2 to 4 pF of stray capacitance, this gives load capacitances near the nominal 20 and 32 pF. The direct connection gives a high load capacitance that puts the crystal very close to series resonance.

Most of the circuit is compactly built on half a Radio Shack 276-159A printed circuit board (see **Figure 2** and **Photo A**). The tester is housed in a Bakelite™ box with a metal front. For reliable measurements, keep leads to S1 and the crystal as short as possible. Use alligator clips as a universal low capacitance crystal socket.

It's easy to test a crystal. Simply clip it in place, hook up the frequency counter, turn on the tester, and flip S1 to find out which load capacitance gives the correct frequency. This procedure also tells you how much the crystal can be "pulled" by changing the capacitance. The 20-pF load gives the highest frequency; the series connection gives the lowest. Overtone crystals will oscillate at the fundamental

TABLE 1

The frequency marked on a crystal isn't always the frequency at which it oscillates. The frequency measured by the crystal tester may be different yet, because the tester doesn't operate in the overtone mode. Here are some kinds of common crystals. F refers to the frequency marked on the crystal.

Type of crystal	Marked frequency (F, MHz)	Operating frequency (MHz)	Measured frequency (MHz)	Load capacitance
General purpose	1 to 20	F	F	Various
General purpose	20 to 60	F	F/3	Usually series
General purpose	55 to 100	F	F/5	Usually series
CB transmit	26.965 to 27.405	F	F/3	Series
CB receive*	26.510 to 26.950	F	F/3	Series
Scanner	30 to 50	F+10.7	(F+10.7)/3	Series
Scanner	140 to 175	(F−10.7)/3	(F−10.7)/9	Series
Scanner	440 to 470	(F−10.7)/9	(F−10.7)/27	Series
Scanner	470 to 500	(F−10.7)/10	(F−10.7)/30	Series

*CB receiving crystals are 455 kHz below the designated channel.

PARTS LIST

Capacitors (ceramic disk or polystyrene, 10 volts, ±20 percent)
C1 ***18 pF***
C2 ***10 pF***
C3 ***470 pF***
C4 ***100 to 150 pF***
C5,C6 ***330 or 470 pF***
Resistors (1/8 watt, ±5 or 10 percent)
R1 ***100,000 ohms***
R2,R3 ***1000 ohms***
R4 ***270 ohms***
Semiconductors
Q1,Q2,Q3 ***2N3904 (or equivalent) NPN silicon transistor***
D1 ***1N34 germanium diode***
LED1 ***Light-emitting diode***
Other
S1 ***SPDT center-off (three position) switch***
S2 ***SPST switch***
B1 ***9-volt battery***
Connector and holder for B1
RCA phono jack for output
Two alligator clips to hold crystal
Circuit board (Radio Shack 276-159A)
Enclosure

FIGURE 3

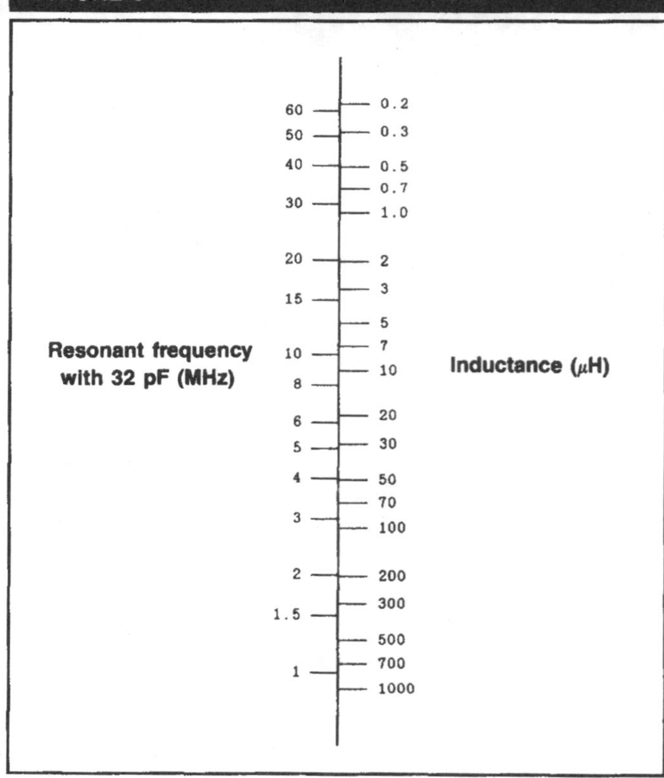

Computer-generated nomograph converts measured frequency to inductance.

frequency. For instance, a 27-MHz CB crystal oscillates at 9 MHz (see **Table 1**). Of course, you can use a crystal whose load capacitance is different from that for which it was ground or use an overtone crystal on its fundamental frequency.*

You can also test coils. Just connect a coil in place of the crystal, set S1 for a 32-pF load, and measure the frequency (see **Photo B**). Now find the inductance using the nomograph in **Figure 3**. You may find it more useful to remember the frequency than the inductance. It's the frequency at which the coil will always resonate with a 32-pF capacitor. Bear in mind that this measurement is inexact because the load isn't precisely 32 pF and the internal capacitance of the coil isn't taken into account. ■

*An overtone crystal oscillates near, but not exactly at, an odd multiple of the fundamental cut. Expect to see a small shift of several kHz when overtone crystals are operated in their fundamental mode. Ed.

Name
Job. No.
Date

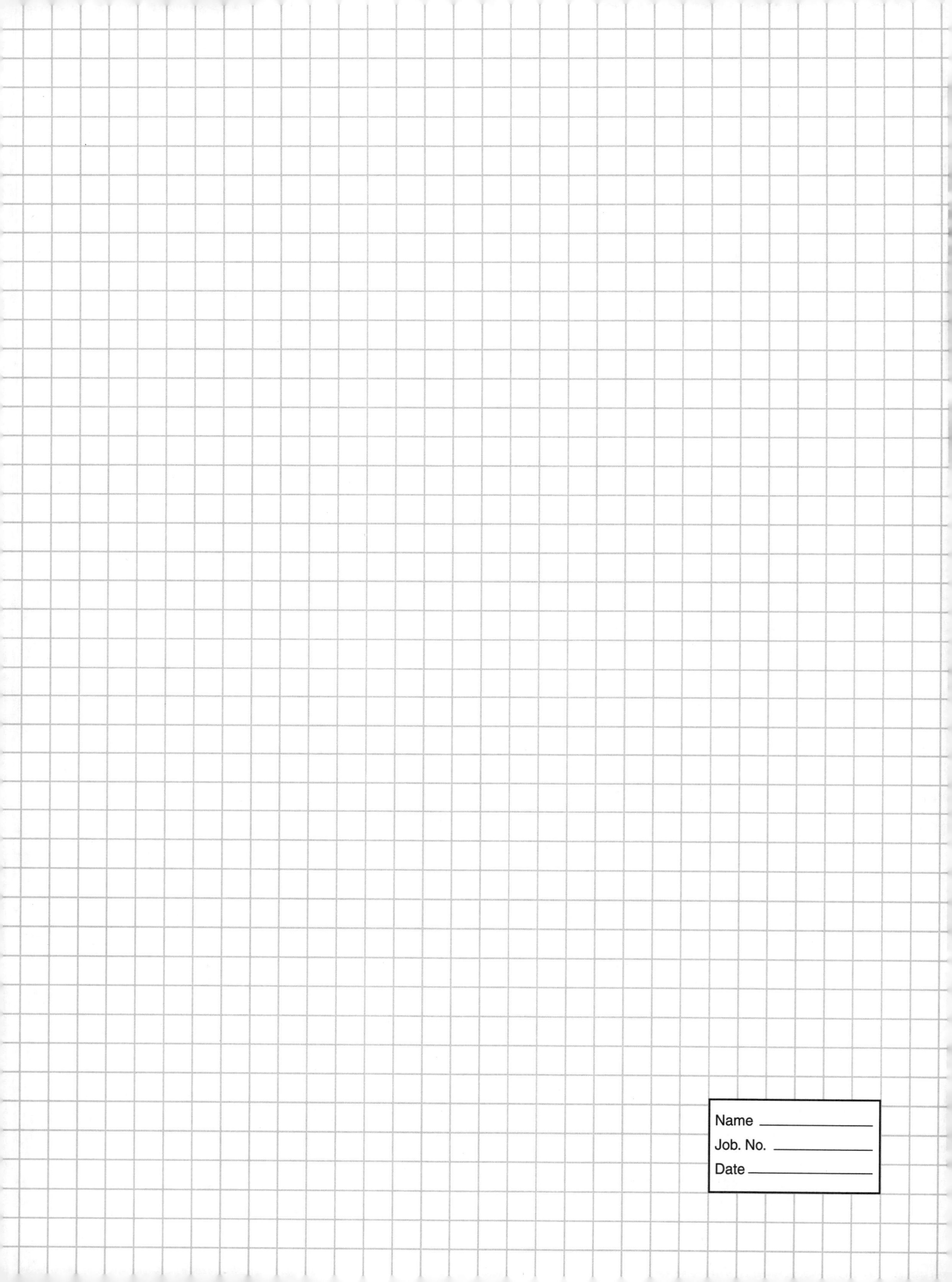
Name
Job. No.
Date